APPIFIED

Culture in the Age of Apps

Jeremy Wade Morris
and Sarah Murray,
Editors

University of Michigan Press
Ann Arbor

Published in the United States of America by the
University of Michigan Press
Manufactured in the United States of America

Printed on acid-free paper
A CIP catalog record for this book is available from the British Library.
Library of Congress Cataloging-in-Publication data has been applied for.
ISBN: 978-0-472-07404-4 (Hardcover : alk paper)
ISBN: 978-0-472-05404-6 (Paper : alk paper)
ISBN: 978-0-472-12435-0 (ebook)

CONTENTS

D. LIFESTYLE/RELATIONSHIPS

E. SOCIAL NETWORKING/COMMUNICATION

F. NEWS/ENTERTAINMENT

Digital materials related to this title can be found on
  www.fulcrum.org at doi.org/10.3998/mpub.9391658

ACKNOWLEDGMENTS

This book is about software that resides in our pockets and offices and homes and cars and nightstands. It is about software that embeds itself into our daily routines, contacts, and communications. But as with most stories of technology, it's not the software that matters as much as it is the people who make, use, play, create, subvert, chat, buy, sell, discover, hurt, trust, deceive, and love with it.

Similarly, this book is about the editors' research interests, but it would be nothing without the wonderful, insightful, and challenging contributions from our thirty-four authors. Our first set of thanks, then, goes to the names listed in the table of contents. When we circulated our call for proposals, we knew that even with a hundred chapters, we'd barely scratch the surface of the app-o-sphere. There are just too many apps to analyze and too many issues they raise. We knew we were asking our contributors to do some heavy lifting, to cover a lot of ground in a relatively short amount of words. We couldn't be more thrilled with how they've taken up ideas about mundane software and digital everyday life and woven them through their respective research interests to truly expand the scope of how we study software and culture. From popular apps to little-known gems, serious programs to novelty jokes, brand new trends to obsolete ones, our contributors have helped us corral and make better sense of this dynamic and ever-changing industry.

Our second set of thanks belongs to the other group of people who made this book a material reality. To Mary Francis for her support pushing this project through to completion and her patience and good humor on countless three-way conference calls. She tirelessly challenged, championed and protected the original vision for *Appified* from day one. To our diligent and understanding production editor, Mary Hashman, Sarah Dougherty, who shepherded this project into the production phase, Pilar Wyman for indexing, Sam Killian in marketing, and everyone else behind the scenes at University of Michigan Press who helped turn the messiness of our early ideas into a physical book readers can hold with pages to turn (or scroll, or swipe).

We'd also like to thank the various audiences and colleagues at ICA, SCMS, and other venues where the editors and contributors have talked about this work publicly over the past couple of years. Your collective criticism, insight, and enthusiastic suggestions have made this collection better and in turn, we hope this text will serve you better in your classes and your research.

Jeremy would like to acknowledge the support he received for this project from the University of Wisconsin-Madison's Office of the Vice Chancellor for Research and Graduate Education. He would also like to thank the students in his Digital Commodities seminar for putting up with early drafts of this project. As always, he owes a debt of thanks to his family—Leanne, Lucas, Rachel, and Justine—who keep him in sync in ways that apps never can.

Sarah would like to acknowledge her writing group comrades—Aswin, Dan, Katherine, Megan—for several rounds of feedback and enduring good cheer. She would also like to thank the students in her Mobile Media Cultures seminars for reading various drafts and offering otherwise inaccessible insights. She's also grateful that Megan's devotion to the weird and wonderful of the apps universe means she'll always know when the International Space Station is passing overhead.

Finally, Jeremy would like to thank Sarah, and Sarah would like to thank Jeremy.

Introduction

Snapchat. Whisper. Yik Yak. Electric Razor Simulator. Vibrator Secret for Women. IAmAMan. I am Important. Candy Crush. Spotify. Evernote. Uber. Boom Boom Soccer. Emoji Keyboard. Is It Vegan? Is It Dark Outside? Is It Tuesday? Is It Love? MapMyFitness. MapMyDogWalk. MyPill Birth Control Reminder and Menstrual Cycle Calendar Tracker. SmartMom. Waiting for Birth. Waiting for Birth Pro–Father's Version. Baby Shaker. Gay-O-Meter Full Version. Tinder. Grindr. AgingBooth. Black People Mingle. iGun Pro–The Original Gun Application.

How else to open discussion on an ecosystem as complex, fascinating, frightening, and culturally significant as apps than with a list? Not only are app stores themselves ordered by lists ("Top Free Apps!" "Best Games!" "New Apps We Love!"), but listing even a small smattering of titles from the millions of apps that now exist reveals the strange encounters and endless choices users face with every visit to an app store. There are apps to help us communicate, travel, sleep, play, and learn. There are apps to help us hide, cheat, share, shop, and save. Some apps keep us connected, informed, delighted, entertained, and safe; others leave us distracted, deceived, disappointed, and addicted. We watch, use, touch, feel, and listen to apps. Apps use, watch, track, follow, monitor, and sell us in return. There are apps that promise to make us better workers, parents, friends, and lovers, and apps whose premise shows the worst sides of what it means to be human. When we look at any single app, it's often hard to conceive how such a trivial and small piece of software could be particularly notable. Viewed from further back, though, from a wide-angled list that apposes each app with like and unlike others, we can start to make connections between all that app stores have to offer--between casual games and productivity suites, between weather apps and dating services, between hidden photo vaults and anonymous activist apps.

The singular insignificance of most of the apps listed above belies the very real collective significance the format represents as an emerging economic market and cultural platform. The apps above are a tiny fraction of the over two million apps available across dozens of app stores (PocketGamer 2016;

Statista 2016a). Apps represent the fastest-growing subsector of the software industry, involving thousands of independent and established developers, from hundreds of countries, and the software they create finds its way into smartphones, mobile devices, cameras, televisions, cars, game consoles, fridges, and other technologies. Media researchers are well versed in theories and methods for understanding the complex ways people make and take meaning from cultural goods like films, songs, video games, and television, but the small and everyday nature of apps as well as their newness as a format has left them understudied and their cultural significance largely unaddressed.

At first glance, after all, apps do seem rather trivial. They are abundant, cheap, or free and often serve limited functions. They can have a rapid rise from obscurity to overnight success, but they fade quickly as different apps emerge, tastes change, or as operating systems update and force obsolescence. Most apps are built to solve mundane, everyday problems: keeping track of one's schedule, waking up, remembering the milk, taking notes, planning workouts. But the quotidian activities they influence and encompass are far from banal: connecting with friends (and strangers), sharing memories (and personally identifying information), making art (and trash), navigating spaces (and reshaping places in the process). Although the sheer number of apps may be overwhelming, as are the range of activities they address, each one offers an opportunity for media and cultural studies scholars to seek out meaning in the mundane. Rather than treat apps as frivolous or incidental, what insights reveal themselves if we take apps seriously, as a specific manifestation of software, as a vector for popular culture, and as a particular aesthetic and functional presentation of code that has expanded the market for software and further integrated it into leisure, commercial, educational, interpersonal, and other spheres of everyday activity?

Appified seeks to answer this question by drawing on the research of emerging and established scholars from a variety of disciplines and backgrounds. Each contributor to this book was tasked with analyzing one individual app and thinking through the issues it raises not only for software as a form of media but also for the relationships between users and their software (and ultimately, users and each other). We began the project with a few key questions in mind: How are apps different from other forms of software? How does the mundane nature of apps reconfigure our ideas of what software can and should do? How do app interfaces affect the experiences of users and the media/cultural content they produce, consume, and interact with? What do different categories of apps (productivity, health and fitness, social, education, communication) say about the ideologies embedded and represented in apps? What digital divides do apps exacerbate? What presumptions do apps and their developers make in the design, functionality, and mode of address of their apps? What opportunities, if any, do apps present for resistance, activism, and subversion through technology? This intentionally broad set of questions is meant to mirror the wide range of apps that populate various app marketplaces and the wide range of analyses that apps make possible. After all, apps, as research objects, cross fields and disciplines.

Apps also pose methodological and theoretical hurdles: How does one research an app that is updated frequently with new features and may only last a few months? (Several of the apps in this collection, indeed, shut down during our production process.) Similarly, how might researchers approach an app built for a specific event, location, or time period? How does a researcher analyze and explain the experience of an app? How do we study the ways that apps are categorized, approved, banned, or featured when these are notoriously mysterious and proprietary processes? Thankfully, our contributors met and surpassed these challenges by thinking about apps both as individual nodes in larger networked media industries and, collectively, as vectors for the production, transmission, and interactions of culture. The result is a volume as ambitious in breadth as it is in depth—a transdisciplinary, geographically, and institutionally diverse collection that aims to make sense of this burgeoning industry and this emerging form of software.

Our modest table of contents of thirty apps is but a drop in the ocean of the app-o-sphere, but we hope *Appified* highlights the range of critical cultural analyses that apps open up when examined as microscopic case studies symptomatic of broader social-technological practices. Ultimately, as the title of this volume suggests, we wanted *Appified* to provide a study of a historically specific moment when an increasing number of everyday activities and routines are being expressed through, carried out by, and experienced as apps. We believe there is much to be learned from this particular moment, one in which cultural products and communication practices are increasingly being appified.

Planet of the Apps

While their colorful, glossy, rounded-edge icons and innovative, personalized features encourage users to be swept up in their newness, apps are more accurately the latest iteration of the software commodity, whose longer history includes formats as varied as punch cards, preinstalled code, boxed software (shrinkware) on floppy discs, hard disks and CD-ROMs, as well as digital bits delivered via direct downloads or online streaming services. Although as recently as the 1960s, software was preinstalled on computers and not technically a distinct commodity (Johnson 2002; Pugh 2002), it was gradually unbundled through the late 1960s and early 1970s (Boudreau 2012; Campbell-Kelly 2003). This led to the emergence of a dedicated software industry, complete with producers, publishers, retailers, and, of course, customers. Corporate and organizational software flourished during this early period (Campbell-Kelly 2003), as did games (de Peuter and Dyer-Witherford 2005; Newman 2017), and by the early 1980s, the term "app" became a common one used as shorthand in tech press articles to describe the various kinds of software applications available (Morris and Elkins 2015). The advent of the web in the early 1990s and its subsequent development through the late 1990s and early 2000s brought a diversity of novel software such as Flash animations, music and video software, email programs, map applications, web games, and more, casting a wide definition around what was considered an app. With the near concurrent rise of mobile phones during this period, software spread from computers and the web to mobile devices like phones, PDAs, and tablets. An early prototype phone from IBM in 1992, for example, offered users access to maps, stocks, and news information (Sager 2012), while more mainstream handset makers like Nokia and Ericsson added small programs—usually in the form of games like *Snake* or *Tetris*—to their cell phones as a way to distinguish their phones from the competition. There were even online stores, like GetJar (founded in 2004), that allowed users to search for and download software onto their phones.

The launch of Apple's iPhone in 2007 and the iOS (Apple's mobile operating system) store in 2008, then, is not so notable for being the first phone to include apps, or the first distribution outlet for mobile software, but rather for crystallizing a certain definition of apps that has come to shape how we think of the relationship between software and ourselves. Web apps still exist and other forms of software are still referred to as apps, but the term "app" now has greater internal discursive coherence and narrower boundaries, in part because of the visual, aesthetic, and technical design of those small and stylish bundles of code that users have been downloading to their phones, tablets, and televisions over the last decade.

After opening up the iPhone to third-party app developers in 2008, the iOS store debuted with close to five hundred apps that garnered over 10 million downloads during its first weekend (Apple 2008). Other tech companies, like Google, Amazon, and Research In Motion (RIM) quickly launched app stores of their own, as did global tech companies and sites like Wandoujia (China), Baidu (China), Anahi (China), Aircel (India), Yandex (Russia), Get-Jar (Lithuania), and SK T-Store (South Korea). While Apple and Google Play remain the market leaders for revenues, app store downloads, and number of users both domestically and abroad, many of the sites listed above are becoming key outlets for the production and circulation of apps. China, for example, where Google Play has little presence, has over two hundred app stores (Shu 2014), the largest of which, Tencent MyApp, counts around 500 million users who, collectively, download over 100 million apps a day (Riaz 2014, Yip 2015). As Nina Li points out (in this volume), these app stores not only demonstrate the shifting nature of software distribution and commodification, but they also serve as definitive cultural guides to mobile media experiences for users, particularly for China's significant population of migrant workers.

The global popularity of apps has made them big business. App Annie, a leading ratings and statistics provider for the app ecosystem whose influence is both described and critiqued in Patrick Vonderau's chapter (this volume), estimates that the gross revenues of global mobile app stores grew to $86 billion in 2017 and may exceed $101 billion in 2020 (App Annie 2018; App Annie 2016). App store downloads grew 60 percent to 175 billion in 2017 and reach 284.3 billion by 2020, up from the 111.2 billion downloads logged in 2015 (App Annie 2018; App Annie 2016). Given that global smartphone penetration is currently around 25 percent and expected to grow to 35 percent by 2020, apps are quickly becoming one of the most frequent ways users interact with software (Anderson 2015; Statista 2016b). Estimates suggest that US smartphone users spend about forty hours a month on apps (Nielsen 2015; Rosoff 2016), and, globally, total time spent in apps on Android phones grew by 63 percent from 2014 to 2015. But while our appetites for apps seems broad—the average number of apps a US smartphone owner has downloaded at any one time is twenty-seven—we are ultimately creatures of habit, with a majority of users employing only four to six apps daily (Smith 2014; Statista 2015). In other words, apps are now a primary platform of their own, sometimes surpassing previous ways of interacting with media, brands, and other cultural content. As Finn Brunton's analysis of WeChat's routinization of financial transactions demonstrates (this volume), apps are insinuating themselves further and further into the patterns of our everyday lives and becoming more habitual and second nature in the process. App

Annie's (2016a) self-serving claims that "apps are eating the Web" may be a bit far-fetched, but it is clear that apps aim to present an alternate version of what it means to be connected and what it means to interact with digital information.

Collectively, this volume argues that apps are a form of software packaging, presentation, distribution, and consumption that significantly shifts users' relationships with software and their understanding of what software does and can do. Not only is the number of apps and the pace at which they have become a primary model for making, circulating, and using software remarkable, but also apps represent a moment of historical rupture in the selling of software. Apps are an aesthetic mode of selling digital objects—one that has been mimicked and ported to a wide range of devices and other media forms. Apps also signify a new logic of production: while there has long been independent and third-party software development, app stores represent a coordinated effort to corral those disparate actors in a way that provides both a (relatively) seamless user experience and a significant amount of outsourced labor, profit, and brand recognition for the app store owners. App stores represent the digitization not just of the software product (which has long been digital) but also the digitization of the distribution chain, a move that has been far more successful than other online means of selling software and one that affords app store providers, as Tarleton Gillespie's (this volume) essay on banned apps suggests, a significant amount of industrial and moral influence over the software users discover and experience.

Apps also represent a different vision for the future of human-software interaction. Jonathan Zittrain (2008) argues that early personal computers and the nascent internet were designed as "generative" technologies—technologies that allowed users to create what they needed or to build easily upon the work of others rather than forcing users to be subservient to a device's or platform's prescribed uses. Drawing on ideas and practices from the free and open-source software movements (see also Kelty 2008) and theories of the Commons, Zittrain defines generativity as "a system's capacity to produce unanticipated change through unfiltered contributions from broad and varied audiences" (2008, 70). Given the control that manufacturers and app store providers exert over devices and software, technologies like the iPhone or apps represent, in this view, highly sterilized and "appliancized" versions of more generative technologies by not allowing users to tinker with the code, customize devices, or run programs or applications of their choosing. Zittrain's critique arrived just before Apple opened up the iOS platform to third-party developers, at a time when the iPhone and its apps were almost entirely closed off to outside tinkering, save for adventurous jailbreakers and Cydia store users. Others have since argued that truly generative systems are more theoretical than actual (Grimmelmann and Ohm 2010; Thierer 2008), that usable systems and their potential for social and cultural generativity are perhaps more important than an openness to technical tinkering or hackability (Burgess 2012), and that the oversight that platform providers exert over third-party developers allows for *more* innovation and creativity *not* less (Snickars and Vonderau 2012). Yet it's hard not to see apps as driving a more restricted and tethered relationship between software, user, and device. While application protocol interfaces (APIs) and software development kits (SDKs) allow for countless apps to be created for mobile platforms, these third-party apps only allow what Zittrain calls "contingent generativity" (2011); they are only customizable at a surface level, they run on devices that allow for very little tampering or tweaking, and they may ultimately limit a user's perception of what agency they could have with a given piece of technology (Gillespie 2006). As Zittrain writes, "Contingently generative apps will appeal to the market, even as they constrict the ability for truly revolutionary applications to come about, and place unprecedented power in the hands of the platform vendor and the governments that will regulate it" (2012, n.p.).

Apps marry mobile hardware and personal software in ways far beyond what computers and boxed software allowed, ways that stretch software's capabilities and how it is popularly understood and used. Apps are not outside the long history of software, but they do present new configurations and encounters of hardware and software (Miltner, this volume); new gestures (Duguay, this volume), movements (Halegoua, this volume), tactile and sensory experiences (Hagood, this volume), new avenues of affect (Keller and Harvey, Chess, this volume) and new forms of bio-control (Rettberg, this volume) that are embedded into the everyday routines and rituals of users.

Mundane Software

We are not the first to recognize the increasing embeddedness of software in everyday life. Since Lev Manovich's call for a move to "software studies" (2001) to some of the earliest work in the field that formed in response (Chun and Keenan 2006; Fuller 2003, 2008; Galloway 2004; Hayles 2005; Manovich 2013; Wardrip-Fruin 2009; Wardrip-Fruin and Montfort 2003), researchers have noted the power of software is "significant but banal" (Kitchin and

Dodge 2011, 10). As Matthew Fuller argues, "Software forges modalities of experience—sensoriums through which the world is made and known" (Fuller 2003, 63). Increasingly "everyware" (Greenfield 2006), software is also becoming troublingly habitual (Chun 2016). Like other new media, software "matter[s] most when [it] seem[s] not to matter at all" (Chun 2016, 1), when it is integrated so precisely and unremarkably into the rhythms of the everyday that its uses and meanings become second nature, instinctive, repetitive action, and common sense. As software becomes habit, it waffles between conscious routine and unconscious action, between the voluntary and the involuntary, the creative and the mechanical, and thus becomes a prime locus of power and ideology (Chun 2016, 8). Today software is literally in the pockets of millions of users worldwide—a fact that early software studies scholars predicted in theory but had not yet encountered in terms of the material format, cultural logic, and commercial industry that apps have inaugurated. Now more than ever, users delegate a vast swath of everyday activities to highly packaged and curated software on mobile devices.

The influence of apps has piqued the interest of software studies scholars (Bogost 2011b; Bratton 2016; Manovich 2013) but their presence on smartphones, tablets, and other portable platforms has also drawn in scholars from mobile media studies (de Souza e Silva and Frith 2012; Farman 2012; Goggin and Wilken 2012; Wilken and Goggin 2015). Moreover, apps now encompass a number of issues that have been key to cultural and media studies, so projects on labor, production, representation, identity, the public sphere, and surveillance now often have to contend with apps. Svitlana Matviyenko and Paul D. Miller's recent collection, *The Imaginary App*, presents apps as "an abbreviated software application—figuratively and literally, linguistically and technically: apps are small programs—pieces of software designed to apply the power of a computing system for a particular purpose" (2014, xvii–xviii). The collection's theoretical pieces, interviews with technologists, and essays about imaginary apps—apps that do not actually exist but work as a form of critique of app culture—offer an important conceptual toolkit for thinking through the realities of mobile and digital culture we try to build on here. Other research on apps, like the special issue of *Fibreculture* on the subject, calls for them to be understood at the intersection of affect and materiality, as "objects that are related to the constitution of subjects, as a component of biopolitical assemblages, and as a means of digital production and consumption" (Mellamphy et al. 2015, 2). Apps' industrial significance has been investigated, given the app is "particularly powerful in its combination of software design and price modeling . . . adding genuinely new meanings to an object [i.e., phone] not originally conceived as a mobile platform for consumers to download data in a standardized format" (Snickars and Vonderau 2012, 3) and the ideologies of specific categories of apps—such as productivity (Gregg 2015a) and casual games (Anable 2013)—have been critiqued. Much remains to be said, however, about the apps themselves—their design, their interfaces, their features, their designers and users, as well as the networks of hardware, software, actors and agents they bring together. Luckily, scholarly research is catching up, with apps now central to several emerging research programs (e.g., Duguay 2017; Gerlitz et al. 2016; Nieborg 2015).

In line with these investigations, *Appified* provides analyses grounded in case studies of individual apps and app stores and places those programs in the wider historical and cultural context of media and cultural studies scholarship, attuned to structures of power, modes of access and representation, and issues of identity. By adopting the modular, single-serve ethos apps themselves represent, we hope the volume builds on the above critiques and presents new methodological (Light, this volume) and theoretical vectors (like Sarah Sharma's focus on temporality in her analysis of TaskRabbit, this volume) for the analysis of this emerging software format.

The collection also further develops the concept of mundane software. "Mundane" is a term to qualify both a particular format of software and the devices that help insinuate that software into our everyday lives. Since for most of software's history, usage was limited to computers, software has not had the chance to inhabit everyday life to the extent it does currently. As recently as the 1980s, for example, computers were still largely specialized machines for specific purposes (accounting, games, calculating, etc.). The rise of multimedia machines in the 1990s and the proliferation of the internet made computers more essential to everyday life and thus made software more important. But the move to mundane software has as much to do with the phones, watches, televisions, cars, gaming consoles, tablets, and other technologies that have opened up their interfaces to app development as it does with the shifting role of software itself. Put simply, one rarely pulls out a laptop while waiting for the bus to kill time with a casual game, or at the grocery store to check on the shared shopping list, or when walking home, scared, late at night to enable a tracking app to ensure they arrive home safely.

The qualifier "mundane" productively extends the notion of software as and in everyday space by considering how the app represents the mobility, ubiquity, and ready accessibility of software. It also places the app within the context of "the everyday" long theorized by cultural studies. Mundane, ordinary, banal, and quotidian: media and cultural studies have consistently taken these qualifiers of everyday routines as central to their fields of study. Whether through Raymond Williams's notion of culture as ordinary (1989), Henry Jenkins's or Janice Radway's efforts to identify pockets of resistance in the popular (Jenkins 2014, 1992; Radway 1984), or Erving Goffman's seminal work on the presentation and management of the self in prosaic social encounters (1959), the study of culture necessitates careful consideration of repeated practices that organize a daily flow. Everyday life is a space and experience that has been colonized by commodities and information technologies but also one of the few arenas where resistance and subversion can authentically originate (Lefebvre et al. 2008). The everyday is a purposeful and known ordinariness, and our ability to imagine and manage that ordinariness as routine is dependent on the media and technologies we incorporate to stifle the "threat of chaos that is the *sine qua non* of social life" (Silverstone 1994, 166–67). The everyday is constructed out of the conditions of possibility for habits to form from the "continuity and reliability of objects and individuals" (Striphas 2009, 10). The patchwork that forms is what Michel de Certeau described as the practice of "making do" (1988), even if affect theorists like Lauren Berlant (2006) and Kathleen Stewart (2007) offer a more cynical take, defining the everyday as the cruel optimism of forward momentum.

By juxtaposing these theories with grounded studies of software, we hope to situate apps as a particular, historical expression of the software commodity that insinuates itself in, and exploits, the mundane in powerful and important ways. Mundane software, then, as Jeremy Morris and Evan Elkins argue, is software that

> spreads out beyond the computer and into a vast range of everyday routines and activity. Mundane software is mundane because it is relatively unremarkable: a to-do list app, a bird-watching guide, an app that mimics the sound of a zipper, etc. This is not to suggest mundane means useless or boring. Mundane software can be necessary, affective, and enjoyable, even if it is for everyday tasks like doing groceries or tracking your exercise. But given the diversity and wide-range of functionality apps provide that seem to defy categorization or unification, we suggest their mundane-ness is precisely what allows for the incorporation of software into a range of everyday practices. (2015, 65)

Mundane signifies a material and aesthetic shift in the form of the software commodity from boxed/shrink-wrapped software that costs between forty and sixty dollars to stylized virtual programs that come freely or cheaply from highly rationalized digital stores. Its flexibility accounts for entertaining distractions (e.g., Temple Run or HQ Trivia), as well as more everyday but essential apps (e.g., the Eat Sleep Baby Tracking app, Rain Rain Sleep app). Mundane software is not a measure of popularity; it can mean programs with hundreds of users (e.g., Stapler Simulator) or those that have attracted millions (e.g., Evernote). It is also not a measure of seriousness; the term is meant to include both novelty apps (e.g., iBeer or Face on Toast) and those more serious in tone (e.g., SMS Lifesavers).

Mundane software is meant to qualify both a particular kind of software—a program that is simple, has a single purpose or limited functionality, is cheap or freely available, is not particularly remarkable in terms of design or content, etc.—and also the ways/places in which it is taken up and used (e.g., during chores, at the store, before bed, waiting for the bus). As Jesper Juul (2010) notes in his discussion of casual games, where "casual" is less about specific generic attributes of a game and more about the practices, interactions, and relationships users have with games and gaming, mundane software is both object and practice. Mundane also helps account for the sheer volume this particular format of software production and distribution encourages. It provides a framework for distinguishing the impact and importance of a particular format of software, at a specific moment in software's longer history. The mundane constitutes at once the daily imperative of making ends meet (Sharma/TaskRabbit), the pressures to participate in societal rituals (Duguay/Tinder), the relentless daily goal of self-improvement (Schull/Lose It!; O'Riordan/FitBit), and the instrumental acts of communication users take part in to find their place in their communities locally or globally (Gajjala and Verma/WhatsApp). We engage in the mundane when we take in news, politics, and trends (Powers/This.Reader; McKelvey/Hillary For America) and when we seek out our favorite sounds (Razlogova/Shazam), images (Rettberg/Snapchat) or videos (Goggin/TubiTV), even if the interactions that occur in these spaces—among users, between users and information, and between self and identity—raise questions that are essential for understanding the role of new media in processes of communication (Shepherd and Cwynar/YikYak).

We use the term "mundane," then, purposefully and provocatively to acknowledge commonality among a disparate set of software programs and in an attempt to bring order to the miscellany that app stores both create and encourage. Despite the number and seemingly endless variety of purposes for apps, mundane helps draw connections between software as potentially discordant as an app to play fantasy sports (Lopez, this volume), an app for participating in a national election as a canvasser or campaign manager (McKelvey, this volume), an instant weather forecasting app like DarkSky, and a walkie-talkie app like Zello that can be used to coordinate both democratic protests and militant ISIS messages. We see potential to expand the scholarly conversation in digital culture if a term like "mundane software" can force connections between, for example, casual games and antiterrorism apps or music-making apps and productivity software (Chess; Elmer and Nasirzadeh; Simon; Murray, this volume). "Mundane" is ultimately a flexible yet useful qualifier for a particular iteration of the software commodity that is marked by its ubiquity, its discardability, and its increasing incorporation into the rhythms, routines, and rituals of daily life through smaller, more mobile devices. It accommodates the popular and the obscure, the trivial and the useful, the fun and the serious, the single-purposed and the multifunctional. Mundane allows us to consider how the ordinary is developed as a market investment (i.e., the rationalized, high-volume production of packaged and marketed apps in Apple and Google stores from a wide variety of developers) while also offering contributors to this volume a frame through which to highlight how apps invite certain uses, how they are distributed and used by various audiences, and how they circulate within the context of daily routines.

The Order of Apps

Dropbox, Waze GPS, Mint Money Manager, Infectious Disease Compendium, Swarm, 7 Minute Workout, Hooch–One Drink a Day, Nigerian Constitution, Dog Supplies, Naughtify–Sexy Adult Emoji for Naughty Couples. This list of apps, like the list that began this chapter, seems unwieldy at first: a jumble of proprietary names, succinct (and strategically vague) descriptions suggesting potential uses and meanings. Each visit to an app store brings us face-to-screen with similarly odd juxtapositions. The relatively low cost of app production and distribution has sparked a plethora of novel ideas for mundane software but also a host of new ways to classify and categorize it: Productivity, Navigation, Finance, Medical, Social Networking, Games, Magazines and Newspapers, Health and Fitness, Food and Dining, Government and Politics, Education, Shopping, Lifestyle, and more. Beyond the categories lay the lists, curated bundles of related content, meant to direct our attention: "New and Updated Apps," "Top Grossing Apps," "Our Indie Picks," "Games Recommended for You," and so on. These tools for organizing and presenting apps highlight different ordering ideologies: objective measures of success, subjective thematic groupings, affective collections by mood or need, algorithmic suggestions from data mining, and editorial recommendations from experts or advertisers. While traditional software retail stores usually classified software in the broadest of swathes, such as Business, Industry, Productivity, Games, and Education (Campbell-Kelly 2003, 208–209), even a casual scroll of today's app categories show just how far into the realms of the everyday software has spread.

Given the bizarre eclecticism across apps, the ubiquity of lists and categories enforce a kind of ontological order. Categories describe the things within them and prescribe how those things should be perceived and experienced (Bourdieu 1991). They appear as a natural ordering of things and present themselves as common sense. But they also, as Michele White notes about eBay's lists and categories, "create a set of objects and relationships between objects, articulate what is recognizable and purchasable [and] structure how things can be viewed" (2011, 8). In other words, the process of categorizing reproduces normative articulations of certain objects, texts, and practices to certain identities and, in the case of app marketplaces, produces visible and less visible apps and digital practices. When faced with disorder, lists are "deployed in order to order" (Young 2013, 502), and it's precisely in this ordering where the power of lists and their subsequent presentation in app stores resides. Perhaps the most notable innovation of app stores, then, is infrastructural: they bring together, package, and present these odd mixings of mundane software in a way that gives them order and coherence and makes them commercially legible. Whereas the software commodity was historically sold through a variety of independent or networked retailers, each with differing levels of quality, reliability, and commerciality, the biggest achievement of most app stores has been the consolidation of a commodity in a way that offers users an easy-to-use, reliable, and secure way to download, install, and use software on their devices.

But the list above also shows the power of lists, at least academically speaking. If the content categories and commercially oriented lists of the app store enforces order and logic and ways of understanding software, our collection seeks to freely associate a number of apps to see what transpires. As Ian Bogost argues, lists can be the "perfect tools to free us from the prison of representation precisely because they are so inexpressive. They decline traditional artifice, instead using mundaneness to offer 'a brief intimation of everything'" (2012, 40). Despite the power of lists and their ability to direct attention toward some goods and not others, to some causes, beliefs and attitudes and not others (i.e., what else could be in "Top Free Apps" other than the top free apps?), lists also "remind us that no matter how fluidly a system may operate, its members nevertheless remain utterly isolated, mutual aliens" (40). Even if lists work subtly to shape how we navigate digital stores, the list also "disrupts being, spilling a heap of unwelcome and incoherent crap at the foot of the reader. In doing so, a tiny part of the expanding universe is revealed through cataloging" (41).

It is in this spirit that our list of chapters—subdivided into categories— seeks to describe, enact, provoke, and expand. The categories we have chosen to structure this book are drawn directly from various app stores. We do this not to merely mimic industry logics, but rather to reflect on them. While the move is partly performative, we also mean to call attention to the power of lists and categories to structure app stores and users' experiences of them, and to the logics of commodification that underpin any act of categorization. App designers and producers, for example, must think in these categories: developers must think about how many other apps exist in the category in which they compete for visibility, and they must engage in marketing and other processes (e.g., app store optimization, as Morris discusses in his chapter) that might help get their app ranked in one of the many lists or bundles that adorns the landing pages of various app stores. Knowing which categories are more or less crowded, how to design an icon that both suits the category and stands out within it, or how to increase ratings and visibility for an app are all part of the business of entering any app store. Placing an app in the productivity category, or calling it a utility, or getting it featured in the "Must Have Apps for the Long Weekend" promotional bundle can make a substantial difference in the future success of an app, even if the process of having an app included in such a grouping is both opaque and proprietary. With software delivered directly from the developer to the customer, there's little need for these artificial designations, but in a unified online retail outlet that aims to present a coherent interface through which to access hundreds of thousands of programs, categories and lists become incredibly important and influential, so app store owners must figure out which categories make the most intuitive sense to consumers and the most financial sense to support. The organization of this volume is meant to call attention to the power these categories enact and to the way they frame how scholars even approach apps in the first place.

However, the need to force mundane software into a manageable and user-friendly interface also means that much of the software sticks out from the categories to which it is assigned. Our chapter organization strategies also hope to call attention to just how damaging acts of categorization can be. To place a rape activism app (see Rentschler's analysis of Hollaback!) or a personal safety app (see Ellcessor's chapter on Companion) in a catch-all category of "Lifestyle"—a category that also houses apps like the home-finding app Zillow, the Consumer Reports Car Buying Guide, dozens of food/cooking apps, and the Houzz Interior Design app—seems to belittle the purpose of the former and relegate/equate the needs they aim to fulfill as nothing more than a lifestyle choice. It is as if the "choice" to get home safely or not be harassed is somehow equivalent to buying a car or decorating a home and should thus be solved through the same mechanism—the app.

Indeed, part of what is remarkable and troubling about mundane software is that it puts in the same category the trivial and the serious, the funny and the frightening. The category "Utilities," for example, contains innocuous apps like calculators and flashlights but also seems to be the category where some registered sex offender–tracker apps reside (see Mowlabocus's chapter). For the majority of the apps in the collection, we have mimicked their placement in the app stores; they are listed under the same headings and categories in which you might find them in the iTunes store or Google/Android marketplaces. With others, we took some creative license and placed them in cat egories where they seemed to connect topically or thematically with the other chapters in their section. In both cases, the intent was to provoke a reflection on the categories themselves. To see an app's purpose reduced to the level of category and the very fabric for understanding this new software is problematic, and it is for these very reasons we call attention to them in this volume.

We open the collection with a section titled **Welcome to the App Store,** which includes four chapters that broadly sketch the app ecosystem, including discussions of the role that app stores and other infrastructural intermediaries play in shaping access to apps and app marketplaces. This section begins with App Annie, an app about apps, and a chapter that reflects on how the app industry measures itself and the role that "ratings" companies play in creating a coherent marketplace discourse. Tencent MyApp, both an app and an app store, demonstrates the diversity and popularity of app culture across China and the substantially different market conditions shaping app distribution in Asia. Also a focus of this opening section are banned apps like Exodus International and a consideration of the power that app store owners (e.g., Apple, Google, Tencent) exert over the content that users discover and download. Finally, we include what we see as a field-defining discussion of a format-specific methodology for approaching mundane software—the walk-through method. Using the case of the discreet encounter app Ashely Madison and building on recent work by Ben Light, Jean Burgess, and Stefanie Duguay (2018), the chapter helps establish a structured yet flexible qualitative mode of studying apps. These chapters are intended to set the economic, theoretical, and methodological frameworks for the rest of the collection.

The next seven sections follow the layout of an app store more closely, beginning with **Productivity/Utilities**, which explores the ideologies behind the drive to be more productive and efficient workers and citizens, ideologies that are often designed into the app from conception. Featured in this section are apps that help users "get things done," often with little thoughtfulness about what the imperatives of efficiency do to notions of labor, time, and care (TaskRabbit) or productivity and shame (Carrot). Meanwhile, an app like See Send, which encourages intuitive reporting of suspicious activity, reduces the racially fraught logics of surveillance into simple snap and swipe gestures. Rounding out this section is the novelty, single-purpose app Is It Tuesday?, an app that is as entertaining as it is insightful in making plain the digital solutionism evident in mundane software. All the chapters in this section look at how the productive and utilitarian aspects of apps are presented as convenient and effective solutions to complex social and cultural issues.

The **Health/Fitness** section considers two popular tracking apps and the historical precedents that contribute to a particular vision of healthiness and happiness that apps subsequently take up as they position themselves as a necessary tool for daily attention to self-betterment. LoseIt and FitBit represent appified continuations of a longer history of debate about how to be healthy and what it means to track and quantify the self through technology in order to achieve socially acceptable, and often very gendered, health and fitness goals. **Lifestyle/Relationships** draws together chapters on intimacy and safety, paying careful attention to the role interfaces and software play in mediating both. Using dating apps (Tinder), personal safety and reporting apps (Companion, Hollaback!), and sex offender tracker apps (Sex Offender Tracker) as cases, the chapters in this section pinpoint in critically excellent yet alarming ways how both the marketing and functionality of apps shape how societal problems are valued (e.g., women's safety as a "lifestyle" issue) and which social problems deserve a technologically solutionist intervention, and of what kind.

The **Social Networking/Communication** category focuses on a variety of messaging and social media applications that appify ordinary, ritual communication practices and in doing so raise questions of community, civility, and commodification in digital technologies. This section looks at the tensions between Yik Yak's initial success with anonymous communication and its business imperatives to authenticate and profile users, Snapchat's emphasis on ephemeral and phatic communication, and WhatsApp's ability to provide diasporic and transnational communities crucial links between private/domestic spaces and the broader publics with which they connect. All of these chapters consider how social apps mediate intimate and affective communication. The chapters on WeChat and Foursquare also point to the extent to which our everyday practices and movements have been colonized by app providers looking to capitalize on the data these mundane activities generate as they attempt to insinuate themselves into all aspects of our daily routines.

News/Entertainment reaches across a number of media aggregation apps to consider how apps influence the distribution of news, images, videos, and political discourse. From trend-spotting apps like This.Reader to the free TV streaming app TubiTV, this section explores the promises content providers see in the appified distribution of media while also considering the economic and cultural reasons why those promises are rarely realized. This section also takes up two apps that contend with liveness and real-time experience: Periscope extends and reconfigures the long-held "scopic regime" of traditional broadcasting into a live-streaming video app, while the gamified political campaign app Hillary for America attempts to draw audiences into campaign participation by making something like the experience of canvassing feel both ordinary and exceptional. Overall, this section shows how apps act as potentially novel venues for message dissemination but are also channels complicated by app infrastructures, expected uses, and longer histories of communication.

The **Music/Sound** category listens in on a variety of sound-based apps to consider how algorithms are shaping the production and circulation of music and music industries and the very timbre and tone of the soundscapes around us. Apps like Shazam and iMaschine 2 remind us of the agency of technologies in shaping our sonic environments: Shazam champions discovery for on-the-go music lovers but also points users to a limited subset of mainstream genres and artists, while iMaschine 2 provides novice and expert musicians with user-friendly tools for creating songs and sounds but may also reign in creativity in the process. This narrowing of the sound of music, or rather the music of sound, is even more explicit in the case of Here: Active Listening, an app that literally reconfigures and personalizes the user's sonic environment. We conclude the volume with the ever-popular **Casual/Games** category, which collects a series of games and leisure-time apps that explore the affects and ethics of fun, sport, downtime, and aspirational make-believe as it becomes appified. Analyzing a keyboard app connected to a popular reality TV competition (RuPaul's Drag Race Keyboard), a fantasy sports gambling app (DraftKings), a game that invites users to participate in the social entrepreneurship of celebrity labor (Kendall & Kylie), and an app that gamifies the caregiving of digital pets (Neko Atsume), the authors in this section address the blurry boundaries between leisure, work, and self-branding; fantasy and legality; and personal communication, play, and corporate marketing.

We hope these distinct categories do not isolate readers from themes that carry across the volume—such as the increasing role of affect and interface design in shaping relationships with software; or the ways neoliberal ideologies of more efficient, self-sufficient, and productive users infiltrate more than just productivity apps; or how the infrastructural and embedded nature of many of these apps are sites of corporate or state power and user resistance. The mundane software analyzed here calls our attention to the modular, microfunctional and highly personal nature of contemporary software and the influence this has on processes of identification and communication. It also reminds us how quickly individual apps fluctuate and change over time, not just in terms of basic features and design but in purpose and meaning as well. The chapters on Yik Yak, Tinder, Foursquare, TubiTV, and others, for example, demonstrate how these apps have changed mission, market position, or otherwise tried to adjust their reputations over the course of their lifecycle. This flexible, iterative, and responsive design is an (often futile) attempt to meet the needs of shifting economies, audiences, and social conventions, with each version update acting as a response to the unpredictable and messy nature of cultural practices like dating, anonymity, public spheres, and everyday movement through space, place, and various taste cultures. The app's relationship to these growing pains reminds us that mundane software is significantly more variable than most cultural goods media scholars are used to addressing.

Ultimately, we want readers to recognize the potential discomfort in categorization, and we hope this discomfort is productive. Labeling a rape activism app as "Lifestyle" or an antiterrorism app as a "Utility" is as concerning as it is problematic and is directly related to the troublesome commercial and cultural challenge of trying to fit a broad collection of apps and their nuanced, specific functionalities into inflexible or vague categories. Lists both recognize connections between the listed objects and generate them. It is our hope that the organization of this volume does precisely that and, in doing so, fosters new dialogues, theories, and critiques of the role mundane software plays and should play in our everyday lives. As more and more interaction with software and networked communication shifts to mobile devices, and as apps begin to proliferate and populate a number of other devices previously absent of software, apps represent not just a fashionable tech trend but a new way of accessing information, experiencing media, mediating commerce, and understanding the self and others. If it's true that, as Intel intimated, the next era of the computer age will be characterized by "computing not computers" (qtd. in Farman 2012, 1), then *Appified* argues that mundane software will play a crucial role in these new forms of computing and, more importantly, in the management and coordination of everyday life.

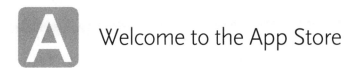 Welcome to the App Store

App Annie
App Industry Rankings and Infrastructure
PATRICK VONDERAU

App: App Annie
Developers: Bjørn Stabell and Bertrand Schmitt (Exoweb/
    App Annie Limited)
Release Date: July 2013 (Current Version: 1.43.1)
Category: Business
Price: Free
Platforms: iOS, Android
Tags: analytics, mobile, big data
Tagline: "Build a Better App Business"
Related Apps: Flurry, Localytics, SensorTower, MoPapp,
    Fiksu

As of 2018 over two billion smartphones were in use globally. The average smartphone user has more than 80 apps on their phone. App store revenue has more than doubled over the past few years to $86 billion (App Annie 2018). In order to navigate this lucrative but overcrowded marketplace, app publishers need to combine the monitoring of their own apps' performance with a granular understanding of their competition. Enter App Annie, a San Francisco–based market research firm that offers a sales tracking and trend analytics web service—and an app that sits on top of its complex web-based platform.

Established in 2010, and now widely considered the industry leader, App Annie bundles services and software that help to chart, track, research, and forecast the app economy, providing data related to app stores, in-app engagement, advertising metrics, and marketing in one single web dashboard. The app, in turn, promises instant access to all of these data on mobile devices. Users of the free version may access only a reduced "top charts" version of App Annie's market stats that consists of a ranked listing of three hundred apps according to country, device, store, operation system, or date. Developers who subscribe to the service get the full app analytics and intelligence. Overall, the usefulness of both the app and the web service consists in providing actionable intelligence—knowledge that is operational for developers, because it helps to position them vis-à-vis the competition and the marketplace, but also for trade journalists and investors. *TechCrunch, Venture Beat, Wall Street Journal,* and other publications use App Annie data by default, echoing its view of a market that is transparent, has very low barriers of entry, and is consequently growing rapidly on a global scale.

This is, in short, how App Annie presents itself and the market and how it aims to "empower the people who change the world through apps" (App Annie 2016). For you and me, the app may hold little interest apart from its listing of top apps and the memorable brand mascot of "Annie" herself, a geeky cartoon-style math teacher who evokes gaming rather than market analytics. There is, however, a bit more to the story of Annie and her app.

First, its history. Who developed it, where, and why? How might a media-historiographical approach help us move beyond the widespread presentist focus on digital media and apps in particular? Second, there are the conceptual concerns about the market that App Annie aims to track. Here, we may follow economic sociology and a strand of research called the anthropology of markets in arguing that App Annie does more than just simply track market activities (Callon 1998; Aspers and Dodd 2015). Its charts, lists, and rankings are fundamental in shaping the market for apps. How does App Annie contribute to describing, organizing, and maintaining the app economy? What should we make of the fact that App Annie both charts the app market and is an app on said market? Ultimately, I argue that App Annie's service does not so much consist in providing a program or access to market data but, instead, in defining a frame for what can be known about markets, or what N. Anand and Richard A. Peterson (2000) call a "market information regime."

The Hidden History of App Annie

Apps do not exist in isolation. Their emergence is inextricably linked to the evolution of the internet from an open web to a controlled distribution channel for proprietary data. App Annie's history begins with a device that prepared for such a transition to centralized market control: the iPhone. Introduced in June 2007, and followed by the launch of the Apple App Store in June 2008, the iPhone quickly became a mass market gaming platform, turning Apple into a gatekeeper over app development (see Gillespie, this volume), similar to the power held by traditional telecommunication networks or film studios, while simultaneously enabling all kinds of software developers, including hackers and self-taught "digital natives," to become entrepreneurs (Consalvo 2011; Flückiger 2011).

In 2000 Norwegian tech entrepreneur Bjørn Stabell cofounded Exoweb, a software company doing outsourcing work for mobile phone makers, in Beijing. Exoweb grew to around ninety people but given its "engineering-centric culture steeped in open source and agile methodologies," it turned out to be neither scalable nor profitable (Bjørn Stabell, email to author, November 13, 2016). In an attempt to change from a services to a product company, a "labs" division was established in 2008. After initial experiments with utility apps and social games for the Chinese market, the division in 2009 developed a series of iPhone games released under the brand Happylatte. Benefiting from the emerging gaming mass market and a new ease of game development facilitated by Apple's SDK and Interface Builder tools, the division had success with *High Noon*, a multiplayer real-time shooter game.

In order to better track their iPhone games' downloads and rankings, Stabell and his team began experimenting with app store marketing techniques in a separate "labs" division. The result of these experiments initially was a free service, hosted on a website run by Exoweb. The appannie.com domain was registered in September 2009, and on March 2, 2010, App Annie went into open beta. Users could register and give the app permission to gather information regarding their own apps' sales and download numbers from various app stores, sharing this data in return for having App Annie giving them a dashboard of their app sales/marketing analytics. "As developers, we always struggled to keep up on our apps in the app store," the team related in App Annie's first blog post. "iTunes Connect just seemed inadequate, and accessing reviews and rankings through the iTunes browser was too slow and limited" (App Annie 2010).

This early history of App Annie is kept out of the company's current marketing materials, and even the initial "gaming-centric" brand design has recently been replaced by a visual aesthetic that is much less recognizable (David 2013; Lopez 2016; Wee 2013). The reason for this is that when App Annie was spun out of Exoweb and refounded by CEO Bertrand Schmitt as a separate company in September 2010, Schmitt built "a new team with its own culture" (Bjørn Stabell, email to author, July 10, 2017). While not an open source company, Exoweb Labs originally had been conceived as a venture builder or startup studio with a culture that was "very friendly and supportive towards open source projects," allowing employees to spend 10 percent of their time learning and contributing to these projects. Open source communities such as those connected to Exoweb grew as social movements in response to the proprietary software industry, based on a different understanding of value creation. Exoweb's community-related mode may have allowed for the explorative part of innovation but lacked capacity to derive a scalable product.

Although App Annie's business model never really changed—Exoweb's related "labs" division was founded to commercialize the service—the cultural shift noted above is indicative of a broader tension between openness and closure, open source and proprietary data, exploration and exploitation that marks app development generally. Research on open source entrepreneurship has described the connection between firms and open source communities as a "parasitic relationship"—that is, one "marked by the exploitation of community labor coupled with actions which ultimately constrict community space" (Tech, Ferdinand, and Dopfer 2016, 143). Although this does not apply to Exoweb or App Annie, turning the product into a venture capital–funded global business included forms of closure: the filing of patents in 2012, a lawsuit against competitor Apptopia in 2014 to protect App Annie's name and brand, and a pricing model that currently charges $15,000 for "mid-market" and $30,000 for "enterprise" subscribers per year (anonymous source, App Annie, September 2016). Today App Annie relies on proprietary data, using an extended network of connected developers and connecting to the back end of the app stores to automatically capture, aggregate, and visualize real-time, raw data—it works with Apple and Google directly.

The app is merely an extension of these long-term dynamics rather than a new or separate development. Although an HTML 5–based solution for iPhone and iPad was already available back in May 2010, an initial version of the app was not launched until 2013. Since then, countless versions have been released, twenty-five of them in 2015 alone. The app is thus the most unstable or adaptable product the App Annie company has on offer. Initially created as a device to freely share market data, it now allows paying subscribers to access an online database while retaining a marketing function targeting all of those developers not yet paying for the service. The app thus somewhat complicates the notion of apps as "software commodities," because in App Annie's case, neither the application program nor access to raw data is what is being sold.

Here we may distinguish between apps as *access points* and apps as *end points*. While some apps offer unique stand-alone functionalities as an end in themselves (think of Doodle Jump or Snapchat), most of them relate to already existing services (such as Facebook or Tumblr), adding or enhancing the functionalities of such services or growing their user base. App Annie has to be seen in the context of a broader shift among businesses toward cloud computing and "software as a service" (SaaS), a licensing and subscription model that promises to lower costs and save time. In this model, hardware is increasingly distributed across different data centers, and software dissolves into "cascades of services that organize access to data and its processing" (Kaldrack and Leeker 2015, 10). As a business-to-business application, App Annie sells precisely such services while also growing its user base.

The Sweet Story of Growth

App Annie pictures the market in the way a cartoonist pictures a character: by extracting selected details and simplifying them in pleasing shapes so that we may be effortlessly engaged by what we are looking at. It has arguably never been easier to access information on a market than by tapping, swiping, and pinching it. The ease of access to data heightens the sense of their immediacy: within seconds, we move between Austria, Costa Rica, and Japan, comparing large aggregates of numbers to the idiosyncrasies of reviewer opinions, scrolling through an app's weekly losses or gains as if they were status updates while still aware of their economic consequences, like a broker contemplating real-time stock indexes. In creating such ease of access to market information, the app shapes our view of this market and of our own mastery in performing in markets. Ranked lists, colored tabulations, and imaginative infographics have become tools in day-to-day business operations and topics of conversation among journalists, investors, and app users. The market they show is transparent, well organized, stable, and scaling globally at a rate faster than the square root of time. As the sales pitch goes, "change is very important" (App Annie TV 2016), at least as long as something remains constant: the growth of the app business itself.

Similar to the relation a cartoon holds to reality, such and other depictions of markets do not represent markets neutrally (cf. Meehan 1986; Bermejo 2009; Napoli 2012). For developers, the app store model is not as positive. In addition to the friction between open source and corporate entrepreneurship, several issues animate the developer community. These issues include the very constraints and conditions imposed by the Apple iOS ecosystem, a system that has shifted the locus of control from production to marketing and distribution (Gillespie, this volume). This includes the problem of discoverability or of being seen at all—if the market constantly grows, so does the competition, and the proportion of apps showcased on an app store's front page decreases constantly. While promoted as a way to monitor your competition (and offering ASO, or app store optimization), App Annie never promised to do away with the winner-takes-all model that constitutes the political economy of the information economy. There's also the issue that apps are frequently copied or cloned, as App Annie's own lawsuit against Apptopia demonstrates, prompted by the latter's improper attempt to market itself as the better solution.

Finally, there's the problem of remuneration. Although top-grossing apps such as Clash of Clans have been estimated to earn between $500,000 and $1.5 million daily (after subtracting Apple's standard 30 percent share), the average monthly salary of an iOS developer is only at $5,200 (Goldsmith 2014, 175). Given that the large majority of apps are never downloaded, today's developer industry is marked by precarity rather than unmitigated success—and benefiting from low barriers of entry are first and foremost those controlling distribution (Nieborg 2016).

In painting the app economy as a frictionless, low barrier of entry endeavor premised on constant growth, App Annie provides exactly the sort of intelligence expected by its subscribers. In relation to the "predictable format" of this and similar market research services, Anand and Peterson have coined the notion of "market information regime" in order to describe how regularly updated information about market activity may become a mechanism by which stakeholders can "make sense of their actions and those of consumers, rivals, and suppliers that make up the field" (2000, 271). App Annie operates on providing this "look and feel" of market information access, whether or not it will help developers to break through. The advent of such market information regimes is probably unavoidable, given that markets are abstract entities that need to be localized—temporal and spatial distances between exchanges have to be bridged, and market actors have to be engaged in exchange itself.

In a similar vein, Hans Kjellberg and Claes-Fredrik Helgesson (2007) speak of "representational practices" as a key element in constituting markets. Representational practices include all forms of activity that contribute depicting markets and how they work, such as the production of statistics, and are as fundamental in shaping markets as the exchange practices themselves. Through representations, economic exchange is turned into pictures, diagrams, and texts, based on prevailing ideas about what and how to measure. Importantly, such representational activities do not necessarily use the most scientifically robust models, as representations always are put to a particular use—for instance, to motivate trading or establish preferable directions for some group of stakeholders.

App Annie has successfully established a frame around the app market, in line with Apple's and Google's attempts to formalize and control software development. "Framing puts the outside world in brackets" (Callon 1998, 15), drawing boundaries between market actors, on the one hand, and the rest of the world, on the other. Such bracketing is, again, unavoidable for those aiming to develop a market, because products need to be qualified, supply and demand need to be constituted, and transactions have to be organized. App Annie's representations of the market are necessary to make apps calculable. Their success lies in not even being recognized as a market-making practice: they are just taken for granted or, in wording preferred by industry stakeholders, "trusted." Becoming the default solution for bracketing market exchange among key actors such as developers, tech journalists, and investors, App Annie defines a frame for what can be known. It sets the epistemic boundaries for market actors and gives them a direction to pursue.

The most obvious instance of market representation created by the App Annie company is that of scale. Its own collections of ranked lists and statistics tells a story of eternal growth, widely echoed among other constituencies. Scale is constituted in basic data used to chart the market—"two billion smartphones globally, 40,000 apps monthly, 41 billion USD annually"—but also in forecasts ("by 2020, 101 billion USD"), standard tools (top charts), metrics (daily use), or to market its own business ("used by 94 percent of the Top 100 publishers"). Thus, despite the volatility in the day-to-day operations of the market, what remains is a generic narrative of a constantly scaling market for apps. It is always the same story, retold with new detail and variation, and ultimately simplified by the usability and interaction design of App Annie. Here, we have to remind ourselves again that scale is neither just a neutral frame for viewing the world, nor that industries scale naturally. As Anna L. Tsing observes, "In speculative enterprises, profit must be imagined before it can be extracted; the possibility of economic performance must be conjured like a spirit to draw an audience of potential investors" (2005, 57). Scale must be brought into being, it is claimed and contested, requiring considerable "data work" (Gregg 2015b). In their aestheticized, cleaned-up, and comprehensive presentation of data, apps are the perfect tool to perform such work.

Not coincidentally, the narrative of scale is inextricably linked to App Annie's own fortune as a business and an app. Growth, of course, is what every start-up has to demonstrate to potential investors, and linking the growth of the app economy to the company measuring that growth has proven to be a highly lucrative investment strategy for App Annie so far. As tech journalist Liz Gannes put it, "Whenever I ask venture capitalists how they find interesting new apps and start-ups, they mention App Annie, so it makes total sense that a few of them would get together to put money into their favorite tool" (2012). App Annie is thus also an actor in a different economy—the financial market. Since July 2011 the company has received six rounds of venture capital funding from investors such as IDG Capital Partners, Greycroft Partners, Sequoia Capital, and Institutional Venture partners, amounting to roughly $157 million, and has made major acquisitions, including competitor Distimo (2014), Mobidia Technology (2015), and AppScotch (2016).

How does the considerable market power held by App Annie relate to the gender and geek-chic of the company's name and brand mascot? Originally designed as a virtual "company spokesperson" to attract male game developers with her warm appearance, the app's "Annie" today stands in a stark contrast to the financialized media business operation behind her. As an aestheticized form of software, the app and its icon create an aura of soft transparency around the project of scale while simultaneously masking App Annie's own hard financial interests. These interests go beyond what mainstream economics describes as a "multi-sided" market (Gawer 2014; Tiwana 2014), because apps and finance do not simply constitute two sides of the same exchange; their products, circulation, value, usage, and form of commodification differ markedly. At the same time, both are obviously also tightly connected. App Annie thus appears as protagonist not so much in a story of scaling but in a story of *stacking* markets, where trading sites are stacked into or on top of each other in often obscure ways (Vonderau 2017).

Conclusion

Tracing App Annie from its gaming-inspired brand and mascot to representational practices that serve a crucial role in constituting the app market itself, there remain two key issues that demand further study. The first concerns a new model for outsourcing software development and its effects on developers, as well as the impact of App Annie's appified service that promises to help developers achieve the unrealistic expectations set by the rhetoric of apps culture. The second issue relates to the politics of market information regimes. While every economy, including the app economy, necessitates epistemic frames that bracket relevant economic actors and operational knowledge from what does not have to be included in exchange, market measurement here also motivates and directs trading. App Annie thus complicates the notion of software as commodity. To grasp this complexity, I have suggested an approach that distinguishes between apps as end points and access points. I have also suggested that we should relate the latter to devices and services rather than just platforms and databases. Apps are always embedded in larger industrial and historical contexts, as App Annie demonstrates—without the iPhone or the shift toward cloud computing and software as a service, the company simply would not exist. As a research strategy, to "follow the service" thus prompts us to move app analyses beyond a focus on social media platforms, acknowledging that "the social" is neither solely in the present nor in the technology.

Ashley Madison
An Introduction to the Walkthrough Method
BEN LIGHT

App: Ashley Madison
Developer: Ruby Life Inc.
Category: Social Networking
Release Date: 2001 (website); 2009 (mobile app) (Current
    Version: 4.4.7)
Price: Free
Platforms: iOS, Android
Tags: hookup app, dating app, cheating, adultery
Tagline: "Find Your Moment" (after July 2016). "Life is Short.
    Have an Affair" (before July 2016).
Related Apps: 3Somer, HaveAFling, HAA, SinMatch

As the introduction to this volume indicates, apps are a challenge to study because, among other reasons, they are usually made technically inaccessible to researchers by developers. This chapter is methodological in nature and illustrates the use of a digital method, the *walkthrough method*, as one way of studying apps in response to the challenges of researcher access to this domain. A fuller exposition of this method can be found in the work of Ben Light, Jean Burgess, and Stefanie Duguay (2018). To illustrate how the method can be used, I use the Ashley Madison app, a dating/hookup app aimed at those seeking to make what the app's promotional materials describe as a "discreet connection" outside the confines of a monogamous relationship.

Ashley Madison aligns with this volume's definition of "mundane software" in two ways. First, the service goes beyond the desktop and out into the world via a mobile device. Second, it seeks to appify and enable the everyday practices of adultery. How we experience the mundane is relative. Not everyone engages in extra-partnership affairs (sanctioned or not by other/s in the relationship), and not everyone uses hookup and dating apps. This has important methodological implications: as app use becomes mundane and recedes into the background of our consciousness and everyday infrastructures, we need to find ways to unpack their taken-for-granted and highly opaque nature. Using the case of Ashley Madison, I briefly apply the walkthrough method to show how it can be used to demonstrate how apps might generate income, attempt to regulate and structure a range of activities and aspects of ourselves, and how we might integrate them into our lives on our own terms, or not.

Walkthroughs are already a popular and important device for users and their consumption practices across a range of digital and non-digital goods (e.g., Grimes 2015, Lee and Hoffman 2015). In such contexts, details are revealed about the artefact in question, creating a narrative of use. Walkthroughs hold the potential to make clear what otherwise might be a taken-for-granted good or service. Beyond use as an analytical tool, walkthroughs usually privilege a mode of use to suit an agenda. For example, a walkthrough of a digital game may show you the best way to win or the best way to cheat, while one concerning a Nutribullet is likely to emphasize its ability to make health drinks rather than chocolate shakes. From a methodological perspective, early walkthrough techniques, originating in software engineering, sought to improve the quality of code and user experience (Fagan 1976). Later, human-computer interaction built on software engineering approaches to create "user walkthroughs" was aimed at creating more usable and useful digital products (e.g., Lewis et al. 1990).

The walkthrough method I employ offers a way to perform a critical reading of apps. It is an empirical method informed by science and technology studies and cultural studies with a focus on interactivity, materiality, text, the human, and the nonhuman. The walkthrough intends to help researchers uncover how an app may present itself to users and how they might engage with it. The method offers a framework that guides the researcher in the targeted analysis of an app by examining elements of the following:

- **the environment of expected use**—including the developer's vision, the app's operating model, and its governance structures;
- **the technical walkthrough**—exploring registration; everyday use; and suspension, closure, and leaving mechanisms through attention to interfaces, functions, content, tone, and symbolic representation;
- and, **assessing evidence of unexpected practices associated with the app**—through the analysis of, for example, associated advice blogs, hacking, resistance, and third-party additions or manipulations of the app.

Researchers can use the walkthrough method in conjunction with appropriate theory in the same way that building theories from other forms of qualitative data might happen. The walkthrough may be performed by a researcher or group of researchers examining the app themselves, or it may involve interviews with users and other stakeholders. The method can reveal imagined and expected modes of use as defined by the app designers as well as those based on user feedback. In this illustrative example of the walkthrough, I will point to various bodies of theory to show how this might shape engagement with the method rather than stick with one or two theoretical angles, as might be the case in a full-length journal article application of the methodology. This chapter is less concerned with any theoretical implications the Ashley Madison app poses and more with showing the utility and versatility of the walkthrough method for researchers.[1]

While I present the walkthrough in a particular order above, there is no requirement for it to proceed in this way. It is also the case that while focusing on one element of the walkthrough, links may be made with others, and a researcher may, for example, shift their attention between examining the environment of expected use and unexpected uses. In the case of my work on Ashley Madison, I became aware of the app and began my study because of unexpected use, so that is where I begin here.

Assessing Evidence of Unexpected Practices Associated with the App

Ashley Madison provides a dating and hookup desktop website and mobile app for people to engage in discreet encounters. In July and August 2015, Ashley Madison dominated the news and online technology press because it had been hacked and close to 37 million account records and other company documents were made public. In the wake of the hack, users filed lawsuits against Avid Life Media, the company behind Ashley Madison (Hackett 2016), partnerships and marriages collapsed, and the fallout from the exposure of identities and data was linked to several suicides (Lamont 2016).

The rationale for the hack, in the words of the Impact Team—the hacker group responsible—was to stop the exploitation of future users: "We did it to stop the next 60 million [users from being exploited]. Avid Life Media is like a drug dealer abusing addicts" (Cox 2015). This unexpected practice, the hack, became the subject of significant media coverage (see especially Newitz 2015). Anna L. Newitz's analyses of the data dumps revealed a significant presence of bots operating within the site and pointed to the deceptive practices of Ashley Madison in terms of the promise of meeting a human being. This finding also had ethical implications in terms of usage of the site—in short, were users being unfaithful when they chatted with a bot? This finding sparked my interest in the app and led me to consider how bots might be present within the app and its environment; it is a good example of how to identify unexpected practices that may become apparent.

The next section documents some of the functionalities of the app. However, it is worth a slight detour first to examine a further form of unexpected practice. Associated with the app was the development of businesses that worked with app data in the wake of the hack. One such example is Trustify, a network of private investigators in the United States that offered two services.[2] First, an Ashley Madison user can pay to find out if their data has been released by the hack. Second, anyone else can check up on someone they think may have used the site (despite the fact that Ashley Madison did not have a validation process in place, so no links could be made between users and their emails). As Trustify's advertising promises:

> If you suspect that you are being cheated-on by your spouse, or that your partner is on Ashley Madison, then you are not alone. The Ashley Madison data breach includes over 32 million users, and millions of Americans are cheated-on by their significant others on a daily basis. Many of these victims of infidelity cite that the

worst part is not knowing if they are being cheated-on. They just want to know the truth. Trustify's Network of Licensed Private Investigators can help you get the truth today. (Trustify 2016)

These two forms of unexpected practices and secondary uses for app data add to the growing library of forms of appropriation of dating and hookup apps. For example, blogs such as *Douchebags of Grindr* seek to publicly shame and call out users in a variety of ways. On Tinder a range of businesses have created profiles to connect with potential customers. My own institution used Tinder to recruit undergraduate students in 2016. Such diversity of use calls for additional theories. In the example of the blog spun off from Grindr experiences, researchers might draw on theories of race; for my institution's use of Tinder, we might instead turn to marketing theory. In the case of Ashley Madison, literatures related to hacker ethics and business ethics could be seen as useful lenses to undertake a further walkthrough-based analysis with a focus on a comparison of the positions of the hackers versus Avid Life Media. However, as I progressed, I decided to focus on the role of the bots as engaging devices purely because they appeared to be a significant actor within the network of associations the hack had revealed (Latour 2005). Examining unexpected uses is helpful as it shows how different individuals and groups may seek to resist, adapt and make use of the app as originally envisaged by the designers.

The Technical Walkthrough

My initial walkthrough of Ashley Madison, conducted in August 2015, was based on the desktop version of the app and was conducted using a Safari browser. For this chapter, I examined the mobile app version in September 2016. I began the walkthrough at the iTunes download page for the app, as I was installing it on an iPhone 6s Plus. The app was categorized as Social Networking and is described as the official Ashley Madison app. Screen shots present the app as a place "Where millions go to find something different, something discreet." This tagline is quite different compared to that on the desktop version I analyzed previously—"the leading married dating service for discreet encounters"—indicating a shift in marketing in the wake of the hack. At the time of the hack, the service reported having over 43.5 million members. In November 2017, the opening page of the desktop site suggested it had over 55.56 million. The hack appears to have been good for business.

The iTunes download page gives a brief description of the functions of the app, indicates it is free to download, and notes that there are optional in-app purchases ranging from $19.99 to $249.99. In-app purchases give users additional functionality such as messaging, gifts, and chat time. The screen shots show the revised black and hot pink branding of the site, much like what I reviewed on the desktop. The images within the screen shots are mainly white women with fully visible faces and one picture of a white man. All of the models in the pictures are what might be considered normatively attractive. One screen shot illustrates the chat function and suggests a woman has sent the first message to a man. This heavy suggestion of female users aligns with data from the hack—particularly that which evidences the role of bots in creating a sense of the significant presence of available straight women for straight men. I could not find any mention of bots or fake profiles on this page; rather the emphasis appears to be very much on meeting "real people."

I installed the app and, once it was opened, was immediately asked if the app could use the location services of my device and send me notifications. At this stage, even though no gender or sexuality data had been entered, a shadowy full-screen image of a smiling woman was presented as a taste of what was to come (I identify as a cisgender gay white man). I was then routed to the registration process, where I selected a user name and password and chose from a compulsory menu of relationship and gender statuses—these included Attached or Single, Female or Male options as well as simply Female or Male without indicating one's relationship status. In each case these options were combined with the term "seeking." For example, Attached Male seeking Single Male was a possible combination. No options for trans identification or more than a one-on-one relationship were present. Discreet encounters via Ashley Madison appear to be relatively normatively configured, leaving no room for gender fluidity or group sexual play, for instance (unless you count the bots multitasking in the background!).

The next point in the registration process asks for typical information associated with dating and hookup apps, such as height, weight, body type, date of birth, and ethnicity. There is also a section titled "limits" that allows users to specify "something short term," "something long term," "cyber affair/erotic chat," "whatever excites me," "anything goes," and "undecided." While this set of options appears to offer some scope for sexual play of different forms, it is moderated by the normativity of the prior section. Through this part of the registration process, we can see how theory regarding gender and sexuality may be helpful for understanding how the app attempts to regulate these in a hookup context. Crucially, this happens not just through registration; it is also signaled in the advertising and marketing that aims to influence the needs and desires of users of the app before they even download it.

The next part of the registration process asks, again, if I would like the app to use my location. This choice is optional, though the "skip" link is, as is often the case in apps, much smaller than the large blue "use my location" button. The user is then invited to upload photographs of themselves with a textual communication allaying any fears of being found out while using the app: "We respect your need for discretion so we've created some tools to keep your identity a secret." These tools are implied to be cropping facilities and the ability to cover your face with an eye mask. When I decided to try not adding a photo, I received a further message on the next screen: "Are you sure you don't want a photo? Men are 37 times more likely to contact you if you have a photo." At this point my preference for men had been recognized and the importance of including a photograph is asserted. I decided to upload a picture to test the implied use of masks as a privacy aid. This led me to a screen where I could upload my photo and obscure my identity by a variety of means (see fig. 1). Here we see how further theory regarding privacy could be used to explore morally charged hookup cultures and one might go back through the walkthrough to reexamine, for example, requests for location and how the tensions of revealing personal information, such as the details of a face, are problematized when combined with other sensitive data.

Once in the app, I browsed further and identified mechanisms for buying credits via the app store to allow me extended interaction possibilities such as sending messages and gifts. There were the usual features to shape search preferences that one might find in many dating and hookup apps, such as height, age, weight, location, and ethnicity. There was also a function to search in relation to the "limits" option from the app—organizing possible connections based on the relationship parameters users specified. I did not explore these functions any further at this point due to limitations of space and focus, but one might expect to find further avenues of investigation. In terms of leaving the app, the only option available is to log out and delete the app. However, it is not possible to fully delete an account through the app; this must be done via the desktop version. Documenting strategies of departure is another useful step in the walkthrough that lends additional clarity to app developers' intentions for user retention and loyalty—what does deactivation look like, how easy is it to leave the app, and is a delete a permanent delete? What digital footprint does a user leave behind?

Interestingly, at no point in the registration process did the app mention

"Obscuring" Identity
While Onboarding

bots or the sign-up terms and conditions, although the latter are accessible from within the app. As I researched the terms and conditions of the site and app, though, bots were mentioned. The technical walkthrough of the app, with its focus on registration, expected everyday use, and closure is a helpful part of app analysis, because it allows researchers to consider how a user may be configured by the app at different stages; the data needed for the app to operate; how default, mandatory, and optional settings work; and notions of the associated tone that pervades the app.

The Environment of Expected Use

Users expect Ashley Madison to provide "a discreet connection." Beyond that, the site operates like many dating apps: users create a profile, search for connections, and communicate with them in a variety of ways. Free user accounts can be created, allowing users to receive winks, send and receive photos, add members to a favorites list, reply to full members, and perform searches. Payment to the site is required to send a custom mail message, initiate a chat session, send a priority message, or send virtual gifts.

Ashley Madison attempts to delegate the morality of engaging in an extra-relationship affair to the user, positioning itself as an intermediary. As I have argued elsewhere (Light 2016b), this resonates with Kylie Jarrett's (2008) commentary on Web 2.0 producers and their strategic denial of authority. But Ashley Madison further delegates the enabling of extra-partnership connections by introducing bots that encourage users to engage. This engagement takes the form of bots inhabiting fake profiles and attempting to make connections with human users by sending them messages or chatting with them (see Light 2016a for an expanded analysis of bot chat). That said, the delegation only goes so far as the app also enables other, less automated forms of connection making. For example, the Traveling Man/Woman functionality of the app, which shows others users when you are traveling in their area on business, encourages users to "pursue a little something on the side." Ashley Madison and its bots work together to provide distance from extra-partnership connections and to enable it. This sets out a very particular vision for the app in comparison to other dating and hookup apps such as Tinder, with its leanings toward authenticity and openness (Duguay 2017), and Squirt, with its focus on enabling public sex (Light 2016a).

I knew bots were present in Ashley Madison from Newitz's analysis of the hacked data, yet they were not explicitly mentioned in the marketing materials, during registration, or presented in everyday uses of the site—until I was approached by them (which I will discuss shortly). I decided to investigate if bots were mentioned in the terms and conditions of the site, as this is where we have come to expect such a disclaimer to be deployed. Data from the terms and conditions revealed the characteristics and ethics of the encounters involving the bots of Ashley Madison and the profiles they work with. In my earlier work (Light 2016a) I demonstrate how the bots are characterized as enhancing the Ashley Madison community (even though they are not named as bots but, rather, "profiles"). For example, the January 2016 terms and conditions stated,

"In order to allow persons who are Guests on our Site to experience the type of communications they can expect as Members, we may create profiles that can interact with them" (see Light 2016b, table 2 for more extracts).

This notion of bots being helpful to the app also carries through to their role in the collection of data about users and the monitoring of user communications to ensure compliance with service terms. As profiles and possible connections, bots are entertainers inscribed with a language of sexually charged playfulness and the ability to send winks, private keys (to additional content about the character associated with the profile), or Ashley Gifts. These bots are central to the in-app economy of Ashley Madison. They seek to have users pay for additional interaction possibilities by contacting them and, indirectly, generate a sense of liveness within the space, because users may be contacted by multiple bot-inhabited fake profiles, encouraging further spending in the belief that there will be someone out there to meet (Light 2016b). In my second walkthrough, the terms of service devoted much less space to clarifying the presence of automation. These are the terms, in totality, updated July 2016:

> Our Site and our Service gives users the opportunity to explore and improve their abilities to interact with others on the Site. However, there is no guarantee you will find a date or a communication partner on our Site or using our Service. Our Site and our Service also is geared to provide you with amusement and entertainment. You agree that some of the features of our Site and our Service are intended to provide entertainment to our users.

While my initial walkthrough of the website did not suggest the presence of bots, nor did my recent app walkthrough, the bots did emerge over time. When I first created my profile on the desktop version of the site, I did so as a gay man looking for other men in the north of the UK. I instantly set my account to private and wrote into the free text on my profile page that I was a researcher just "here to see how the site works." I also noted that I did not have institutional ethical approval to talk with users (hence why I locked down my account). However, over the first three months of opening my account I received seventy-five system messages from female profiles, geographically located in many different countries, and despite my profile being hidden from user view. The bots were there but not very clever. Exploring the environment of expected use of an app is of value in determining what the app developer wishes to gain from users, how it expects them to use it, and what is not allowed. This environment also provides indications of the funding and resource models associated with an app as well as the market or population it seeks to engage.

Conclusion

I have illustrated one method by which we might study apps and locate our findings about them in broader theoretical and empirical narratives of the digital. In this example, for instance, my walkthrough has suggested links to ethics, hacking, gender, sexuality, privacy, race, bots, and business. It might also point to theories on algorithms, criminology, data mining, or adultery. My illustration is necessarily partial and incomplete as, indeed, any walkthrough will be, since it is impossible to document everything and quite unhelpful to do so. The walkthrough method is based on the principle of building up an interesting set of associations that tell a very particular story of mediation in much the same way as a good case study might. There is no easy or single answer as to how to start building a story around an app. In the case of Ashley Madison, I was guided by interesting stories in the press regarding bots and the positioning of the site in highly gendered terms. Therefore, my walkthrough reflects this point of entry and my application of the method to more narrowed, specified research inquiries.

However, the walkthrough method does point the researcher to particular forms of, and environments of, mediation. Looking for what is expected, what is unexpected, and the detail of the technical walkthrough will lead a researcher in particular ways in contrast to, for example, interviewing someone about how they use an app. Indeed, one might argue for an engagement with both approaches, to develop multiple accounts—theory- or researcher-driven as well as user experience. There are challenges, though, as the most interesting mediators might not reveal themselves in an obvious fashion. I was drawn to Ashley Madison only because of mass media interest, and then to the bots because of Newitz's reporting. This leaves us with a struggle to find the missing, the hidden, and the non where apps are concerned. This struggle is further compounded by the proprietary and locked-down nature of apps—a more general problematic of the increasing commercial interests in other digital forms such as the web and streaming services. Further, apps cannot be seen to operate necessarily in isolation. As I have shown here, apps may have mobile and desktop versions each with different affordances. While the walkthrough has limitations, as all methods do, it provides researchers with a new way of studying locked-down media and the ability to combine it fruitfully with other methods such as interviews, ethnographies, and other digital methods. Moreover, the walkthrough method, while focusing on apps, offers potential for the study of other sites of digital mediation too.

Notes

1. For more detailed uses of the method, with theoretical applications, see Duguay 2016, 2017, and Light 2016a, 2016b.

2. I thank Dr. Molly Dragiewicz at the Queensland University of Technology for making me aware of Trustify.

Tencent MyApp (*Yingyong Bao*)
Android App Stores and the Appification of Everything
LUZHOU NINA LI

Developer: Tencent
Release Date: December 2013
Category: App Store
Price: Free
Platform: Android
Tags: app marketplaces, digital distribution, migrant labor
Related Apps: Baidu, Qihoo, Wandoujia

Tencent MyApp (*yingyong bao*), a dominant Android app store developed by one of China's largest internet companies, Tencent, has been instrumental in shaping China's recent economic and social life. Tencent's MyApp is itself an app but also an app store that distributes and sells apps. Using this dual role that Tencent plays in a vast and rapidly changing digital landscape, this chapter considers how Chinese app stores and their infrastructures shape communication and culture, specifically with respect to migrant and creative workers in China.

Readers may wonder why it is not Google Play but instead local players that have come to dominate the Android app distribution market in China. So let's begin with a general discussion of the app industry in China to see how Tencent MyApp has risen to domination in the global geography of technology. App distribution in most countries is straightforward: Apple's App Store for iPhones and Google Play for Android devices. In China, however, while Apple's revenue grew to $18.37 billion in the first fiscal quarter of 2016 and accounted for close to a quarter of its global revenues (Matney 2016), Google has gained little value from Chinese Android phone users. This is primarily because of the political wrangle between Google and the Chinese state that eventually led to the company's forced decision to withdraw from the Chinese market in 2010. The Google Play store did not prove popular among China's Android users before 2010 (Millward 2014), and yet its withdrawal created more space for local players building on top of the Android operating system to flourish. When Google left China, there were only 35 million smartphones manufactured in the country each year, but the number soared to 390 million by 2014. By March 2017, 87 percent were Android phones (Kantar Worldpanel 2017). The explosive growth of the smartphone manufacturing industry consequently created a lucrative third-party Android app market where tech giants, manufacturers, and telecom carriers all rushed to compete for market shares.

This diversity, albeit existing only for a short period, shows how the intervention of the Chinese state, which eventually led to Google's withdrawal, produced an indigenous technological space and disrupted the global geography of ICT (information and communication technologies) development, thereby countering the global technological order defined by transnational corporations. Yet the palpable trend toward market consolidation in the past few years, which took place in almost every sector of the Chinese internet, including Android app distribution, has brought about another hegemonic regime domestically. In July 2013 the search leader Baidu acquired one of China's most popular third-party stores, 91 Wireless, for $1.9 billion (Yee and Carsten 2013). On the heels of this most expensive deal in China's internet history, Tencent revamped its app store in December 2013. MyApp, or *yingyong bao*—literally, "treasures of apps" or "application treasures"—incorporated Tencent's seven major PC and mobile app distribution platforms into MyApp (hereafter, Tencent MyApp) (Sun 2014). Monopolization was inevitable. By 2015, Baidu, Tencent, and Qihoo already held 80 percent of the market in app distribution (Mao 2015). The once popular indie stores developed by small firms such as SnapPea (*wandou jia*) and Wooden Ants (*mu mayi*) gradually lost ground and were acquired by larger companies, putting an end to the fragmented and more diversified app distribution system. Alibaba's merging of SnapPea into its mobile business unit in 2016, for example, highlighted the worries hovering over many Chinese start-ups: how can an independent app store developer survive in a market when massive tech oligopolies (Baidu, Alibaba, and Tencent, also known collectively as BAT) are starting to compete in their areas of expertise?

BAT's foray into the app distribution market grew from the recognition that the app store is the first point of entry to the mobile internet where most end users discover, download, and pay for apps, thereby making it a control point or "bottleneck" in the mobile industry (Pon 2015). This recognition, however, came quite late. Tencent's initial mobile deployment strategies, perhaps those of others as well, relied on an assumed logic that the mobile internet is merely an extension of the desktop. By this logic, the app store was only one among Tencent's many other platforms in distributing apps. Other than the app store, Tencent also relied on its mobile properties such as the mobile QQ browser to distribute software, because browsers/search engines had been considered the first point of contact in the desktop era and were used for web-based software distribution. Replicating its desktop structuring, Tencent was hoping to effortlessly secure its mobile dominance by reproducing all of its star desktop products in the mobile context, from QQ (Tencent's popular web-based instant messaging software) to the QQ browser. However, this logic quickly proved unproductive, especially with the rapid and unexpected rise of WeChat, Tencent's messaging app that was entirely native to mobile platforms (see Brunton, this volume). The strategic upgrade of MyApp as a distribution platform, then, took place in this context. The rise of WeChat pushed Tencent to reconsider its mobile strategies and unify its app distribution under a single portal: the Tencent MyApp app store.

App stores are a portal to a vast amount of mobile content and services. Browsers and search engines, even in their mobile forms, can only lead to traditional online content, whereas apps can lead to content and practices that are otherwise unavailable. The strategic importance of app stores to tech companies thus lies not only in the changes these marketplaces represent to previous forms of software distribution but also, and more importantly, in the ways app stores affect the control and access to content and services in the mobile context. The rise of app stores, then, further extends the notion of software as commodity (Morris and Elkins 2015). It reorganizes the relationship between technology platforms and third-party app developers, governing creative labor in particular ways.

Moreover, the access that app stores control bears critical implications for their role in guiding and defining people's everyday mobile phone use. In most cases, they reproduce themselves as a continuation of familiar, long-established consumer spaces in the digital context. My research of young migrant workers' mobile experience in urban China particularly supports this observation. In the following sections, I show how Tencent MyApp, as both an app and a marketplace, has become an important technological site that articulates two of China's most unsettled social groups—migrant and creative laborers—thereby participating in the governance of a particular part of economic and social life in China that is now mobilized and connected not only through apps but through app stores as well.

Android Store and Migrant Workers: An Information Gateway

In 2014, while conducting research in Shanghai on the video-streaming industry in China, I saw numerous migrant workers using mobile phones to watch online video, a reality that was never reflected in my conversations with industry practitioners who viewed highly educated and highly paid young professionals as their target audiences. Out of curiosity, I went to a local restaurant with a sociologist who had worked there for months researching the gendered experience of migrant restaurant workers and waiters. With the permission of the restaurant manager, I spent two weeks observing mobile phone use by workers and talking with them about their experiences. My original focus was on video, but the stories I heard from workers, such as their initial experience with exorbitant bills resulting from streaming video over a cellular data network instead of Wi-Fi, made me realize that the ease with which workers navigated video apps and their seemingly smooth mobile experiences were hard won. Indeed, the shift from internet access via a desktop or traditional cell phone to a smartphone required a steep learning curve.

The restaurant where I conducted my fieldwork is located in one of Shanghai's southern suburbs. Perhaps due to geographical proximity, many of the workers I spoke with came from the same place—Anhui, a province adjacent to Shanghai. The way in which they got their jobs through kinship networks probably also accounts for this concentration. Almost all workers I spoke with were Android phone users except one young male worker who used an iPhone 4. The phone brand workers used ranged from well-known manufacturers such as Huawei and Xiaomi to lesser known ones such as OPPO and Cool-Pad. Price was a primary concern for workers in choosing Android phones instead of iPhones, which appear to be more popular among China's middle-class population. A Lenovo phone user told me, "My phone cost me about RMB 1,400 (about US $200). It is much cheaper than iPhones."[1] Most of the workers were not first-time smartphone owners, and they changed phones frequently. One young male worker was using his fourth smartphone. The mass production in the device manufacturing industry produces phones at low cost and of mediocre quality, which makes a frequent phone change both affordable and necessary.

Workers' working hours were organized around the restaurant's business, which mainly covered lunches and dinners. Within this time frame, workers had two hours of break in the afternoon, which normally ran from 1 or 2 PM to 3 or 4 PM. During the break, workers gathered in the banquet hall located on the second floor of the restaurant. Some took naps, but most of them chatted, played mobile games, and watched online videos they had downloaded earlier via a Wi-Fi connection. Wi-Fi was introduced to the restaurant during the time of my fieldwork, but before this development, workers had already found various ways to access Wi-Fi in the workplace, in the places they lived, and in public spaces.

As Android phone users, all of the workers had Android stores installed on their phones, although a frequent phone change often also led to changes in stores. Most phones had preinstalled Android stores and bloatware built in by manufacturers,[2] but workers would still choose to install third-party marketplaces like Tencent MyApp, Baidu 91 Mobile Assistant, and Qihoo 360 Mobile Assistant. Thanks to the aggressive public presence of intermediary institutions such as mobile phone retailers and repair stores, many workers obtained their initial instructions for phone use and recommendations of software selection from this informal network. These stores, along with street kiosks providing downloading and rooting services (e.g., printing and photocopying stores), thus constitute an important institution in the technological diffusion among migrant workers. Their middle-class peers, on the contrary, might have more stable, systematic ways of developing digital literacy (e.g., from schools). The development of Tencent MyApp gained momentum in 2014 through a series of commercial campaigns. A Xiaomi phone user who had Tencent MyApp explained that the salesmen at the phone shop had recommended and installed the app on his phone.

The gateway nature of Android stores was well articulated by migrant workers. One worker showed me his phone and told me, "You need to download an assistant before you can download other stuff." The term "assistant" he used was certainly derived from the naming of popular Android stores such as Baidu's 91 Mobile Assistant and Qihoo's 360 Mobile Assistant, but it was also evoked as a general term to describe the perceived function of Android stores as a category of totality. In this case it was made clear that Android stores were lived and felt as a guiding portal in workers' mobile experience. A conversation with another Tencent MyApp user reiterates this point. When our conversation randomly moved to the pre-smartphone experience, the user recalled, "[Using the non-smartphone] you can also get online, but it didn't have this software [app store]. It only had *yidong mengwang*." The *yidong mengwang*, which literally translates to "Mobile Dream Net," was a mobile internet service portal developed by China's telecomm carrier China Mobile based on SMS (short message service) and WAP (wireless application protocol) platforms before the massive rise of smartphones. It partnered with mobile content and service providers to deliver multimedia products ranging from music to games to its subscribers. The service was then promoted with handsets from manufacturers such as Motorola, Nokia, and Ericsson (Bourne 2001), and as a portal to a vast amount of entertainment and information, its gateway function was obvious. In this context, the way the worker put Tencent MyApp together with *yidong mengwang* reflected a mental structure in which a gateway quality was assigned to the former.

As an information gateway, Android stores like Tencent MyApp guided and defined workers' phone use to a large extent. From conversations with many of the workers, I learned that the "Must Have" (*zhuangji bibei*) list recommended by Tencent MyApp and others was the one most workers would rely on when they first began using a phone. As the phone use became routinized, workers usually further relied on lists of top downloaded apps from different categories in the store to discover new apps. As one user said, "I download apps to try them out." The most frequently downloaded apps by the workers were music, online video, gaming, and other entertainment-oriented content. Parkour games were particularly popular among workers during the time of my fieldwork as was reading online novels.

The light-entertainment orientation of the app store provision somehow echoes the consumerist logic that inspires the functional and technical design of Android stores like Tencent MyApp. New technologies, as Raymond Williams's (1974/2003) reading of television suggests, are always simultaneously a combination and development of earlier forms of sociocultural activities and an adaptation of received forms that leads to changes that are not in any obvious way derivative and can therefore be seen as innovating forms of the technology itself. Thinking of the app store as a new cultural form in this particular case, however, it is more productive to emphasize the continuation aspect of its development. As Bryan Pon (2015) notes, app stores can be conceptualized as a realization of essential market functions—economic exchanges between buyers and sellers along with supporting institutions that make this exchange possible—being remediated in the digital context. As such, the information-seeking behavior is substituted by online search, and in a similar way, the face-to-face relationship building as a precursor to trust is replaced by online recommendations.

An organized store catalog format in app marketplaces, for example, is accomplished through categorization, and the ways in which apps are named, classified, and categorized often have clear consumerist allusion. In Tencent MyApp, for example, the app MTIME, which features a non-mainstream cinematic culture-and-taste public, is classified into the less politicized category of "Entertainment." Similarly, the social app Douban, which has a reputation for hosting dialogue about alternative cultural politics and social issues, was placed with ride-sharing and Groupon-type apps in "Life." Therefore, much like department stores and shopping malls, the design of the app store bears an underlying structure of consumerism that might be challenged by users in everyday life or in more serious political/ideological contexts, or might not.

Tencent MyApp, Creatives, and Mass Entrepreneurship

Tencent MyApp developed rapidly in 2014. By the end of the year, the store reportedly had a daily distribution volume of 110 million downloads (Riaz 2014). Tencent's strategy in developing its store was an integral part of its larger blueprint to migrate its open platform to the mobile context. The company opened up in mid-2011, allowing third-party software developers to access its huge user base across its online properties, which were mostly PC-based products before 2014 (Xiang 2014). In 2013 Tencent's Open Platform evolved to include offline entrepreneurship incubation projects involving various levels of local governments, high-tech industrial zones, financial agencies, and venture capital investors. It established its first incubation base in Beijing in November 2013, aiming to provided tech start-ups with one-stop service support in project provision, training, investment solicitation, legal backup, and monetization (Qu and Wu 2013).

This hybrid online/offline entrepreneurship model was extended to the mobile environment in 2014, with Tencent MyApp being positioned as the core of its mobile open platform deployment (Wen 2014). Sitting at the center of Tencent's mobile platform plan, Tencent MyApp was not only a distribution platform but also a techno-economic institution of entrepreneurial development that positioned upstream app developers on the production/distribution end and millions of users, including migrant workers, on the consumption end. Tencent's incubation project has only accelerated since its move into mobile. By April 2015 the company had established more than twenty incubation bases in major cities across the country, attracting over five million entrepreneurs (IT home 2015). In the same year, Tencent announced an upgraded incubation program named "Mass Entrepreneurship Space" (*zhongchuang kongjian*), which opened incubation centers in twenty-five cities with a total space of five hundred thousand square meters (He 2015).

The new program directly corresponded to Chinese premier Li Keqiang's call for "mass entrepreneurship and innovation" to increase employment and drive China's economy in early 2015. The state-endorsed mass entrepreneurship program particularly emphasized an inclusion of young creatives and college students. I see this particular approach to unemployment and social instability in a declining welfare system as a form of neoliberal-oriented governmentality. In a much more unprotected exposure to the market (i.e., entrepreneurship), individuals were mobilized to be responsible for their own success. The participation of tech giants like Tencent gave the mass entrepreneurship program a more appealing persona than state-sponsored projects, because successful high-tech entrepreneurs "operate in the popular imagination as models of achievement for the aspiring young" (McGuigan 2014).

The state-driven strategy of mass entrepreneurship created conditions for "variegated governance" in which large tech companies such as Tencent acted like a corporate government in managing populations and social spaces (Ong 2006). This much is now clear: stimulating employment on the one end (creative labor) and providing entertainment on the other (migrant labor), Tencent MyApp seems to pacify two of China's social groups of which the Chinese state is most wary. Of course, this overly deterministic conclusion does not account for all the ongoing social, economic, and technological changes taking place in contemporary China, not to mention frequent mergers and tensions among local and transnational corporate players. Yet there are always pockets of resistance and subversion, as evidenced from two stories from my fieldwork. The first is about an app called Omnipotent Keys (*wanneng yaoshi*), an app that allowed workers to hack urban Wi-Fi hotspots to get free internet access.

This app consistently stayed on the "Must Have" list in many app stores, including Tencent MyApp. A couple of workers told me they used the app to obtain Wi-Fi access in their dorms, though the app sometimes failed to crack more complicated Wi-Fi passwords. This app gave workers a sense of active, if limited, control over their technologies as well as a sense of mobility.

The second story concerns a twenty-seven-year-old worker who slowly discovered documentaries, a less popular category in video apps, that he downloaded from app stores and quickly became obsessed with the genre. The app store brought him to serious topics on history, war, and politics, which he was eager to discuss with people who shared his interests. These ordinary technological and cultural experiences have the potential to bring incremental changes and material consequences to the various ways app marketplaces will continue to act as cultural intermediaries that condition our entrance to mobile, appified worlds, particularly in the vast and rapidly shifting landscape of Chinese digital culture. It is precisely this tension that Tencent MyApp embodies as an *app store* that has heavily shaped mobile use and economic tech development in China, but also as an *app* that affords migrant laborers and everyday users access to popular entertainment, media, and literature.

Notes

1. Workers' names were removed in the essay for purposes of anonymity.

2. Preinstalled apps have been a key source of income for China's smartphone manufacturers to compensate for the low cost of mobile devices. The relationship between software developers and device manufacturers has evolved into a gray profit chain involving phone retailers, agencies, and rooting service providers at different levels who all participate in app preinstallation to make a profit (Wu 2015). Bloatware reduces the quality of smartphones and jeopardizes user experience. Workers complained about the slow speed of the machine and a long list of hard-to-navigate apps that they couldn't delete. Brian Larkin (2004) argues that contemporary scholars of technology have put in much effort in understanding the phenomenological and cognitive effects of technology when it is working at its optimum. What is less discussed, however, is how technology influences through its failure as much as through its successes. In China, migrant workers' experience of smartphones preloaded with bloatware shows a very similar kind of failure as the one described in Larkin's analysis of Nigeria's digital video infrastructure, so the impact to app users should be subject to similar critical scrutiny.

Exodus International
Banned Apps, App Stores, and the Politics of Visibility

TARLETON GILLESPIE

App: Exodus International
Developer: Exodus International
Release Date: March 8, 2011
Category: Lifestyle
Price: Free
Platform: iOS
Tags: gay conversion, religious freedom, hate speech,
     banned
Tagline: "Change is possible."
Related Apps: Manhattan Declaration

The Exodus International app was available in Apple's App Store for only three weeks. That it appeared at all and that it was gone so quickly both call for explanation. The app itself was unremarkable in its design, largely duplicating the event listings, videos, and articles available on the organization's website. More remarkable was the organization itself: Exodus International advocated for a controversial "gay conversion" therapy, in which those experiencing homosexual proclivities and eager to resist them could be counseled or reconditioned so that they might "grow into heterosexuality" (app text, qtd. in Jackson 2011). This approach, presented as a positive therapeutic option, especially for threatened families urgently seeking "healing," framed homosexuality as an unwanted condition to be rectified. Gay rights activists, concerned with the psychological harm of such treatments and the organization's regressive attitude toward the LGBTQ community, challenged Apple, arguing that by approving the app and making it available, Apple was endorsing the organization. This forced Apple to make a decision: remove the app, or not?

This particular app, and the responses to it, can serve as a lens for asking a broader question, raised clearly by Gerard Goggin: we need to know "what kinds of activities, projects, aims, groups and individuals may access apps—

and upon what terms, and subject to what social, and power, relations they may do so. These are basic cultural, political questions we ask of all our media systems" (Goggin 2011a, 150). To answer such a question requires attending not just to apps but to the providers behind them and the infrastructures for organizing and distributing apps—in order to appreciate the power they exert over apps, developers, and the field of possible tools that users encounter. The fact that Apple can highlight some apps and reject others entirely helps reveal the form and consequence of their power (Gillespie 2015).

Exodus International is also a reminder that apps can be political—not just in their content but as claims for political visibility in and of themselves. Because mobile providers, especially Apple, decide which apps will reach users, they are faced with choices that are unavoidably political, with real effects for the people involved and the issues they advocate for. While Apple promises a fair and impartial review process, in decisions like the one they faced about the Exodus International app there is no choice they could make that is not political. As a consequence, they are increasingly being taken to task for how they curate the public space of apps.

Apple Rejects Lots of Apps for Lots of Reasons

Exodus International can serve as a stand-in here for almost any of the controversial apps that Apple has rejected or removed over the years (Hestres 2013). As soon as Apple launched the App Store and opened the iPhone to third-party developers, it began rejecting or removing apps from it. Some were merely crass, such as Baby Shaker (a drawing of a baby who cries until you shake it into silence), Pull My Finger (one of dozens of apps that provide fart noises), and Slasher (a picture of a knife with a horror-movie scream when you slash your phone up and down). Apple removed I Am Rich, which cost $999 and did nothing but show a picture of a shimmery jewel. Other rejections seemed motivated more by Apple's economic self-interest: some podcast apps, web browsers, and email clients too closely duplicated the functionality of (i.e., competed with) Apple's own apps; some incorporated programming languages or methods Apple wanted to avoid; some simply came from competitors like Google.

Other apps ran afoul of content prohibitions for profanity, nudity, and violence: apps from established performers, including South Park and Nine Inch Nails, or the graphic novels *Murderdrome*, *Zesty*, and *Sex Criminals*. Some decisions were questioned for being overly prudish, as when Apple requested adjustments be made to a comic of Joyce's *Ulysses* (exposed breasts), a comic of Wilde's *The Importance of Being Earnest* (men kissing), the Ninjawords dictionary app (common profanities), or the Eucalyptus ebook reader (the Kama Sutra). Apple was also criticized for briefly removing the UK's *The Sun* newspaper and Germany's *Stern* fashion magazine until nude images (already published in their print versions) were excised. They banned Bang with Friends, a Tinder-like hookup app aimed specifically at facilitating sexual encounters, but later accepted the same app when it was renamed Down. Apple briefly removed third-party Reddit apps, until they ensured they would not provide access to the dirtier parts of the site.

In May 2010 questions about Apple's oversight grew when more than five thousand sexually suggestive apps were removed from the iTunes store in a single sweep. Developers complained about the opacity of the review process, noting that the Playboy, Hustler, and Sports Illustrated's Swimsuit Issue apps, which seemed no less explicit, were allowed to remain; others felt that purge of sexually suggestive apps was capricious given that users could access every pornographic corner of the open web through the built-in Safari browser.

Perhaps most troubling was the rejection or removal of apps with political content, including Bushisms, MyShoe, and Freedom Time (parodying President Bush); Obama Trampoline and H149 (parodying President Obama); and NewsToons, a collection of political caricatures by Mark Fiore that had already appeared in major newspapers—and had earned him a Pulitzer Prize. All of these were initially rejected for making fun of public figures, a rule Apple has since dropped. Apple removed the Wikileaks app after Wikileaks posted the massive cache of leaked diplomatic cables, because it was no longer in compliance with the law—though the legality of Wikileaks' actions was far from settled, and Apple did not remove the app Truthseeker, a third-party app that also provided the Wikileaks documents. They approved but later removed Metadata+ and Ephemeral+, apps that tracked US drone strikes, for "objectionable content." Apple has been cautious when a legal or public issue is changing—for instance, initially rejecting cannabis apps like Ease and Mass-Roots before later loosening their policy. After the racially motivated shooting in South Carolina in 2015, Apple (like other platforms) removed all apps that featured the Confederate flag, then reinstated some that used the flag in historical or educational ways.

Apple has also rejected a number of "serious games" apps, which explore publicly relevant or politically challenging issues in game form: Sweatshop

HD highlighted exploitative labor conditions; Phone Story and Permanent Save State criticized Apple's manufacturing partner Foxconn for its working conditions; Smuggle Truck dealt with the hazards of illegally crossing the Mexico-US border; Joyful Executions criticized the North Korean dictatorship; Endgame Syria highlighted the ongoing Syrian civil war; Ferguson Firsthand offered a virtual reality walkthrough of the Ferguson protests. Several ran afoul of a rule that, as far as I can tell, has no corollary in traditional media or in other digital environments: under Apple's rules on violence, "'Enemies' within the context of a game cannot solely target a specific race, culture, a real government or corporation, or any other real entity." Designed to complement Apple's rules against hate speech, the rule managed to catch apps with educational rather than xenophobic aims in its dragnet.

Of course, the cases just mentioned are only the ones that made it into the press. They are hardly a comprehensive or even representative account of all the apps Apple has rejected. Many more are rejected for noncontroversial reasons: they don't work, they misuse Apple's trademarks, or they don't interface correctly with the hardware. Others are such a clear violation, not just of Apple's rules but of law and morality, that their rejection isn't contested publicly. Even in cases where developers were surprised or frustrated by a rejection, many say nothing: many developers reasonably feel the need to maintain a good working relationship with Apple, as success in the App Store may be essential for their livelihood.

Rejected apps represent a shadow of the world of available apps (Brunton 2013), a reminder that the market for apps is curated on not just technical and economic criteria but political and moral ones as well. As Luis Hestres argues, "What types of content Apple allows developers to make available through apps, as well as the processes that govern such decisions, deserve closer scrutiny because of the consequences these decisions and processes could have on the ability of both developers and users to express themselves politically, socially, and culturally" (Hestres 2013, 1268). To understand apps we must also understand how and why they are curated.

Apple Has Built an Infrastructural Bottleneck to Manage App Distribution

Looking closely at app rejections, especially when the criteria is not functionality or quality, sheds a different light on debates about Apple and the model of software distribution it has championed. Some heralded Apple's opening of the iPhone to third-party apps, not only for massively expanding the range of offerings for users, but for helping amateur developers more successfully compete with major software companies (Banks 2012; Johnson 2010; Jordan 2015). Others criticized Apple for imposing a "closed" system that exerts control over every phase of the software development process (Zittrain 2008; Doctorow 2012; Nieborg 2015). Jonathan Zittrain, for example, argued that Apple's infrastructure for the iPhone is antithetical to the "generative" openness of the personal computer, or even the Android mobile OS (operating system) platform. Others disputed Zittrain's analysis, suggesting that there is no such thing as a completely generative system, that there are (and must always be) some constraints to how open a system can be; so the more important question is, In what ways is it open and closed, and with what effect? (Grimmelmann and Ohm 2010; Thierer 2008). Pelle Snickars suggested that Apple's management of the app development process created the conditions for a flourishing of creativity and innovation, "a gated community of code" (Snickars 2012, 166) in the best sense; a less governed approach to app distribution would have been no more democratic, just more susceptible to a flood of malware, spam, and porn. And Goggin noted that it was Apple that managed to open "the vice-like grip that carriers had held on mobile software" (Goggin 2011a, 153) before the iPhone.

When we bring the removal of apps into the picture, it is clear that Apple is trying to have it both ways: an enormous range of software, lightly governed, all distributed through their marketplace. The fact that apps are "tethered" not only means Apple can upgrade or kill an app, but it also makes Apple obligatory: developers must work exclusively through Apple, design on their terms, and submit to their review, and users can get only the apps that Apple approves and distributes. Apple's ability to curate apps depends on a technical and institutional infrastructure designed to keep Apple's guiding hand along the path of an app's development, and to position the App Store as a bottleneck through which apps must pass before they reach users. It is a different arrangement than older models of software distribution and requires specific technical, economic, and legal arrangements to be in place.

First, the software development kit (SDK) needed for designing apps compliant with iOS devices is available only from Apple. To get it, developers must sign a contract that not only spells out the economics of their relationship but also requires them to submit their software to Apple for review. All of this allows Apple to guide the development of and "certify" apps—as functionally sound, and as editorially acceptable.

Second, the App Store is the only place where Apple users can get apps.

Unlike Google's Android phones, iPhones as configured can install software only through the App Store. Some users have circumvented these digital protection systems, "jailbreaking" their iPhone (Rogers 2013), which then allows them to load uncertified apps from anywhere. But Apple regularly closes vulnerabilities exploited by jailbreakers when they next upgrade the iOS. For nearly all iPhone users, the App Store is the only game in town. App developers must pass through this bottleneck, and users must wait at its exit for approved apps to emerge.

At this precise bottleneck sits Apple's review process. Every app must pass through a review by Apple's team before appearing in the store. This not only means Apple can filter out apps they deem inappropriate; it also allows Apple to ensure they work properly, prevent spam and malware, prohibit specific software configurations, insist on interface design consistencies across apps, and exclude apps that might undercut Apple's own services. The App Store bottleneck also means Apple can control the financial side of this exchange: how prices are set, how updates are delivered, how customer feedback and reviews are received and responded to (Morris and Elkins 2015; Nieborg 2015). And they can direct how apps are presented to users: how apps are categorized, how the collection of apps can be navigated, and how specific apps are introduced and highlighted. Apple has constructed the curatorial review, the clearinghouse, and the marketplace—and gets to set the terms for how they all work.

These are the kind of regulations that are endemic to running a "multi-sided platform" (Boudreau and Hagiu 2011), where the platform's economic interests depend on coordinating an exchange between more than one audience—in this case app developers and iPhone users. Apple had already begun to establish this logic with the iTunes music store: making their store the only point of exchange proved an effective way for Apple to insinuate itself into the music business and challenge the power of both record labels and online sellers. Now their specific control over this multi-sided platform helps Apple ensure that, for example, any further purchases between user and developer must be "in-app," rather than the app directing users to an external website: Apple takes 30 percent of all in-app purchases. Extracting rent always requires some form of control to keep participants where that rent can be extracted (Elaluf-Calderwood et al. 2011; Tiwana et al. 2010).

This is not the only way to run an app store. Android users can download apps from Play, Google's official app store, but developers are also able to provide their apps independently and can even create Android app markets of their own. Users who can't find the pornographic apps they want, for example, can go to Mikandi to get their fill. Google retains the ability to kill apps loaded on Android phones, though this power seems to be reserved for removing malware or privacy breaches. Still, both platforms share a sensibility that they are to some degree responsible for the apps that run atop them. Both depart from the logic of shareware, for example, where developers can produce tools, in whatever programming language they like, and circulate them through the web or informal networks to interested users. It is a model of software distribution that thinks of apps as "for" the platform, as a destination if not as a commission. And this logic is not specific to apps: this store/review/bottleneck structure is being deployed elsewhere, first and foremost by Apple, which has extended this infrastructure from iPhone to iPad to AppleTV to the Apple Watch, as well as to the MacOS for desktops and laptops (Morris and El-kins 2015, 77). Microsoft has developed a similar arrangement, now using the Microsoft Store as the central means for installing software onto Windows OS.

This Gives Apple the Power of a Gatekeeper to Render Apps Visible or Invisible

That Apple even faced a decision about Exodus International was only because they had built themselves into the position of gatekeeper (Barzilai-Nahon 2008). Overseeing the distribution of apps, not just in the particular way Apple does but in any way at all, means the apps that users see are the product, to some degree, of that oversight. It is a position Apple wants and enjoys, for a number of reasons, but one that from time to time can be politically precarious.

First, requiring review must come with a matching promise to move apps through that review process expediently. For small-scale developers, a delay or an unanticipated rejection can be costly, even devastating. This puts an enormous burden on Apple's small review team, tasked with reviewing a growing queue of apps carefully and quickly enough that the bottleneck doesn't become a logjam. Second, Apple must articulate guidelines that are clear enough for developers to follow, consistent enough that they can be fairly applied by reviewers, imaginative enough that they can anticipate apps that have not yet been developed, reasonable enough that they can be defended under public scrutiny, and comprehensive enough to discourage policy makers from intervening. Third, oversight can quickly creep—from setting prices, to imposing standards, to upholding quality, to enforcing legal mandates, to making moral judgments. Fourth, gatekeepers can find themselves saddled with some sense of responsibility, to users and to the public at large. And, fifth, when what lies beyond the gate is the public, Apple is bestowed with the power to grant some apps public visibility and to render others invisible.

As in the case of Exodus International, Apple's power to grant or withhold visibility resides most fundamentally in the review process. But beyond this, Apple has crafted additional layers of visibility into its App Store, which different apps enjoy to different degrees. Apps are categorized, recommended, sequenced into browsing flows, ranked by sales, identified as "trending," featured on the front page, included in an Apple ad. All of these choreograph an app's visibility, both to possible users and to the public at large.

From Apple's perspective, these are efforts to "surface" apps that will interest users and keep them using their iPhones. App developers compete for this visibility, most often and obviously for economic reasons. Visibility generates attention and downloads, which generate visibility; this means apps face either a vicious or a virtuous cycle: the rich get richer while unpopular apps tend to stay unpopular (Lotan 2015b). Visibility in the app stores will often make the difference between an app becoming wildly successful or disappearing entirely into the archive (Lotan 2015a).

This Power to Grant Visibility Turns the App Store into a Venue for a Political Contestation

Platforms that offer not just access to users but public visibility, like App Stores, draw those wanting to be publicly visible for other than economic reasons: activists, media outlets, experts, newsmakers, fans, crackpots, and ideologues. And with the prize of political visibility come contests over the right to and the political significance of that visibility.

The Exodus International app didn't do much more than the organization's website, or do it in a way that took distinct advantage of the mobile interface. That means, even without an app, those who wanted to access Exodus International's material through their iPhone could do so through the Safari web browser Apple provides. Exodus International built the app in the pursuit of visibility: not just to reach more people in a new and convenient way but to claim the right to a spot on a public platform. And while critics certainly wanted the app gone to prevent new users from discovering it, more than that they objected to this claim of public visibility. Challenging the app, and Apple, was a way to challenge Exodus International, to undermine the organization's public legitimacy. But while critics' target was the organization, their critique was of Apple too. They saw Apple as not just enabling Exodus International but tacitly endorsing it; at the very least, the move by critics to claim that Apple was tacitly endorsing the app was a strategic move in and of itself.

Apple was forced to then weigh, on the one hand, Exodus International's right to provide its services and offer its perspective through an app; and on the other, critics' claim that in accepting the app Apple was endorsing an offensive position and insulting the gay community. This was a decision in which both choices were political, and both would appear political. Keeping the app risked offending the gay community and inviting other radical organizations to submit apps themselves; removing it risked offending evangelicals and highlighting Apple's willingness to squelch political and religious speech. Because Apple wants to review apps for quality, remove apps they fear will offend, and remain within the obligations of the law, they find themselves in a position where review is sometimes an unavoidably political intervention.

Critics and activists have taken advantage of this, calling Apple to task for contentious apps like Exodus International. In fact, Hestres worries that susceptibility to public pressure is the real problem: "There is no reason why citizens should be any more comfortable with Apple's susceptibility to public pressures than they would be if Internet Service Providers (ISPs) and Web hosting companies were subjected to similar demands" (Hestres 2013, 1274). App stores, and social media platforms more broadly, are increasingly becoming sites of controversy—not just as a means of distributing political speech but as places to contest public meaning and legitimacy. Calling on Apple to remove an app is a powerful way to debate the appropriate contours of public visibility and the power of platforms (Clark et al. 2014). It also highlights the importance of examining Apple's power to curate. Users are not just debating politics via social media but debating the politics of social media as well, the very contours of public discourse as a by-product of the workings and interventions of private providers like Apple.

Some were troubled both by Apple's decision to remove the app and by their very ability to make decisions like this. Discussing both Exodus International and Manhattan Declaration, Rebecca MacKinnon argued:

> Though these two apps are appalling to anybody who believes in gay rights, and though the public expression of homosexuality is offensive to some conservative

Christians, the free speech rights of people with both sets of beliefs are protected by the First Amendment. . . . as long as the open Web remains the primary
platform for Internet content, it is technically possible for controversial groups to
reach an audience—as long as they reside in a country where free speech is protected constitutionally and by the legal system. But if the bulk of the potential audience abandons the open Web for closed, app-based systems, then political activists
are increasingly hostage to the whims of corporate self-governance. (MacKinnon
2012, 129)

Of course, Exodus International and its critics did avail themselves of the web
to make their point. But MacKinnon is right to sound the alarm in case we
were to be reduced to a single speech environment structured with the kind of
corporate oversight as the App Store.

Still, I want to consider Apple's role, even if they are arbiter of just one
venue of speech. Though both sides of this political issue could speak freely
elsewhere, Apple's power to curate some of them into the App Store and some
of them out has political ramifications, once Apple sets itself as the gateway to
public visibility and, by association, a stamp of public legitimacy. This concern
grows as a platform grows, becomes more public, and becomes more dominant in the field (Ananny 2015). Apple might prefer to distance itself entirely
from whatever apps their developers make available. Social media platforms,
content sharing sites, and online markets do this all the time: merely a conduit, safe harbor, the right to intervene but not the responsibility, buyer be-
ware (Gillespie 2017). Since Apple does not commission apps or pay for their
production, the traditional reasons we hold intermediaries responsible are unavailable. Yet in very real ways Apple has inserted itself into the middle of this
exchange between app developer and user, by asserting their right to pick and
choose apps and building an infrastructure for doing so. In addition, the public tends to see Apple as the provider of apps, both practically, because that's
where users download and pay for them, and experientially, because apps feel
like such an integral part of the phone or tablet. This puts app providers in a
profoundly difficult and unavoidable political position.

B Productivity/Utilities

TaskRabbit
The Gig Economy and
Finding Time to Care Less

SARAH SHARMA

App: TaskRabbit
Developer: Task Rabbit Inc.
Release Date: July 2011 (Current Version: 4.5.9)
Category: Lifestyle
Price: Free
Platforms: iOS, Android
Tags: gig economy, digital labor, mobile media economies
Tagline: "We do chores. You live life."
Related Apps: GigWalk, EasyShift, Uber, Lyft, ShiftGig, Takl,
    Freelancer

TaskRabbit was founded in 2008 by Leah Busque, a former IBM software developer, after she grew increasingly frustrated that she had no time to buy dog food. Busque realized how helpful it would be if she had a neighbor or a friend to rely on. With the guiding ethos of "neighbors helping neighbors," she started TaskRabbit—a mobile app and online company that connects freelance laborers to local "demand." She needed a friend, so she started a company. I will return to this enterprising equation later.

The company's punchy tagline for this lifestyle app reads, "We Do Chores, You Live Life." The weblog *LifeHacker*'s promotion of TaskRabbit gets to the heart of the matter:

> We all have stuff on our to-do lists that we're either too busy to finish, or let's be honest—we just don't feel like doing. That's where TaskRabbit comes in. Whether it's getting groceries, putting together furniture, or picking up a Craigslist purchase, TaskRabbit's network of reliable do-ers will take it off your hands. (Miller 2013)

So, yes, let's be honest. Dog food just isn't that much fun to buy. It would be rather repugnant to ask a friend or neighbor to run this errand just because you didn't want to. As the old adage goes, "You are the company you keep," so you probably also only hang with others like you, people who live life and only do things they love. Buying dog food isn't one of them, and it probably isn't for your friend either. TaskRabbit can get the job done, but it can also save you from depending on others outside of economic exchanges. The app will connect you directly to an amenable subset of people for whom choring isn't so boring.

Prospective taskrabbits, or taskers, go through the company's careful protocol, which includes a criminal background check and online interview. TaskRabbit helps people "find jobs they love, at rates they choose, and a schedule that fits their life" ("Become a Tasker," TaskRabbit.com). Paraphrasing from the website: *TaskRabbit has enabled Tatiana to have the flexibility to work around the schedule of being a full-time mom. TaskRabbit has given Chris peace of mind in knowing he can still pay rent if his music gigs aren't paying all of the bills. And Jason just loves delivering good service with a smile, loves the variety of taking on different tasks and meeting great people, and finds much pleasure in seeing clients react to a job well done.*

After getting hired by the company and downloading the appropriate software, a taskrabbit's day is punctuated by notifications whenever there is a job in their vicinity. Several taskrabbits bid on the same job. From there, the person who hires them manages and organizes the assignment (date, time, cost). The task is electronically billed, which, as one client of TaskRabbit claims on the website, "greases an otherwise awkward transaction." The platform's four most popular types of service: handyman, house cleaning, moving, and personal assistants. With all chores taken care of, TaskRabbit's promotional video promises, "you can get on with the life you should be living." In the video this *living life* culminates in a fun afternoon of whitewater rafting. As for the taskrabbit anthropomorphized into a cute bunny, it merely hops along to the next task.

As with all such companies currently vying for success in the gig economy, the descriptors used to explain the labor TaskRabbit reorchestrates with the use of their app are very telling. TaskRabbit describes the work as "freelancing" and "micro-entrepreneurial," giving the workers some sense of an identity. Yet in the same breath the company's marketing rhetoric refers to the labor these freelancing entrepreneurs undertake as random and the menial chores involved hardly worth the time of anyone when given the choice. There is one app, one ad campaign, and one website. The attempt to empower people while devaluing their actual labor would seem like an oxymoronic advertising technique or corporate strategy doomed to failure. But, sadly, it isn't. What TaskRabbit offers in exchange for the labor it demeans is the ability to control time. And with the ability to control time, it offers both the short-term worker and the client the choice to not care about the task at hand while having the effect of diminishing the care about the labor that goes into reproducing the social order. And as I will explore below, this company strategy depends upon deepening the sexual division of labor.

The political potential of care as an organizing concept for social change is increasingly evacuated of its power with the rise of companies such as Task-Rabbit. The app expands the conceptual boundaries of what actually consti-tutes care labor, and the work of social reproduction in general, while it de-values this labor as a useless pursuit of time that can be undertaken by anyone else any other time. TaskRabbit makes claims upon time or free time as that which should be free of sustaining and maintaining life while it expands the realm of what should constitute this life deeper into the consumptive realms of superfluous privileged needs and demands. The success of TaskRabbit rep-resents an impediment to a feminist post-work and post-gender future.

Freed Time

TaskRabbit extends already established forms of outsourced care work in the home, whether of gardeners, cleaners, drivers, or milk delivery. It also replac-es the traditionally feminized secretarial support, or office care work, with its rabbits doing coffee delivery, lunch orders, photocopying, scanning, and quick jaunts to the post office. TaskRabbit outsources and turns into gigs the labor associated with social reproduction work—the type of labor historically devalued as work women do out of love for free or for little pay. But it is not only women doing this labor for TaskRabbit; it is the new precariat (Stand-ing 2011). Like other workers in the sharing economy, taskrabbits tend to be male, ethnically similar to the US workforce. Most workers use this type of employment within the sharing economy as supplemental income. In the case of TaskRabbit, it is estimated that only 15 percent work for the company on a full-time basis. Roughly 70 percent have at least a bachelor's degree, which also indicates the possibility that this work is seen as transitional or supple-mentary. There is no doubt that the success of the app is fueled by the growing insecure labor conditions in which people find themselves—being without stable work for some coupled with the ever expansive maintenance of a consumptive lifestyle for others. Yet what also underlies all of this is the fact that TaskRabbit can depend on the discursive currency and illusory promise of the novelty of freed time: a new temporal experience of being able to control when you work.

The *New York Times* recently ran an exposé on workers in the sharing economy focusing in on TaskRabbit (Singer, August 16 2014). In this instance the taskrabbit profiled was a single mother who had spent almost sixteen hours doing menial work that included assembling furniture and gardening but then also driving for Uber in between. Interestingly, the discourse here is about the hard manual labor she undertook for the day. At the end of the day she had worked well beyond an acceptable workday. Along with two hundred dollars, she also had a backache and a tension headache to show for it. Recognizing that it was a good day in terms of cash but not a sustainable livelihood, she expressed that the ultimate payoff was that at least *she can control her own time*.

Like the other taskrabbits profiled on the app's website, the ability to control time refers to the immediate moment, the contours of the exact day. But temporally speaking, uncertainty looms for anyone employed by the company. There is no guarantee of a good working day tomorrow or any secure future. The taskrabbits may plug in and plug out when they want. They may determine the length of their working day as well as their own geography. As a type of labor that is downloaded on demand, they might not know the exact contours of their day or even what they will be doing the next hour, but their labor gets to feel like a choice when there is clearly no choice but to work.

Having control over one's time in the absence of security and with the promise of laboring under the temporal parameters of the day as one chooses suddenly seems like an acceptable trade-off. Being able to control the minutes and minor details of the task justifies otherwise precarious working conditions. Even if the boss is decentralized and diffuse, incarnate in every transaction, the relationship is fleeting and more palatable than other types of workplace domination and exploitation. Alienation is experienced as fleeting, nothing one must identify with, as the undignified moments of this type of labor pass quickly. This is much like an Uber driver in London who tells me he enjoys the work because it lets him be in control of his time. He makes this proclamation right after saying he has spent the last thirty minutes quite bored and waiting in his car for a fare.

The mantra "I control my own time" is said while entirely recalibrating to the time demands of others and while doing tasks that others have devalued as less valuable pursuits of their own time. "I control my own time" is also a statement espoused by those controlling the time of others by outsourcing all of their tasks. For this privileged, consumption-fueled population that depends upon gig workers, the promise of free and malleable time depends on a blanket relegation of "chores" to the domain of *not living life*. The work of social reproduction quickly falls into that domain. Between the taskrabbit and the client, no one cares, and neither does the company. What is increasingly on the line, however, are the gendered politics of care.

No One Cares

The mantra of newly freed time also does capitalism and patriarchy's important work of normalizing alienation deeper and deeper into the realm of care. Free time is the trade-off for caring less, but it has a gendered component that affects how social reproduction is culturally understood. Most profiles for TaskRabbit's workers tend to focus on the young male workforce with a new gendered spin on the type of labor they are undertaking. An article in *Money* profiles three young male taskrabbits who each espouse the shared company mantra of more free time to pursue their own unique lives, such as being able to live on a boat. But the focus on their unique lives is accompanied with a list of their craziest TaskRabbit jobs (Zimmerman 2015). These "crazy jobs" included being a "stroller bouncer" at a school to tell "nannies and mothers" where to park strollers, seventy dollars an hour for an all-day T-shirt-folding gig for a start-up company whose delivery from the factory came unfolded, and a late-night task of helping a distressed woman move from her apartment. The underlying assumption in the article is that it is men who are undertaking this "crazy" work—work oriented toward the feminization of labor and managing women. Their work is seen as helping women—whether by rescuing them, directing them, or undertaking their unfinished factory labor at a significantly higher wage.

Feminist Marxist labor debates in the 1970s—specifically the Wages for Housework campaign—recovered reproductive labor as a category of women's generalized experience (Dalla Costa and James 1973; Federici 2012). "Reproductive labor" refers to producing labor power—transforming raw materials and commodities to maintain the worker daily as well as the future workforce—feeding, clothing, education, socialization, and caring. These early feminist interventions can be misread for maintaining heterosexual ideologies within the labor struggle as well as in the home, charged with essentialism and maintaining a rigid gender binary. But in fact this theory does quite the opposite: it denaturalizes care and explains how the process of naturalizing care as feminized is precisely how it continues to be devalued.

This focus on social reproduction, we must remember, was a corrective to the misogynistic but also racist elements of Marxist approaches to the economy and class struggle. The focus on social reproduction and wages for housework was meant not to domesticate women but to provoke new ways of living, laboring, and loving that placed care at the center, including attention to forced care and the intimate and complex relationships between caregivers and receivers of care. Moreover, by making social reproduction central, it opened up an inclusive category; it does not necessarily refer to heterosexual or biological regeneration of human life but how labor power is reproduced and how life is sustained and regenerated.

Politicizing the power of social reproduction requires only the fundamental recognition that we were all born, that we will all die, that all lives depend on reproduction, and that this reproductive work has historically been undertaken by women for little or no pay. Care labor is banal and mundane; it is also joyful, dirty, gross, hard, disgusting, demeaning, empowering, boring, sexy, sickening, creative, physical, mental, voluntary, spirited, disheartening and heartening, fulfilling and unfulfilling, skilled and unskilled. This type of labor is increasingly taken on by gig workers but in temporary and fleeting arrangements so that there are no ties that bind. There is no rouse of love in the air. They are doing it out of the promise of time control, freed time. Doing care labor for time and not out of love is the new element in the equation. But the tasks are increasingly devalued.

Care labor is increasingly on the popular agenda as an inequitable form of affective gendered labor. The Sunday sections of liberal newspapers have columns arguing that thank-you cards, play dates, snack sharing, potluck contributions, bake sales, gourmet daily meals, pressed shirts, and stain removal are unequally distributed in the home. It is many of these same tasks that are outsourced to TaskRabbit. The politics of the gig economy are mostly discussed for who it benefits as a form of labor and whether or not it provides the means for new forms of organizing and worker solidarity (Olma 2014; Scholz 2016). When platform cooperativism and organizing are on the table, these taskrabbit workers are not mobilized under the banner of care as a grounds for solidarity; instead they are united by the fact that they are workers. But another conversation TaskRabbit prompts is a deeper consideration of the roots of this labor as embedded within the sexual division of labor related to care—one that the gig economy now extends and perpetuates but with newly precarious bodies. TaskRabbit represents a significant impediment to politicizing the politics of care labor, because it further abstracts and negates recognition of mutual dependency while expanding what might be understood across the public imaginary as care.

We might start to consider, too, how the app maintains a normative gendered divide even when its jobs are undertaken by young men. This occurs in at least three different ways. First, in a heteronormative household, male refusal to participate requires the reorchestration of care labor to other bodies. There is no single middle manager or boss accountable to workplace exploitation, bullying, and abuse. It becomes diffuse. In sum, within the TaskRabbit imaginary everyone has time and no one has to care. There are more and more products/tasks/gigs, and being able to orchestrate labor for someone else— "I'll get you a driver, I'll get an Uber for you, I'll send you a TaskRabbit"—is a reorchestration of care by way of sending someone a gig. But this isn't all I mean when I suggest that the category of care is expanding. This market is expanding the sphere of care labor. It is not that care labor has expanded into the market or that care is monetized; it is that the sphere of what constitutes care is expanding and becoming muddled with frivolity and privilege.

Second, the types of affective/care labor that circulate as "needs" actually revolve little around the sustenance of life, and thus TaskRabbit sustains and maintains privilege mistaken as care labor. This starts to provoke an urgent questioning of whether or not a boundary and parameter should be drawn around what is care and what is social reproduction. Delivering a single piping-hot crepe or cup of coffee to someone's work desk is beyond sustenance, beyond reproduction. Reorchestrating care labor to different undervalued bodies does not reveal a new cultural value now placed on care; it merely divides it up, distributes it differently while continuing to devalue it within the sphere of a useless pursuit of time. This signals a greater societal exit out of relations of care while simultaneously expanding the sphere of what constitutes the sphere of social reproduction. Left aside is the real work of determining what is needed to sustain our lives.

Third, in almost every instance of a taskrabbit worker profile, the labor undertaken is demeaned through a gendered power dynamic. For example, in a *New Yorker* article about TaskRabbit, the author states, "If you're a Tasker, it might be disappointing to discover that you're expected to clean up after the birthday party rather than, say, pick up the cake. But doing the dishes may be better than nothing" (Raphel 2014). A culture so pursuant of free time denies this mutual dependency while it rests on gendered and classed preconceptions of what is a meaningful versus useless pursuit of time. It is the promise of free,

malleable time that keeps the worker working and the tasker tasking even under duress or troublesome work conditions. A politics of care requires cultural
recognition of mutual dependency, but this is precisely what is negated with
TaskRabbit. In fact, the preferred word in Silicon Valley for this type of labor
is "distributed." In some ways TaskRabbit is seen as a feminist solution—this
is an old debate. It can alleviate forced care by redistributing it. TaskRabbit
reorchestrates the labor of care and adds additional affective layers to it. For
example, in an article celebrating the convenience of the app, the author ex-
claims, "Shout-out to my sister's boyfriend Ramsey for TaskRabbit-ing Nyquil
to my mom when she was sick" (Lee 2015). Shout-out to Ramsey for using the
app or shout-out to the Rabbit for doing the job, I wonder?

I think it is important to also point out that a focus on social reproduction
is meant to call into question the restrictive institutional ordering of everyday life into discrete spheres of home and work—provoking the necessity to
multiply the spheres that constitute life, care, and intimacy beyond these two
mutually oppressive poles of home and work (Weeks 2011). It includes the
notion that we should be encouraged to care for those who fall out of the
realm of our normative lives (Haraway 2016). Making time for one another,
whether stranger or friend, starts to seem like a radical concept. Neighbors
helping neighbors is the premise for this start-up. TaskRabbit maintains the
institutional ordering of everyday life by incorporating more bodies into the
domestic sphere and the workplace. It does not free care from the home; it
enslaves more populations to maintaining the home and extends and expands
the sphere of social reproduction while deflating the political potential of care
as grounds for solidarity.

Conclusion: Time to Care

A more radical perspective related to time would come by way of a feminist
refusal as to what is cared about, what is free time oriented toward? Feminist
calls for refusal of work are not about free time or just refusal of capital but
a refusal related to patriarchy and capitalism while placing care at the center.
Kathi Weeks's *The Problem with Work* (2011) elaborates a refusal via a living
wage that does not return women to the sphere of reproduction with the
promise of more time for the family or more care work but, rather, time to
explore new intimacies and connections that break away from the normal-

izing and confining spheres of social reproduction. A shortened work week isn't dedicated to expanding the sphere of gendered labor within the home or redistributing labor. In other words, this post-work paradigm does not include hiring a taskrabbit either.

TaskRabbit represents a narrowing in the cultural imaginary of what to do with one's time while expanding what you can get others to do for you. It is no wonder the animated mother in the TaskRabbit promotional video is split into two before she is freed by a taskrabbit and able to go rafting. This split into two reveals the expanding sphere of social reproduction. On one hand the case needs to be made for regulation, livable wages, unionizing, and worker's rights, but on the other we might need to think about how participation in this economy—the ordering up of labor on demand for mostly superfluous, extravagant, and superficial wants—makes difficult politicizing the sphere of social reproduction. Perhaps the ultimate question isn't what to do with one's free time, but what are the political parameters of care and how does a normative attachment to the sexual division of labor impinge upon what the dream of free time is supposed to deliver. TaskRabbit represents a narrow cultural imaginary of time and what needs to be cared for and about.

Carrot
Productivity Apps and the Gamification of Shame

SARAH MURRAY

App: Carrot
Developer: Grailr LLC
Release Date: January 2013 (Current Version: 5.2.4)
Category: Productivity
Price: $2.99 each, $11.99 bundled suite
Platform: iOS
Tags: social technologies, shame, affect, productivity, self-surveillance
Tagline: "the A.I. construct with a heart of weapons-grade plutonium"
Related Apps: BetterMe; Wake N Shake; Zombies, Run!; DroughtShame

Accomplish Something . . . Anything

"Greetings, lazy human." This is how Carrot To-Do, described at the time of its release as the most sadistic productivity app on the market, introduces itself upon launch. Created in 2013 by screenwriter turned independent developer Brian Mueller, the app is presented in promotional materials as an "A.I. construct with a heart of weapons-grade plutonium" (MeetCarrot.com).

Carrot To-Do is the lead in an aggressive suite of five goal-oriented apps, including a fitness tracker, a talking calorie counter, a mean-spirited alarm clock, and a talking weather forecast robot. Purchasable as individual apps or as a bundle, all are anchored by "Carrot," a constructed sentience with a monotone female voice and a dark sense of humor. The premise of the suite is simple: users are more efficient when their app is their tough-love taskmaster. Accomplish something—check a task off your to-do list, stay under your calorie count, get up early, or open the app for a quick check-in—and you "level up" and are rewarded with a carrot (e.g., a digital kitten to feed and care for, badges, stars, jokes, and, occasionally, offers to download another app). Fail to complete goals, and the apps experience a mood change expressed in a fresh batch of insults intended to playfully tap in to a user's sense of self-worth. As the app store description notes of Carrot To-Do, the original app in the suite and the focus of this chapter, "You keep this sadistic AI construct happy by getting things done in real life" (MeetCarrot.com).

We have long been preoccupied with machines that talk back, evident by how much scholarly work and popular press on communications technologies open with Hal, the social computer made famous in 1968's *2001: A Space Odyssey*. But what might it mean when your app doesn't just talk back, but *talks back*—through shame, disgust, disparagement, sarcasm, and, occasionally, public humiliation? As software continues to immerse itself in the everyday, Carrot's use of shame subtly reconfigures how, when, where, and with whom powerful feelings of humiliation, stress, or foolishness are shared. This should prompt us to consider our increasing reliance upon, and comfort with, the social roles we assign our technologies, particularly in a contemporary context already teeming with the violent culture of internet shaming (e.g., Wikileaks' launch in 2006; "The Fappening" in 2014; fatal shaming based on false information on Facebook and Whatsapp in 2018). Carrot's use of negative reinforcement to stand out in a saturated world of productivity demonstrates one way mundane software is a site of shifting relations among users, their digital worlds, and the social imaginary. Through a process of digital shaming, the emotional invocations of everyday social interactions are transferred into the simulated life-ness of an app and its disciplined user.

That Carrot To-do (hereafter, Carrot) is symptomatic of the contemporary neoliberal atmosphere of personal accountability, and its corollary of self-discipline and self-motivation is a given. Like most productivity apps, Carrot slides effortlessly into the envisioned user's daily routine with its swipe-friendly interface and cheeky tone, making a lighthearted game of one's overscheduled day. The app's singular functionality (Morris and Elkins 2015) as a list maker and its minimalist design (for which it has been praised) are malleable enough for the often incongruous momentums of aspirational goal setting and the daily banality of just getting by. Moreover, because shaming is a process that is increasingly found in the hands of sentient-simulated devices, Carrot's functionality is anchored in an infrastructure designed almost exclusively for human-computer interaction.

With its hyperbolic nature, Carrot offers itself up as an easy object of study; through its excess it makes plain the pervasive nature of both apps and the primary logic that undergirds them, an insistent, ordinary productivity that is maintained, in part, through the transferability of affect. Carrot directs

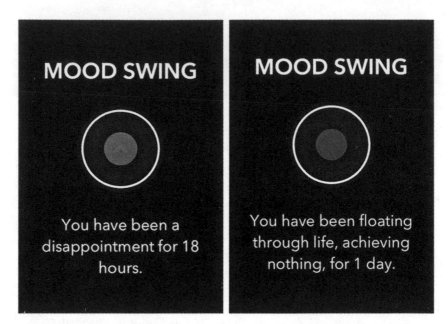

Carrot's Shaming Interface

and redirects the cultural boundaries of shame as it shifts from an affective encounter anchored in interpersonal social relations to a generative symptom of intimate technological practice. This emergent relationship among shame, users, and their technologies is visible in three overlapping characteristics of Carrot: (1) its imperative for constant user interaction and the subsequent production of intimacy, (2) its use of shame as a dismissible comedic device that grounds the app's gamified reward system, and (3) the simulated normative modes of gender and sexuality that support the transfer of shame from interpersonal dynamic to human-app interaction.

I Don't Need (or Understand) Excuses . . . Just Get it Done

Although the app's marketing materials are replete with references to humanity and life-ness—it has "a heart of weapons-grade plutonium" and plenty of feelings—Carrot's constructed sentience is quite limited; the app is designed to react to time- and list-based movements by the user but little else. While the app experiences simulated "mood swings" based on the user's productivity, it doesn't sense your mood. In other words, Carrot responds when you check things off your list or when you haven't opened the app in a while, but the app has no way of perceiving factors like depression, being overly busy, or other contextual life factors for why you might be slacking off or why you might need a motivational boost. The fact that Carrot's designed sentience is simulated rather than organically produced is key to understanding how shame is operationalized. Put differently, Carrot makes no attempt to anticipate or react to users' needs or emotions, and shame is similarly evacuated of the human rhythms of action and reaction.

Carrot's look and feel are filled with intertextual resonance that builds on popular representations of modern machinic beauty as minimalist, functionally driven, and affectless. The app borrows heavily from the visual and tonal depiction of *Space Odyssey*'s computer Hal with an emotionless voice and spherical "ocular sensor" that are nearly identical to those of the central processing unit at Hal's core. The ambivalence of automated sentience as an imperfect form of "life capture" (Turkle 1995, 299–302) becomes all too clear when Carrot chides you for accidentally touching the area of the screen where "her" ocular sensor resides. Promotional videos and text, as well as the voice within the app, gender Carrot as an irreverent, monotone woman. The familiarity of the snarky tone, synonymous with digital assistant "machine speak," calls to mind the history of tempering Siri for mainstream use (Bosker 2013) and the subsequent pleasure of poking and prodding her for signs of life-ness. Carrot's primary function to playfully discredit its user's worth is communicated through purposefully affectless vocal delivery (e.g., "Your tasks are bad and you should feel bad"). The app's gamified utility is similarly one-dimensional and inflexible: do something, get rewarded with snark; don't do something, get rewarded with snark (e.g., "Congratulations! You're not a waste of space after all!"). Overall, Carrot's aesthetic and interface are indifferent and unforgiving, approachable and intuitive without intuiting, and they posit spiritless, sarcastic reproach as a primary technological affordance of our personal digital tools. And yet Carrot's mode of relationality is built on shame, a profoundly social and human characteristic.

Shame and Social Interaction

Shame is a powerful component of lived social relations. As Katherine Sender writes, "Existing in the eyes of the other, however painfully, is central to the experience of shame" (2012, 82). C. H. Cooley's "looking glass self" (1902), which became the epistemological basis for much of microsociology and social psychology, contends that we develop, monitor, and adjust the self in the reflections of others. So even if shame is ultimately a relation of the self *to* the self, it is motivated by the external discipline of others who are emergent in deeply entrenched interpersonal culture. It follows that shame might be understood as the experienced consequence of this social monitoring, a momentary process of de-subjectification (Vallury 2013), a "halt to looking" (Tompkins, Sedgwick, and Frank 1995, 134). Shame is what is experienced "when bodies cannot or will not fit the place" (Probyn 2004). Although much debate has encircled the question of whether shame should be characterized as cognition-emotion or affective-embodied, most assessments of shame describe it as a temporary experience of out-of-placeness. Thus, shame functions both as an expectation of participation in social normativity and a visceral reminder of one's failure to do so. As Elspeth Probyn argues, shame is a highly personal experience, while it also clearly demonstrates our dependence on the material and habitual world.

Shame is also consistently theorized as *productive*. For Probyn, shame might be considered productively as a rerouting of the body and its habitus—that is, if shame knocks us momentarily out of step, it also has the potential to be profoundly transformative. For some, shame is subversive, productive in its visceral reminder of one's noninclusion in normativity and the hopefulness of being out of step, being different; for others, shame functions as discipline, as a productive moment of self-reflexivity, an opportunity to reassess and consider making a change for the better or toward the more normative (Munt 2008; Halperin and Traub 2009; Halberstam 2005).

What, then, is the productivity of shame when shame is the literal mechanism of productivity? Put another way, what is clarified by *how* shame is deployed in the space of an individualized, microfunctional (Morris and Elkins 2015) to-do list app? Apps bank their functionality on the routinized and the aspirational. Carrot is an app that combines these expectations of the everyday with a persistent yet lighthearted, jokey, even "mean girl" style reminder that one is out of place. In doing so, Carrot's mode of digital shaming suggests one way that apps become positioned as mundane software. Taking on complex social attributes of human life and pushing toward a deeper strand of neoliberalism, Carrot may encourage a "post-shame" sensibility in which one of the heaviest emotional invocations of interpersonal culture can be "tried on" as a novelty of productivity.

"It's just you and me, together forever."

Carrot encourages constant interaction. In fact, the principal "to do" requires users to keep the app in regular rotation by actively building up and tearing down an interminable list of tasks. Neglect to open the app with enough frequency, and Carrot gets nastier with her put-downs. Of course this imperative to engagement is not unique to Carrot. Push notifications—automated reminders from your device to engage with downloaded apps—are a valuable economic functionality of mobile software. Keeping up with *any* app is commoditized as productivity in and of itself, linked to revenue opportunities such as in-app purchases and premium upgrades. Given that so many apps are free, repeated and constant user engagement is necessary for profit; Carrot just makes this relationship more transparent by blatantly demanding frequent use and getting "hurt feelings" when ignored. The app's demand for a user's frequent return is both endearing and annoying, like playing a game with a petulant child who will cheat to win, but also an example of our "ensnarement in the loop of drive," the repetitive circuits of activity that Jodi Dean characterizes as the communicative capitalism of modern digital life (2010, 43).

The app's automation, masquerading as Carrot's demand for attention, is linked to a context-less, time-based evaluation of the user's progress. Following short periods of nonuse, Carrot's push notifications generate messages that connect the user's lack of in-app activity to worthlessness: "I am so, so disappointed in you right now." Upon opening Carrot after longer stretches of inactivity (i.e., a day or two), the first message is one of disparagement: "You have achieved nothing for two days!" While this constant nagging and clingy nature is automated, it is also paired with a simulated reflexivity: "Am I creeping you out?" asks Carrot following a series of pushed messages including, "I can't believe you left me alone for that long!" and "You kept saying my name in your sleep last night." Reminders are not just designed productivity but also the designed intimacy of ever present technological nearness, a perpetual awareness of the physical and emotional proximity between user and app.

Also intimate is how well Carrot the AI purports to know you. List-making apps have built popular sharing components into their functionality (e.g., Wunderlist, Evernote, Any.do). Carrot opts for the primary shared component to be the knowledge of failed productivity that is passed back and forth between the user and the app via constant monitoring of the list's progress. The app is not only programmed to offer persistent reminders based on calculation of time-based productivity (or lack thereof), but it is also designed to be the primary holder of your successes and failures. Carrot knows when you mess up and is in charge of communicating those failures to the outside world—that is, Carrot selectively shares your progress on social media with your permission. However, this sharing is offered infrequently, difficult to navigate within the app, and more likely to be offered in light of your *lack* of progress rather than your achievements (e.g., "Shall I just go ahead and tell everyone how lazy you are?"). Thus, as Carrot acts as the broker between you and social media, this management of your social accountability is a persistent reminder from the app that indeed, it's "just you and me together, forever."

Mark Andrejevic's (2006) notion of lateral surveillance is a helpful way of understanding how contemporary digital media practices include not only the user's agreement to be watched but also a willingness to watch back or, rather, watch forward. For Andrejevic, lateral surveillance is a user's willingness to participate in a digital world where the cost of interactivity is not only the implicit agreement to be socially monitored but to monitor others as well. Productivity apps shift the social nature of lateral surveillance toward an inverse lateral monitoring: users give permission to be surveyed by the app in order to monitor their goals, tasks, and to-dos. In the case of Carrot, this is taken one step further. The promise of interactivity on which lateral surveillance rides ties the intimacy of surveillance directly to the productivity of shame. Thus, when Carrot congratulates you for swiping a task off your list with "You're the worst you that you can be"; when the app pushes out a notification that says, "You snore when you sleep"; and when users reopen the app to find Carrot's needy note reading, "You, you abandoned me," shame is the route through which Carrot not only reminds users of their failed productivity but also demonstrates its proximity to the user. Failure to check something off your to-do list becomes synonymous with being a bad boyfriend, equal parts humorous, endearing, and alarming. And yet the gravity of the user's failed out-of-placeness is immediately neutralized with humor.

This subtle shift in lateral surveillance from something that describes the social monitoring of each other to something that describes constant monitored interaction between a user and their device is notable precisely because shame is the axis on which this transition takes place. Carrot knows what you have or haven't achieved more than anyone else and is designed to do nothing but hold and recirculate that knowledge of your out-of-placeness. Through Carrot's designed technological intimacy, a user's productivity also generates a measurable relationship between the user and their app, and in turn, an evaluative assessment of how well the user has integrated productivity software into their everyday life.

"Let's play a game. We'll call it Don't Suck at Life."

In addition to establishing technological intimacy through humiliation, Carrot uses a shame-based gamified structure to incentivize the traditional to-do list. What the user "wins" is a humorous opportunity to be shamed. The app extends the satisfaction of digitally checking a task off your list by allowing users to earn approximately forty points per completed task and to "level up" when a dozen tasks have been completed. As each task is completed, Carrot chimes in with flowery rhetoric presented in a dry, monotone congratulatory message, such as "Dazzling," or "Glorious," or "You're a winner." A reward is offered at each level. Most often the rewards are more shaming mechanisms. For example, pass level three and Carrot awards you with a home screen icon notification that constantly reminds you of tasks unfinished: "The number will hover above my ocular sensor, silently mocking you [upgrade_002]." When a user's digital kitten is unlocked at level ten—a heavily promoted "carrot" in press reviews and descriptions of the app—it is gifted with a number of caveats, including the expectation that users must level up to earn food to care for their starving cat.

In Carrot the "gamification" of the to-do list grind—the playful, mildly competitive approach to a non-gaming context like productivity—is offered through the humorous possibilities of the reward of shame as a tongue-in-cheek game. Not only is it funny when your app says, "Hey, meatbag," but this casual disparagement serves as your trophy for a job well done. Shame is not mobilized as a disciplinary measure of social interaction but as a simulated nag that mirrors and makes light of one's failed productivity (e.g., "I just calculated which came first, the chicken or the egg. What have you done today?").

Orit Halpern, following anthropologist Rosalind Morris, sees the mechanics of shame as embedded in digital affordances, as a sort of post-shame media governance, or a shame*lessness*. She writes, "In the ability to always already feed the image of ourselves to ourselves in the near future, we make it impossible to feel shame or remorse" (Halpern 2015, 228). This affectless reflexive futurity that Halpern proposes might be considered in light of what it means to assign the task of shaming users to an AI construct with a mean streak. In Carrot, gamified teasing—you check off tasks in order to *earn the right to be teased* by an app with jokes about your uselessness—is arguably a form of shamelessness. The app enlists simulated sentience to ensure that shame functions as a humorous and lighthearted mode of keeping software in your everyday routine, and in doing so, it makes visible the changing nature of shame's productivity. Yet because the app's scripts are both meant as a joke and exactly punitive, Carrot's mode of shamelessness has the potential to undermine the seriousness with which shame pervades digital culture in both disappointing and hopeful ways. Carrot's messy use of shame highlights critical questions that the app's market success should prompt for scholars: For whom is shame fun and lighthearted? For whom is shame pleasurable? In utilizing shame as a joke to force productivity, Carrot may also downplay the actual harm and pervasiveness of shame in digital culture. Or it may reduce shame's power.

"She'll Reward You. But be careful! You do not want to make her upset."

Finally, by gendering, sexualizing, and stressing the app's "beautiful, gorgeous, gesture-based" interface, marketing materials call attention to lifelike but one-dimensional qualities of the anthropomorphized app. If, as Jeremy Morris writes (this volume), novelty apps "puncture the solutionist veneer" of mundane software's political economy, then Carrot reconstructs a solutionist veneer of enlightened sexism (Douglas 2010). The app is explicitly referred to as "she" in press and promotional materials. Many of the programmed dialogues are highly sexualized, suggestive, and stereotypical. For example, Carrot's disparagements during bouts of nonuse, called "mood swings," are thinly veiled and long-held notions of gendered hysteria and women's "nagging." Similarly, to use the app frequently is to be a "glutton for punishment. I guess you just like it rough." At the same time, Carrot never experiences tonal fluctuation due to emotional variance; instead, it is the app's actions and expectations (i.e., the process of reminding you of your halted productivity) where mood is imparted.

The tension between Carrot's beautiful interface and its affective strategy of shaming lies in how, specifically, these gendered and sexualized attributes are applied. Identity retains a one-dimensionality via the use of reductive applications of gender and sexuality. The prevalence of gender and sexuality-based online shaming has been acknowledged by many scholars, but in this case the application of identity is not the site of shame but its transfer. So instead of users being, for example, "slut-shamed," harassed, discriminated, or shut out of Carrot, through the anthropomorphization of a mobile software application, women are signified as the reductive communicators of shame. Through repetitive use, the image of a nagging, irritable woman with an endless "honey do" list is mapped onto a more humanlike social process of sharing shame with your app. Put differently, Carrot fosters a complicated and intimate relationship between you and your app at the expense of an oversimplified "identity prosthesis" built out of gendered, sexualized, and racial stereotypes (Nakamura 2013, 5). As developer Mueller explained in an interview, the original AI construct of a snooty cartoon dog didn't work as well as the idea of the app being like a person constantly nagging you. Carrot, he noted, was inspired by the women in his life, "But it's only because Carrot cares" (Dormehl 2015, n.p.).

Conclusion

Shame is both public and private, personal and social, intimate and performed. We are accustomed to the ways communication technologies mediate social interaction. After all, the purported role of social media is to do just that: help us socialize with one another. A mobile software application that determines, calls attention to, and is responsible for one's out-of-placeness—the appification of shaming—is an observable shift from media assisting us in navigating the social world around us to *being* the individualized social world in which shame is shared. When Carrot talks back to us, when she reminds us we are worthless and that we haven't achieved as much as we'd liked, self-improvement becomes synonymous with sharing something as messy and intimate as shame. While the app can and does subject users to public humiliation, more significant is how shame becomes the ethos of the relationship between users and Carrot's constructed AI.

This is not to suggest that Carrot represents a dystopic future wherein we choose intimate relationships with apps over people. Instead, productivity apps as excessive as Carrot prompt us to consider how the integration of shame into mobile software and everyday computation poses a reconfiguration of shame as being *less* productive in social interaction, demoted to an interaction of designed sentience. In fact, if shame is most often characterized as a humiliating experience born of social monitoring and normativity, then the meaning of transferring shame to a technological practice between users and their media—perhaps the transition from shamefulness to shamelessness—should be further explored as a mode of productivity that quietly puts shame in its less-than-human place.

See Send
Antiterrorism Apps and
Suspicious Activity Reporting

GREG ELMER AND BAHAR NASIRZADEH

App: See Send
Developer: My Mobile Witness Inc.
Release Date: January 2013 (Current Version: 4.5)
Category: Utilities
Price: Free
Platforms: iOS and Android
Tags: lateral surveillance, antiterrorism
Tagline: "If you SEE suspicious activity, SEND a photo or
    note."
Related Apps: Nextdoor, BART Watch, Outbreaks Near Me

Activated by a simple press of a fingertip, the mobile app—or application—is seemingly poised to solve whatever problem the smartphone or tablet user can imagine. Moreover, this simple act of touch on the fingerprint sensor of a personal mobile device, in lieu of a PIN or password, seamlessly works to "authenticate" the purchaser within the mobile media app. The implementation of such an application into mundane and everyday practices of media technologies entails everyday transaction, transportation, and communication. The need now for both finger and fingerprint is particularly germane for security-based apps in our post-9/11 insecure world—a world where threats, risks, and dangers must be efficiently identified and eliminated in a timely fashion. The mobile phone user's fingerprint thus serves not only to securely purchase such an app; it also promises to become the medium through which viewing, identifying, and reporting threats becomes an intuitive act via the seamless integration of threat reports into the networked communications infrastructure of police and law enforcement agencies. Of particular interest for this chapter is the See Send mobile software app, which encourages users to remain vigilant in the face of terrorism and to visually report—using their smartphone cameras—possible risks and threats that surround them.

This chapter questions the intuitive logic of such apps and cites a host of other like-minded apps, such as Nextdoor and ACLU Mobile Justice, to argue that intuitive threats can mean very different things for different groups and communities in Western societies.

Intuitive Design

With the rise of the graphics-enabled World Wide Web browser in the mid-1990s, interface designers, in conjunction with advertising and search engine companies, began to experiment with navigational tools, signposts, and content templates in an effort to manage the attention of internet users. In this new visual economy, what Richard Grusin (2000) aptly called "desktop real estate," *intuitive design* became king. Above all else, the web interface, followed by the tablet, and most recently graphics-enabled smartphones, would need to be easy to use—hence "when a user sees it, they [should] know exactly what to do" (Laja 2017). The actions of the user, in other words, require little to no contemplation. Through the interface, action becomes instinctive.

Such techniques of intuitive action are, of course, not restricted to front-end interface design. The See Send mobile app, developed by various security agencies in the United States, also invokes the intuitive practices of users to report "suspicious activity"—an undefined term on the app website and interface—arguably, in order to "help in the fight against terrorism" (Apple Inc. and Google Play 2018). Thus, users should presumably already know how to use the technology and what kinds of activity it should be capturing and reporting to security agencies.

The See Send app, an offshoot of the US Homeland Security "See Something, Say Something" campaign, invokes a form of lateral surveillance among the US citizenry (Andrejevic 2005). But in using the mobile phone camera, it inverts the gaze from the known subject of the selfie (the self, sometimes with friends) to the unknown and potentially dangerous Other. Initially developed by the New York City Transit Authority, the campaign seeks to raise "public awareness of the indicators of terrorism and terrorism-related crime, as well as the importance of reporting suspicious activity to state and local law enforcement" (US Department of Homeland Security 2017). Similar campaigns, calling upon citizens to report suspicious behavior, have been deployed in times of armed conflict and heightened periods of civil unrest and terrorist activity. Joshua Reeves (2012) has argued that such distributed forms of policing can be traced back to at least the conquest of Roman Britain, where communities actively participated in reporting suspicious and potential criminal activity.

Unlike other times and technologies, today's Homeland Security mobile app, aptly condensed to the abbreviated "See Send," folds the intuitive moment of suspicion into the act of reporting. The app's intuitive design thus enables intuitive-based reporting. In other words, the user-friendliness of the app folds or integrates the smartphone's photographic capabilities and its access to networked communications into a seamless reporting device. A suspicious person or thing can be visually captured, geotagged, time-stamped, and sent almost instantaneously to the Department of Homeland Security for analysis.

Yet what is it that makes the object of the app and smartphone camera intuitively suspicious and dangerous? How does intuition serve to identify and report "indicators of terrorism"? We argue that the See Send app negates the cultural and subjective components of intuition, and we consider how other apps—such as the ACLU's police reporting app that captures potentially illegal police activity—can highlight the dangerous unspoken preconceptions of suspicious behavior that the See Send app motivates.

Intuitive Thought-Action

As an unconscious process, intuition typically refers to a mental state that enacts a spontaneous judgment, an immediate perception, or a "gut-level" feeling about a future event. In this sense, intuition mirrors the larger post-9/11 doctrine of preemption, where action is privileged over contemplation, reason, and debate (Elmer and Opel 2008). These familiar everyday practices of media technologies—enabled by the swipes of fingertips—allow for unconstrained connectivity and seamless interpersonal interaction through writing text messages, taking pictures, geotagging, and then sharing them with friends and family members via mobile networked devices, generating and maintaining affective anticipation of connectivity, mobility, and flow of social networks of "always-on" media. Nigel Thrift (2004) consequently refers to such user practices as a "technological unconscious," while Katherine Hayles offers the competing "technological nonconscious"— in short, "the everyday habits initiated, regulated, and disciplined by multiple strata of technological devices and inventions, ranging from an artefact as ordinary as a wristwatch to the extensive and pervasive effects of the World Wide Web" (2006, 138). As Hayles explains, everyday unconscious use of technology functions "through somatic responses, haptic feedback, gestural interactions, and a wide variety of other cognitive activities that are habitual and repetitive and that therefore fall below the threshold of conscious awareness" (140). Accordingly, everyday interactions with media—what the subject does to media, and simultaneously what media do to the subject—establish an affective relationship between media and the body, what Grusin terms "the affective life of media" (2010, 90–118). Grusin describes the reality of media regimes in the post-9/11 environment as "the ontological aspect of premediation, the extension of media forms, practices, and technologies into the future so that the future will always already have been remediated" (51). Consequently, "premediation" in the post-9/11 world of media power is concerned with the "liveness of futurity," or the constant media visualizations of futurity. It is the way heterogeneous media networks function to ensure the future has already been premediated before its emergence into the present, through which the media actively participates in the mechanism of securitization. This is to prevent any future catastrophic event, such as 9/11, by mobilizing the public's affectivity of anticipation, thus "generating and maintaining a low level of anxiety" (46).

If we look to social psychology, however, intuitive technological practices are not always strictly defined as subconscious or, moreover, lacking in reason. Over the past century, many theorists in social psychology and cognitive science have attempted to define and make sense of the nature of intuition. Of these recent theorists, arguably the best known is David G. Myers (2002) and his mapping of the relationship between the dual processing of rational thinking and intuitive inclination, between what constitutes objective and subjective truths. Building on the concept of dual processing, Myers argues that thinking, memory, and attitudes all operate on these two levels of knowing—conscious and unconscious. He defines intuition as "our capacity for direct knowledge, for immediate insight without observation or reason" (1) that is perceived as fast, easy, and effortless. That said, Myers insists that intuition is not divorced from reason or conscious thought; rather, the intuitive decision is informed by this dual processing.

By contrast, Jean E. Pretz and Kathryn Sentman Totz (2007) present a new measure of intuition, suggesting three distinct aspects of the nature of intuition: affective, heuristic, and holistic, which taken as a whole can speak to the importance of social and mediated discourse, an essential part of post-9/11 security doctrine and culture (cf. Grusin 2000). The mediation of fear by mass media outlets, the fact of politicians shamelessly naming the Muslim Other as universal culprit, and the inability to determine from fiction and rumor from reality in a networked communications environment all speak to Pretz and Totz's broader social definition of intuition (2007, 1248–49), and its mobilization in the See Send app. Thus, while intuition at first glance would seem to denote a lack of thinking and reasoning, we find that it is inexplicably linked to both conscious and social worlds—to the stories that we tell ourselves and others. In short, we bring our experiences and biases to the intuitive act, such as the act of reporting a suspicious activity.

The See Send App

Following the November 13, 2015 attacks in Paris and the subsequent state of heightened security, the state of New York unveiled a new antiterrorism mobile application, called See Send (also known as "See Something, Send Something"), as part of a fast-growing interest in sweeping nationwide suspicious activity reporting (SAR) techniques and technologies. Besides New York State, the use of SAR application was extended to five other states (Colorado, Louisiana, Ohio, Pennsylvania, and Virginia), and is gradually reaching the whole nation. Developed by My Mobile Witness—an app developer company that has collaborated with law enforcement agencies since 2008—the new crowdsourcing mobile application expands on and parallels lateral surveillance trends deployed and promoted by the US Department of Homeland Security's national initiative. Originally introduced in 2010, the campaign invited citizens to be alert, vigilant agents in helping to prevent and fight against terrorism by reporting activities they find suspicious to local law enforcement through Terrorism Tips Hotline. Following this logic, the See Send campaign now benefits from the seamless and familiar digital experience of a social networking application and allows users to notify the appropriate state division of Homeland Security or major city intelligence center of suspicious objects, people, and circumstances that may be linked to terrorism through a simple and fast process of sending geotagged pictures or text messages from their mobile device. This resembles the familiar and ordinary participatory practices of media technologies that make up the technological unconscious, or nonconscious. Pending processing, analyzing, verifying, and sorting of the submitted

SAR information, if relevant, the appropriate law enforcement agency will be notified for security measures (New York State 2016).

One feature See Send offers is "Watch For," in which users can find tips and information about what to look for and when to submit suspicious activity. It also provides eight signs of terrorism, ranging from individuals eliciting information, to photographing and monitoring activities, to acquiring items such as weaponry, maps, and computer. In this sense, a list of identifiable behaviors and activities, which may or may not lead to some sort of event of traumatic nature in the near future, becomes a qualified indicator of potential risk and threat. As Brian Massumi argues, "Fear is the anticipatory reality in the present of a threatening future . . . the felt reality of the nonexistent, loomingly present as the affective fact of the matter" (2010, 54). This perceived reality indeed legitimates preemptive action, that these activities must be identified and prevented prior to their actualization. Citizens are encouraged to reconsider and reevaluate these practices of taking pictures, gathering information, asking questions, and shopping that make up ordinary routine acts among, for instance, tourists, students, researchers, shoppers, and the government itself. Citizens, therefore, are required to be alert and vigilant about these activities in terms of what they can instantly and intuitively find "out of the ordinary" within their daily lives.

In this sense, individual intuitive knowledge plays a significant role in guiding an identification process through which activities and specific individuals and collectivities may be linked to terrorism. Since the 9/11 attacks, and forcefully emphasized under the Trump administration, the figure of the terrorist has been mobilized to conform to specific (racialized) bodies. Subsequently, a new category has been created to cast out those who are associated with inherent difference and on the basis of their appearance and their Muslim identity, a category that pervades beyond the troubling bureaucratic practices of political and legal institutions and into the mundane routines of everyday life. As Sherene Razack argues, "Through [the] various media spectacles and legal campaigns, the pundits, politicians, lawyers, and journalists warn of a deadly clash of civilization between a medievalist Islam and a modern, enlightened West, and declare the urgent need for the West to defend itself against the Islamist threat" (2008, 5). In this sense, it can be argued that the figure of "Muslim-looking" peoples and attributes become the constant focus of American anxiety and hence the target of racialized biases in today's post-9/11 environment.

Apps and Their Potential Roles in Social Biases

In addition to the state-sponsored "war on terror" campaign, the intensification of these interactive lateral surveillance practices based on intuitive insights is also evident in the deployment of crowdsourcing applications in other more local or domestic contexts. Apps such as Nextdoor, BART Watch, Sex Offender Tracker (see Mowlabocus, this volume) and Outbreaks Near Me, respectively, range from sharing information about local suspicious behavior with neighbors, to notifying regional police of suspicious activity on public transit, to tracking sex offenders, to reporting disease outbreak information. For instance, the popular free social networking application Nextdoor—used in more than one-hundred and sixty-nine thousand neighborhoods across the United States—works as a localized crowdsourcing application, encouraging users to stay connected with their community and neighborhood by sharing the latest local news, events, and services with those residing nearby. One of the features Nextdoor offers is the crime alert, allowing users to learn about local crimes and suspicious activities and share information about individuals and activities they find suspicious within their community. The Nextdoor administrators explain that the app allows users to "share information about crime and safety issues" while asking them to refrain from posting content of a discriminatory nature (Nextdoor 2016).

The usage of the mobile application, however, has been different from what the Nextdoor administrators intended. Critics have pointed to the possible prejudice facilitated by the mobile application, because some content posted by the users contains racial profiling and embedded racialized rhetoric. Drawing on various posts of such nature, Pendarvis Harshaw writes, "Rather than bridging gaps between neighbors, Nextdoor can become a forum for paranoid racialism—the equivalent of the nosy Neighborhood Watch appointee in a gated community" (2015). Nextdoor, then, is just one of many interactive crowdsourcing safety tools that has been criticized for its potential racialized bias. Apps like SketchFactor, for example, which was launched in 2014 and later banned, claimed to offer safety for users through a crowdsourced mapping feature, which suggested the safest routes for travel, geotagged "sketchy" areas, and allowed users to share their experiences. As the *Washington Post* reported, SketchFactor relied on inaccurate data based on crowdsourcing user experiences and information in addition to public crime data, ranking areas on the scale of one to five based on their relative "sketchiness" (Dewey 2014). Not surprisingly, the app came under fire for promoting racialized biases and socioeconomic judgments.

ACLU App and Reporting Police Misconduct

Contrary to the interactive lateral surveillance practices in the commercial and state contexts that encourage citizens to be vigilant and report against some specific fellow citizens, Nextdoor and SketchFactor seek to offer technological solutions to maintain discipline within the confines of familiar and intuitive spaces and places—known communities where users are unlikely to meet Others or unknown cultures and practices. Such intuitive apps, in other words, all require an intuitive space and a set of external dynamics—the object of the smartphone's camera must be recognizable as being out of the ordinary. Yet we must also recognize that potential risks are also, sadly, *ordinary* occurrences for many marginalized and racialized communities. Intuitive action for ordinary occurrences begs the question "Ordinary for whom?" That is to say that users' intuitive reactions and use of apps such as See Send or Nextdoor are wholly dependent upon their own experiences of *being in and out of the ordinary*.

Yet another mode of intuitive technicity and lateral surveillance was developed through the ACLU's initiative Mobile Justice, an app launched in 2014 that was eventually made available in eighteen states in the United States. A similar app, the Stop and Frisk Watch developed in New York state by a former Occupy Wall Street protester, allows users to monitor law enforcement and report unlawful encounters and misconduct, including "racial profiling, over-policing, and military-style policing tactics," to the regional ACLU office (ACLU 2015, 2016). Both apps similarly invite users to video record potential police abuse incidents they have witnessed or experienced using their mobile device, include as much information as possible about the incident, and then send the video to their local ACLU office. The mobile application also offers the "Witness" feature, which uses geolocation to provide users with police stop reports. In addition, the "Know Your Rights" feature provides resources in terms of fundamental legal rights, concerning primarily vulnerable groups and communities, ranging from "criminal law reform," to "free speech," to "immigrants' rights." Even as apps like ACLU Mobile Justice attempt to harness intuitive action to the affordances of mobile reporting in a way that intends to protect citizens from abusive or overbearing governing bodies, they risk participating in the same troubling articulation of premediated fear of certain bodies and movements to the mundane place of swipes and snaps in the digital everyday.

In other words, what the ACLU Mobile Justice and Nextdoor technologies demonstrate is that state or citizen appeals to report unusual or out-of-the-ordinary behavior fail to recognize uncommon and contrasting fields of experience in light of cultural, racial, ethnic, and socioeconomic contexts and communities. The See Send and Nextdoor apps further facilitate such assumptions through technological intuition and habit: point the camera toward the Other, swipe a finger to send the image. In contrast, ACLU Mobile Justice may prompt a mode of reporting that bears the burden of a longer history of lack of response and a failure to seriously take certain bodies as being threatened. Put differently, the lateral surveillance built into mobile reporting apps may not be able to adequately capture how intuition shifts based on who is reporting to whom. The question remains whether intuitive practices and technologies may lower the threshold for reporting events to such a degree that they start to redefine and make mundane—across communities and backgrounds—potential threats and the troubling technological impulse to act without thinking.

Is It Tuesday?
Novelty Apps and Digital Solutionism
JEREMY MORRIS

App: Is It Tuesday?
Developer: Kenneth Friedman
Release Date: May 2012 (Current Version: 5.0.1)
Category: Productivity
Price: Free
Platform: iOS
Tags: microfunctionality, mundane, ratings and reviews, app
   store optimization
Tagline: "Tells you everything you need to know about
   Tuesday"
Related Apps: Is It Dark Outside?, Is It My Birthday?, Is It
   Christmas?

If it didn't tell us so much about the political economy of app stores and the evolving nature of the software commodity, Is It Tuesday? would be little more than an entertaining punch line—an obscure side note in a sea of mundane software. But like other seemingly trivial software, the app's existence, reception, and features highlight precisely what's "new" about delivering software as an easily consumable, mobile, and mundane commodity through app stores that are shaped by ratings, reviews, and other algorithms.

The app's purpose is as straightforward as its name: helping users find out whether or not the current day is, in fact, Tuesday. Opening the app launches a black screen with a single sentence in orange letters: "Is It Tuesday?" Tapping the device reveals the answer to the question, which, on six of seven days of the week, is no. The answer on Tuesday, on the other hand, is an emphatic yes, complete with an exclamation mark and animated fireworks. Each tap of the screen also reveals how many times a user has checked whether or not it is Tuesday since they first acquired the app, as well as a "Global Tuesday Counter," which tallies, purportedly, the number of times all users of the app have launched this perplexing inquiry. If the counter is accurate, the question

"Is It Tuesday?" has been asked and answered nearly 4 million times (as of March 2018).

Swiping left from the main screen reveals the app's sub-features: a timer that counts down to Tuesday, a "Tuesday Fact" screen (e.g., "The actress Tuesday Weld was born on a Saturday" or "Remember: Tuesday the 10th occurs just as frequently as Friday the 13th"), and a calendar that calculates whether or not you were born on a Tuesday. Created by Massachusetts Institute of Technology engineering/computer science student Kenneth Friedman when he was still in high school, the app was introduced on May 21, 2012. It has been updated over twenty times since its launch to fix bugs and add new features (such as posting about the app to Twitter or Facebook). The app has, at last count, 233 customer reviews in the iTunes app store, close to a five-star rating in almost all markets where it is available, and has been featured on dozens of websites. Beyond simply a joke-in-code that went a little viral, the app's popularity in ratings and reviews and its limited feature set highlight how software is undergoing a microfunctional turn and how the very distribution of digital media shapes the consumption process.

Every Solution Has a Problem

Whether taken at face value as a helpful app for users who genuinely have few other ways to verify the day of the week in which they are existing or as clever parody, Is It Tuesday? is remarkably focused as a work of software conception and design. The app presents the existence of a problem (i.e., I wonder if today is Tuesday?) and provides a stylish, mobile, and dedicated solution. It is a joke, but one that highlights a trend that can be seen in so many of the apps that populate app stores—what I have elsewhere called "microfunctionality": single or limited-purpose software that splits up and breaks down multistep activities into smaller, more focused and specific components (see also Morris and Elkins 2015). This is perhaps one of the most notable changes in the software commodity since the arrival of app stores and the delivery of software as apps.

Apps are merely the latest expressions in a much longer history of selling software. Originally an integrated service provided by computer hardware manufacturers in the 1950s and 1960s, software was "unbundled" and sold as a separate stand-alone product in the late 1960s and early 1970s. The history of selling software after this can, at least for the purposes of this brief chapter, be distilled into three major periods: (1) the 1970s to the mid-1990s, when software was sold generally as boxed disk drives or on CD-ROMs (aka shrink-ware) through retail stores or via mail order; (2) the mid-1990s to the mid-2000s, when software was sold as direct downloads to computers via online outlets; and, (3) the late 2000s to the present, when software migrates to mobile devices and circulates as "apps" via unified app stores from multifaceted technology firms (e.g., Apple, Google, Microsoft).

In this history the earliest forms of software, outside of games, were enterprise programs, spreadsheets, and other business-oriented software that evolved quickly from programs for calculating and organizing information in databases to multipurpose knowledge management tools. A boom in software development took place in the late 1970s and through the 1980s as computers became more "personal" through programs such as word processors, home accounting software, and typing tutorials. Software historian Martin Campbell-Kelly notes that from 1975 to 1983, there was a huge increase in the number of software programs available, culminating in thirty-five thousand programs from about three thousand vendors in 1983, most of which were available between fifty and five hundred dollars as disks in shrink-wrapped boxes (2003, 208). As software migrated online in the mid and late 1990s, the method of distribution may have changed, but the idea of software as multipurpose and complex still comprised the majority of software exchanges (e.g., Lotus 1-2-3, Microsoft Word, Photoshop). While the emergence of Flash animation programs, casual web browser–based games, and other web applications were technical and cultural precursors to today's apps, it wasn't until the rise of mobile devices and app stores that the software commodity truly became appified.

Not only was the price of apps notably different from the software that had come before it (i.e., largely free, but with a significant number of apps between $0.99 and $4.99), but the potential purposes of software were being reimagined. As Ian Bogost notes about apps, "The days of the software office suite are giving way to a new era of individual units, each purpose-built for a specific function . . . or just as often, for no function at all" (Bogost 2011b). Although there is certainly no shortage of complex, multifunction software suites in app form (e.g., Microsoft's Office suite, Omni Group apps), app stores seem to have unearthed a format and ecosystem that is an ideal design for the creation and circulation of programs designed to meet minor, micro, and mundane tasks. Freed from having to justify the twenty- to fifty-dollar price tag that historically accompanied software suites, apps encourage turning routine and mundane needs into problems solvable through software. This microfunctionality begets what Evgeny Morozov calls "solutionism": "an intellectual pathology that recognizes problems as problems based on just one criterion: whether they are 'solvable' with a nice and clean technological solution at our disposal" (2013).

Much of Is It Tuesday?'s appeal is how the solutionism implicit in most apps is so explicitly on display; it's hard to know who, if anyone, is sincerely in need of a dedicated app to confirm it is Tuesday, but "given Silicon Valley's digital hammers, all problems start looking like nails, and all solutions like apps" (Morozov 2013). It is certainly true that part of the reason for the proliferation of applications (millions have launched since app stores emerged in 2008) may be because the barriers to produce, package, and market software can be lower than through the traditional software retail chain, opening up app development and distribution to a greater number of participants. But the explosion of new apps is also because of the fine-grained level at which software can be generated to both create and solve specific problems or needs. Just as Moore's law is less a scientific principle and more a collective, industry-agreed-upon goal (Sterne 2007, 20; Auletta 2009, 52), Apple's slogan, "There's an app for that," becomes less a description of current affairs and more a challenge for the future design of solutionist software.

Apps, in this sense, are less like the boxed software from which they descend and more like gumballs or toys in the dispenser machines at malls or grocery store exits: trivial, cheap, plentiful, easily consumable, and designed for impulse purchases. The history of automated vending machines dates perhaps as far back as holy water dispensers from 270 BC, but the machines really took off in the late 1800s when they began selling gum, stamps, tobacco, and other mundane commodities. This new form of vending was "promptly dismissed as an amusing novelty with scant practical value" but by the mid-1900s "a new future was envisioned for the vending machine" that saw it as playing a dominant role in retailing all kinds of products (Segrave 2002, 7). Some of the earliest gum ball machines in the US were placed on the elevated platforms of New York City's rail stations, and vending machines proliferated over the next century to reside in workplaces, malls, retail stores, and public parks. Just as making candy, toys, and other trivialities available in this new format reshaped retailing practices, retail spaces, and the very format of the commodity that could be sold, the conversion of software into apps reconfigures software's potential purposes, as well as the infrastructure around its creation and circulation. Whereas vending machines were revolutionary point-of-purchase devices, mobile app stores make every point a possible point of purchase. Instead of gum and trinkets, though, app stores sell solvable problems.

Every App Pales in Comparison

The delivery and distribution of software in a new format, the app, enables the creation of new kinds of software. It also allows for new interactions around software that shape and influence the consumption process in interesting ways. In particular, the integration of reviews and ratings into the discovery and distribution process unites the previously separate moments of product acquisition and discourse about the software product. Just as Is It Tuesday? cleverly calls our attention to the microfunctionality and solutionism of all apps, the reviews for the app parody the very genre of app reviews and call attention to the larger political economy governing app stores.

Previous versions of the software commodity (i.e., shrinkware) also had reviews and rating mechanisms to guide consumers through the purchase process, usually through magazines like *MacWorld*, *PC World*, *PC Magazine*, and *Computer World*. The move to online distribution of software, as with many other digitized media commodities, brought critical discourse about products into the very same environment as the point of purchase or acquisition, such as on sites like Download.com/CNET Downloads or other early software repositories. App stores like iTunes, Google Play, and Microsoft, however, give added weight to these reviews and ratings, not only by displaying them prominently alongside an app as a user considers acquiring it but also by integrating reviews and ratings into the algorithms that determine how an app is displayed and discovered in the first place.

The 233 reviews for Is It Tuesday? are almost wholly positive, from the most basic signs of support ("Wonderful app, couldn't live without it"—TuckerTuesday, November 29 2013) to more elaborate displays of appreciation ("This app is my favorite by a landslide. This app taught me how to love"—Michelin Men QS, October 16, 2012, or "It helped me find the cure for cancer"—Ivan_72, May 21, 2012). While there are those who truly seem befuddled by the app and give it a one-star rating ("I don't need to know if it's ***ing Tuesday! Get a life!"—Videogamecheatsultra, January 29, 2013), even those who dismiss it rate it a perfect five stars ("This app is quite literally thee [*sic*] most stupid and pointless thing I have ever downloaded, however, I check it multiple times a week and am in love with this app. Good work!"—

BrutalNightfall—June 2, 2013). Developer Friedman acknowledges that some early reviews likely came from his friends, but he quickly saw the app expand beyond his social circle, attracting reviews from around the world. The app's most negative reviews appeared after a brief software update when Friedman had included firework animations on days other than Tuesday, angering some of his loyal user base (email to author, August 6, 2015).

Positive or negative, the app reviews are a performance, a knowing wink from users that they are in on the triviality of the app's very existence. However, they also suggest that users are in on the promises that are so often made by app publishers that any of life's biggest problems can be solved through an app. There are apps that claim to help users lose weight, gain muscle, save money, save time, create, learn, relax, connect with family and friends, or find love or sex or both. Reviews of apps that meet these needs reinforce these claims (e.g., a review for Tinder proclaims, "Thanks to this app I found an amazing boyfriend"; another review for the calendar/to-do app 24me states, "I cannot tell you what I would do without this app! It's perfect for everyday planning! If you are looking for a planning app, LOOK NO LONGER!"). So when reviewers of Is It Tuesday? write, "This app SAVED my life. My house was scheduled to be bulldozed on Tuesday. Luckily enough, I had this app to remind me when it was Tuesday. I even started dating now that I can know what day it is" (Thorg10, June 11, 2012) or "My poor grandma takes all of her important pills on Tuesday (she's very arthritic) but I would always bring them to her on Monday or Wednesday. It was a disaster. Now I don't have to worry at all" (#Franswagg, September 20, 2014), they are performing an extension of Is It Tuesday?'s premise as well as critiquing the solutionist mind-set that underpins apps more generally.

The reviews for Is It Tuesday? also highlight just how microfunctional software has become. One reviewer, for example, notes that Is It Tuesday? "perfectly embodies Apple's spirit of simplicity and focus. Most software developers would create a bloated app that does too many things, like telling you days of the week other than Tuesday. But Kenneth Friedman knows how to do one thing and do it well. A++++; would buy again" (AriX, January 28, 2015). Another review, titled "An Excellent Example of the Genre," extols how Is It Tuesday? stands out among the other "far drabber" and "wanting" Is It . . . ? apps in the store (willia4, May 22, 2012). Both reviews underscore that even seemingly simple software like a calendar app that keeps tracks of your daily, weekly, and monthly dates and appointments can be broken down into even more focused and micro components (i.e., focusing only on Tuesday). Just as music genres have splintered and fragmented in an endless quest for marketing niches, software, as app, has broadly expanded the varieties and genres available while providing extreme levels of specificity for the task(s) to which it can be put.

Even if reviews like those above are dismissed as fakes or as organic reviews from caring friends helping to boost the app's profile, these performances matter for the user's experience of the app. Not only can reviews and ratings shape a user's initial choice to acquire or download an app, but they also feed into the app's placement and visibility within the app store and whether an app will get featured on the various landing pages or "Top Apps" rankings that populate most app stores (e.g., "Best New Apps," "Top Free Apps," "Best Back to School Apps"). Although Apple does not publicly share data on how it weighs its ranking algorithm, developers anecdotally believe the formula looks something like this: "Ranking = (# of installs weighted for the past few hours) + (# of installs weighted for the past few days) + Reviews (star rating + number of reviews) + Engagement (# of times app opened etc.) + Sales ($)" (Edwards 2014).

Google is similarly secretive about its app store rankings, but developers report similar measures affecting their apps' visibility (Pratt 2013). Given the overwhelming number of applications in the major app stores, it should not be surprising that an entire sub-industry has sprouted dedicated to gaming the ranking system through user ratings and reviews. Just as a robust and shady search engine optimization sector has grown around web search in the past two decades, there is now a burgeoning business in app store optimization (ASO) that promises developers the highest visibility for their applications. ASO tactics include innocuous activities such as providing advice on logo design, keyword choice, in-app retention strategies, and other aspects of an app's display page, to more questionable practices such as paying for bulk downloads to increase download statistics or using public relations firms to plant fake, glowing reviews of an app (see, for example, Miguel Helft's [2010] review of the Federal Trade Commission's ruling against Reverb Communications, whose employees wrote deceptive positive reviews for clients' apps without disclosing to other users that they were being paid to do so).

Although there is little hard data on how reviews and ratings affect an app's circulation, there is certainly enough anecdotal discussion to fuel a growing market for ASO services. For Friedman, the reviews of Is It Tuesday? "definitely help spread the app" (email to author, August 6, 2015). Beyond introducing his app to new users though, the reviews also contribute to the character of the overall experience of the app: "People that read the comments before downloading get a laugh even before opening the app. I really think it adds to the humor of the app from start to finish" (email to author, 2015). In other words, app reviews may be performances, but they are performances with consequences; they shape how an app is received and understood but also whether or not it is even discovered in the first place.

Conclusion

Just as earlier reformattings of the software commodity provided "a moment when contemporary computer technology created a business opportunity for a new mode of software delivery" (Campbell-Kelly 2003, 3), the switch to apps and app stores has reconfigured the possibilities for the software product. The relative accessibility of app stores, at least compared to previous iterations of the software retail chain, has helped developers with any idea—big or small, trivial or essential—publish and circulate their software. It is safe to say an app like Is It Tuesday? might not exist, or at least exist as widely, in the shrink-wrap retail software regime that drove software sales in the 1970s, '80s, and '90s. Today's app stores not only put new kinds of software on display in new ways, but they also make critical discourse about software part of the discovery and acquisition process. The reviews and ratings for Is It Tuesday? spoof similar such reviews for other apps but also for countless other websites and outlets (e.g. Yelp, Amazon). Digital stores factor the *quantitative* aspects of reviews and rating systems into the complicated algorithms they use to gauge the popularity of their products, and to help determine where and how frequently any given app appears in an app store. The *qualitative* components of reviews, on the other hand, shape the decisions users make about acquiring an app at the point of purchase, as well as the experience of the app before (and after) using it.

The rise of cheap or free apps in app stores linked to mobile phones and tablets has also helped to insinuate software into an ever increasing number of our daily activities. Software has become mundane: it is no longer a shrink-wrapped box you trudge to a retail store to buy but, instead, a common and integral part of the next generation of computational devices. It can (and is still) used to accomplish complex and multifaceted tasks, but it is also increasingly simple, limited in its functionality, and aimed toward achieving rather everyday, routine or otherwise unremarkable tasks. The need Is It Tuesday?

fills, at least on its surface, is as mundane and microfunctional as it gets. But the ways in which the app, and its users, comment on the format of the software commodity and the mechanics of distributing digital goods in digital stores make it far from a trivial piece of software (or, rather, they highlight just how meaningful the trivial can be).

Technology has long been a solutionist enterprise. Whether telephones and tanks, calculators and cement, water dams and washing machines, almost every new technology has brought with it promises about the individual and social problems it will solve. But apps signal the arrival of an environment in which software is vying to solve even the most mundane, routine, and microscopic problems. From knowing what day it is, to remembering your grocery list, to turning your phone into a flashlight or mirror, app developers are on a collective quest to seek out problems they can address through software.

Of course, there is an endless range of problems to meet, from social, cultural, or technical ones, such as increasing access to the internet for rural or low-income communities or increasing literacy rates among certain communities and demographics, to more attainable ones, such as finding a user a ride to work or providing a workout plan for the week. The problem with solutionism is that it privileges problems that are solvable, not necessarily those that are socially, morally, or ethically in need of being addressed. Is It Tuesday?, in both its extreme microfunctionality and its performative reviews and ratings, briefly calls apps' solutionist tendencies into question. Addressing the role that apps more generally play in upholding solutionism, however, appears be a problem for another day.

C Health/Fitness

LoseIt!
Calorie Tracking and the
Discipline of Consumption
NATASHA SCHÜLL

App: LoseIt!
Developer: FitNow
Release Date: July 2010 (Current Version: 9.1.11)
Category: Health and Fitness
Price: Free / Premium $39.99/yr
Platforms: Android/Blackberry/iPhone
Tags: diet, agency, self-regulation
Tagline: "Helps you stay in your calorie budget"
Related Apps: MyFitnessPal, MyPlate, MyDietCoach,
     Lifesum, FatSecret, CalorieKing

LoseIt! appifies weight management—and, by extension, health—through features that mitigate the tedium of tracking and budgeting caloric intake. The premise is that users will make better decisions about what and how much to eat when equipped with a digital assistant to keep track of the cumulative caloric consequences of these mundane choices. This review is concerned with the assumptions about human agency and the technological mediation of health that inhere in the app's design logic, marketing appeals, functional affordances, and user practices.

I approach these assumptions in a genealogical spirit, stepping back in time to explore a trio of artifactual antecedents to the LoseIt! app: a sixteenth-century "static chair" for home use, the Victorian bathroom scale, and portable calorie counting tools. Each, in its respective historical context, participates in a nervous social debate over proper self-conduct in relation to changing norms of health consumption and the risks of dependency on external devices. A consideration of these earlier instruments and debates lays the groundwork for an examination of contemporary weight management apps, allowing us to better grasp what is—and is not—novel about their design features, the demands they make on users, the associations they draw between mundane comportment and health, and the ways they fail.

Body in the Balance: The Static Chair

For over thirty years, the sixteenth-century Venetian physician Santorio meticulously weighed his body, all he ate and drank, and the bodily waste he
produced. He began after learning, to his astonishment, that for every eight
pounds of food and drink he consumed he evacuated only three.[1] To account
for the difference, he drew on the classical theory of "insensible perspiration,"
or the wafting of moisture through the skin's pores. Because it was imper-
ceptible, argued Santorio in his 1614 *Ars de Statica Medicina*, the only way to
determine—and manage—these worrisome quanta was via rigorous weighing.[2]

The doctrine of static medicine derives from the Hippocratic notion of
bodily health as an essentially preservative endeavor: a body that weighs the
same amount every day is a healthy body. "If there daily be an Addition of
what is wanting, and a Subtraction of what abounds, in due Quantity and
Quality, lost Health may be restor'd, and the present preserved," wrote Santorio (Dacome 2001, 474).[3] To support this self-balancing act, he devised several
tools, including a table and bed that doubled as scales and a contraption he
called the "static chair" that hung from the beams of his home (see fig. 1);
seated here, he took all of his meals.[4]

For over a century, Santorio's methods inspired self-experimenters around
the world—from Benjamin Franklin to King Frederick of Prussia—to build
their own static chairs in the quest for the equilibrium of health (Dacome
2012, 385).[5] Thus reads an entry from the "tedious calculations" made by the
South Carolina physician John Lining in the 1740s:

> eat 10 6/8 Ounces of roasted Lamb, Bread and Shallots, drank 40 Ounces of
> Punch, and used no Exercise; in these Two Hours made 3 3/8 Ounces of Urine
> and, being exposed to the Wind, perspired only 12 Ounces, though I sweated a
> little all the Time. (qtd. in Kuriyama 2008, 417)

Confinement to a chair was a small price to pay for access to the critical variable of "insensible" matter that, daily, departed Lining's body and demanded
replenishment in equal measure.

More than a tool of accounting, the static chair was also a tool of behavioral regulation, notes Lucia Dacome (2001, 467); it "mechanically enforced the
control of ingestion." Santorio advised would-be weighers that, prior to sitting
in the chair to eat, they place at the opposite end of the hanging beam a weight
equivalent to that of the food and drink they wished to consume so that once

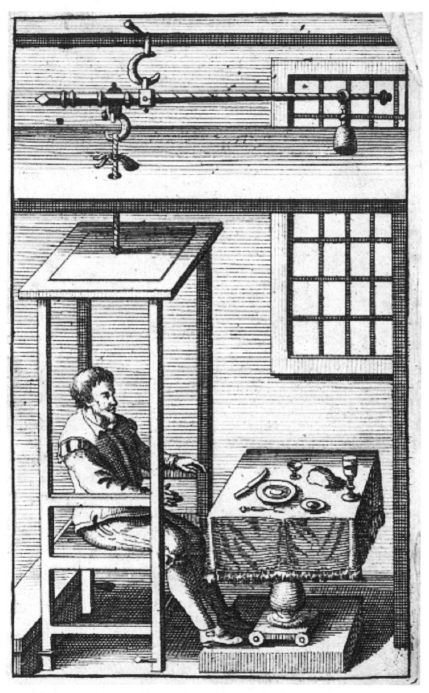

Frontispiece, *De Statica Medicina*, 1615

the meal had been consumed, the seat would drop below the level of the table, "sanctioning the end of the meal"—as depicted in the much replicated frontispiece of a seated figure in the chair, his hand extended toward an unfinished meal now out of reach (Dacome 2012, 381).

A century after the publication of Santorio's treatise, medical authorities began to object to the power his weight-tracking system vested in the chair itself. They worried that dependency on the chair—both as a tool of measurement and an enforcer of conduct—would atrophy users' natural sense of when and how much to ingest, ultimately depriving them of agency. In 1712 Scottish medical writer and apothecary John Quincy cautioned that adherents of the chair should avoid "the Exactness of the Sanctorian Calculations," lest they become obsessed with "how often they must weigh themselves, and whether they ought to drink and eat by the Ounce" (quoted in Dacome 2001, 478). Excessive self-scrutiny and submission to the dictates of the chair, warned the famous doctor George Cheyne, could pervert both one's appetite and one's capacity for self-regulation. "I do not dine and sup by the clock but by my chair," read a satirical letter in *The Spectator* penned by the political essayist Joseph Addison. The protagonist confessed that he had become slave to his static chair in an effort never to deviate from his ideal weight (of 200 pounds). "I used to study, eat, drink, and sleep in it; insomuch that I may be said, for these last three years, to have lived in a pair of scales." He concluded grimly that he found himself "in a sick and languishing condition."

At stake in Addison's satire was the "sick and languishing" condition of society in a context of rapid commercial expansion and rising consumption that posed new risks to the health of citizens and demanded of them new kinds of vigilance over their bodies. In league with anti-luxury campaigners, medical practitioners regarded "the over-consuming body as a body that was bound to perish, and warned that an over-consuming society was a society affected by a terminal disease" (Dacome 2001, 490). Healthy consumption was to be achieved by cultivating inner restraint, not by recourse to external technology. Along these lines, Quincy advocated "self-control rather than weighing on the Sanctorian chair" (480).

Truth Telling: The Bathroom Scale

Given that the static chair was not an object of widespread use, these heated warnings may seem out of proportion with its historical reality. But the stakes were high when one considers that the act of self-weighing was becoming "part of well-bred people's daily practices of well-being" (Dacome 2001, 490) in the burgeoning consumer society. By the mid-1700s it had become fashionable for customers at certain London coffee shops (including women, by century's end) to periodically record their weight in store ledgers, and soon scales began to spread from learned and aristocratic homes to bourgeois European households.

As scales became more mundane, the way they supported health changed: instead of helping people keep their weight steady, they helped keep people honest about the consequences of their consumption habits. Weight was no longer a number to stabilize but rather an index of self-discipline, with the scale serving as a sort of lie detector or "materialized conscience," as Hillel Schwartz observes in his historical account of dietetics, *Never Satisfied* (1986, 17). If in the static tradition weighing had been performed to track the discharge of excessive, potentially polluting elements that threatened to *destabilize* weight (Kuriyama 2008, 433), now it was performed to track the intake of excessive, potentially polluting elements that threatened to *increase* weight.

As fatness became increasingly suspect in the Victorian era and "people began to accept the notion that the body when weighed told the truth about the self" (Schwartz 1986, 147), coin-operated penny scales came to occupy quotidian venues—railroad stations, pharmacies, fairgrounds—where they publicly chimed, sang, and flashed users' weights. Over time they became more muted and discreet and, eventually, moved into the personal sphere of the home bathroom, where unclothed bodies could be more precisely and routinely weighed (165). Stepping on the scale became a private reckoning with the effects of dietary choices on one's health and the fitness of one's figure, to be practiced "from the cradle to the grave," as one scale maker advertised (169; Crawford 2015). Anticipating the solutionism of present-day app developers, scale makers rebuked those naysayers who insisted that "dieting must be an act of internal regulation" (Schwartz 1986, 96) and sanctioned the daily use of self-gauging technology as an ethical way to live in consumer society.

Budgeting Diet: Calorie Counting

Reinforcing the transfer of cultural anxiety from the toxic residues of excrement to the superfluities of consumption, a new variable in the nexus of health and weight soon rose to dominance: the calorie, understood as a unit of energy contained in food and expended through bodily movement. "As a

habit of seeing through and beyond food, calorie counting gave dieters a power which no scale could equal" (Schwartz 1986, 177). In the 1930s, pocket-size
calorie-counting notebooks and calculators gave material form to this new
scale. Precursors to today's digital apps, these gadgets facilitated self-discipline
by keeping track of the accretion of small acts of consumption throughout the
day. As Santorio's chair measured the exit of insensible evaporation out of the
body, calorie-counting systems measured the entry of insensible energy value
into the body.

The focus on calories also shifted the temporality of one's attention: instead of assessing consumption choices retroactively as scales did, with calorie
counting "dieters could truly anticipate the future: so many extra, invisible
calories would mean so much extra, invisible fat," writes Schwartz. "Counting calories was like compounding interest; it assumed an appreciation of
promises and futures—how long a walk it would take to burn off a chunk
of chocolate, how many flights of stairs to climb off a piece of pie" (134).
Calorie-counting notebooks and portable calculators—and, now, apps—are
aspirational rather than static; they allow dieters to calculate the risk of present consumption choices against their desired future weight.[6] To this day, the
calorie has remained a coin of the realm of health—which has itself become a
financialized "regime of anticipation" (Adams et al. 2009) characterized by an
unwavering actuarial stance.

As we will see, what digital apps bring to the task of health is not simply
an easier means of recording and tracking energy intake but the capacity to
automatically sense and factor in its expenditure, and to inform users of their
input/output ratios at any given point in the day. Driven by always-on algorithms, weight management apps seek to automate the vigilance required of
citizens who wish to be fit and healthy, notifying them in real time so that they
can adjust their behavior in accordance with weight goals.

Logging to Lose: The App

Upon first launching LoseIt! (whose icon depicts the face of the humble, analog step scale), users supply the following information: current weight and
height, sex and age, goal weight, and how quickly they wish to lose weight.
The app then calculates a customized daily calorie allowance that appears
henceforth atop the screen alongside a tally of calories consumed, calories
expended through exercise, and calories remaining in one's "budget." Rather

like a stock market ticker, this set of values changes as users move their bodies and log items they have consumed.

Although the app automates many aspects of weight management, the act of logging itself (initiated by pressing "Log" on the app's dashboard) is one that users must diligently perform—although most often they need not manually enter the caloric value of meals as most food items can be found by searching the app's vast, constantly updated database. For instance, "egg" returns 1,308 results that can be narrowed with further descriptors. The database includes most restaurant chains (a cheeseburger from Applebee's is 970 calories; clam chowder from Panera is 630) as well as some relatively obscure dining venues (the quinoa-hijiki salad with tofu from Dojos in Manhattan's East Village clocks in at 300 calories). Name-brand, packaged items can be logged simply by scanning their barcodes. The app learns over time to present up front the items most frequently logged by a user.

To remind users to log in the first place—something calorie-counting technologies of the past could not do—LoseIt! texts them at set intervals with such messages as "An empty log is a lonely log"; "You bite it—you write it!"; "Time to weigh in!"; and "Take another 200 steps for bonus exercise points." More nudging happens through the "Social" feature: here app users can link their accounts and motivate each other to complete goals, join groups that follow certain diets, or participate in challenges (from the simple "log something every day" to the daunting "eat vegan for three months").

Beside "Log" and "Social" in the lower dashboard appears "My Day," where users can visualize their status with respect to assorted values: color wheels and bars indicate if they are under, over, or within their allotted daily and weekly budgets for calories (see fig. 2); a pie chart reveals the proportions of fat, carbohydrates, and protein in their consumption of nutrients; and "Steps" indicates paces for a given day—data synchronized in real time with smartphone step counters or other fitness devices. "This is how I know as the day goes along how I'm doing," says one user of the "My Day" feature. The ability to constantly check one's status evokes the obsessiveness of Santorian weighing, yet the object is not to preserve one's current weight but to reach a future weight.

This aspirational temporality is most salient in LoseIt!'s "Goals" feature, which draws on users' current weight (entered manually or automatically synched from their wireless scales) to graph losses and upticks over time, representing the future as a dotted line that converges with their desired weight—a point that moves farther or closer depending on how fast they wish to lose

LoseIt!'s calorie budgeting screen (left) and "Goals" feature (right), representing future weight on a graph

weight and how rigorously they stick to their caloric budgets. With this feedback, the app imports future consequences into the present, allowing users to adjust as they go, making trade-offs between food and exercise. "I use the data to slow down my eating during the day or make better food choices if I am nearing my daily caloric limit," notes one user review; "I might have a lighter lunch if I see I had a high-calorie breakfast, or go for an extra run." Instead of using ropes and pulleys to maneuver users out of physical reach of their unfinished meals, the app provides information and sends notifications; it offers itself as a digital compass for navigating the confounding, tempting, and sometimes toxic landscape of everyday consumer choices, promising to supplement users' shortsighted perspective with a continuous, informatic gaze able to compute the consequential aggregate of their small daily actions.

LoseIt! has been downloaded millions of times, with users rating the app an average 4.5 out of 5 stars. It consistently places in the top ten "health and fitness" apps on both Google and Apple stores. "More than 60,000,000 pounds and counting!" boasts the LoseIt! website, presenting a live counter that ticks up by approximately one pound per second. Despite minor differences in features and functionality, LoseIt! and its competitors—including MyFitnessPal, the longest-running and current leader in the market—share a common logic of health in which we must be as conscious of caloric intake as possible, in real time. If we do not log calories as they are consumed, we risk consuming too much, thereby delaying the attainment of our aspirational weight.

Dangers of Dependency: User Testimonials

Sitting on her couch with her laptop and iPhone, Krista told me how she had come to be in the top one percent of users for a popular calorie-tracking app. In her mid-thirties, after meeting a romantic partner who liked rich Italian food, Krista gained seventeen pounds; she gained another eight when her mother died. "I decided I needed a little counter," she recalled. Krista had never done Weight Watchers or followed a diet program, but she came across the Loseit! app and discovered that she found it compelling to log her calories throughout the day. "It became an addiction a little bit, I had to do it every day. It got really *granular* for me."

Krista pulled up the app on her laptop and reviewed data from a random day a few months back. Breakfast: Greek yogurt, blueberries, and chia seeds. Lunch: beet greens, a small baked potato, and great northern beans. Dinner: more beet greens and beans, fried with egg whites in olive oil. "It was pretty much a perfect day," she commented, "not just the calories but the nutritional ratios." The satisfaction Krista took in accumulating sequential days of caloric "perfection" motivated her to modulate her diet and lose weight. "I describe myself as an *incrementalist*—I pay attention to small bits and how they add up." She went on to draw an analogy between calorie tracking and quantified trading, her professional field: "It's like machine learning on yourself. It's recursive because you're trying to achieve these numbers and ratios and you learn through trial and error, so the app feeds into how you behave over time. It has this algorithmic component to it." What Krista describes is not the atrophying of natural appetite that Cheyne feared in the 1700s but, rather, a subtle shift in her appetite: "It starts external and objective and then it becomes internalized—so you can just look at your plate and intuitively know the calorie value."

Countering this benevolent narrative of habit change, some charge that weight management apps insinuate themselves so thoroughly into users' routines that they wrest agency from users and enslave them in obsessive loops of logging (Zomorodi 2016). The questions Addison raised in 1711 in his cautionary caricature of the overzealous self-weigher are just as relevant in today's appified context: What are the risks of ceding calculative and regulative control to an external device? Might one's will wane? Do these devices promote or endanger health? User testimonials posted to the iTunes app store, Amazon.com, and other online review sites speak to these questions. One woman who lost forty pounds with the app's assistance comments, "I probably don't need to use it anymore because I know when I am going over my calories but I just can't stop using the app. It's my little helper. I need it to keep me on track."

Some users speak of this need as an "addiction," using the term in a lighthearted, positive fashion:

- I couldn't have lost these 14 lbs. without the app, and now I'm addicted to it (instead of food!).
- I love this little app. It is so fun and addicting to add food and exercise then watch the calories go up or down.
- Am I the only one addicted to this? I find myself checking in several times a day.
- Nope, not the only one! Just recently started using the app again, and I'm on the darn computer all the time.
- Definitely not the only one. I am very addicted to this now but I figure there are worse things to be addicted to!

Others speak of addiction in a darker sense—not to describe the pleasure of checking and entering numbers but a reluctant enslavement to the device and its routines. "This app has me in my seat too much—and that's what I have to change in the first place," writes one poster, calling to mind Addison's satire of the man who could not unseat himself from his Santorian weighing chair. One former user confessed that he spent the majority of his days compulsively recording everything that passed his lips—"even the 'calories' of vitamin tablets, even glasses of water. . . . At the end of the day it would tell me how much I would weigh in five weeks time if my net calories were the same as that day. I was glued to the app—I panicked if it stopped working."

Another writes that use of calorie counting apps by those with eating disorders is so common that deleting such apps becomes, for many, a milestone of recovery. Echoing Anthony Giddens's (1991, 105) argument that eating disorders are "extreme versions of the control of bodily regimes which has now become generic to circumstances of day-to-day life," a reviewer of LoseIt! observes that the app is not merely an accessory to pathological eating but also its trigger and accelerator: "Even if you don't have an eating disorder when you initially download this app, it can so easily become addictive and get out of control. It's not rational or okay in any way to be congratulated on a starvation diet by an app masquerading as a top class 'health and well-being' app." She goes on: "There are some mobile phones that come with MyFitnessPal built in and you *cannot delete the app*. How dangerous is that?" What is appified here is not health, she suggests, but pathology, echoing the fears around dependency and the inducement of a "sick and languishing" state that were articulated around weighing practices in the 1700s.

Conclusion

And yet, in the contemporary world there are subtle differences in the ways that health and the project of selfhood are conceived. Health, once understood as a baseline state temporarily interrupted by anomalous moments of illness, has been recast as a perpetually insecure state that depends on constant assessment and intervention (Dumit 2012; Lupton 2013). In the era of so-called lifestyle disease (e.g., chronic conditions such as obesity, type 2 diabetes, heart disease), we are all potentially sick and must cultivate "continuous reflexive attention" (Giddens 1991, 102) to our quotidian decisions: what to eat, how much to move, when and how long to sleep. At the same time that well-being is construed as deriving from mundane market choices, consumers are understood to constitutionally lack the knowledge, foresight, or inner will to reliably make the *right* choices.[7] Everyday health and fitness apps like LoseIt! present themselves as a partial solution to this double bind of contemporary self-regulation, offering us a way to fulfill the cultural demand for self-management while delegating a portion of the tedious, nebulous labor involved in meeting that demand.

Notes

1. That vapors escaped the human body was not a revelation, but the magnitude was shocking—and troubling, given the importance of systemic equilibrium in the humoral tradition (Kuriyama 2008, 416; see also Renbourn 1960).

2. Santorio, a friend and colleague of Galileo, is known for introducing the quantitative approach to medicine and for inventing various measurement instruments, including

a precursor to the modern thermometer as well as the pulsilogium—a pulse-rate meter considered to be the first machine of precision in medical history.

3 . Female bodies were thought unable to reach this ideal of health, being inherently unstable and always fluctuating (to a large degree due to menstruation)—never static (Dacome 2012, 475).

4 . Until Santorio, scales had been instruments of commerce, used to weigh items such as animals, coins, gems, and grains (Dacome 2001, 468). While human figures appear on scales in ancient depictions of the "weighing of souls" and in various historical cases of witch weighing, it is only with the static chair that physical health, rather than salvation, came to be at stake in weighing (469; Schwartz 1986, 11).

5 . *Ars de statica medicina* (1615) was reprinted some forty times and was translated into English (1676), Italian (1704), French (1722), and German (1736).

6 . In the 1970s the Weight Watchers program attempted to make food choices easier by assigning numerical value to foods and offering prepackaged meals of set caloric value.

7 . While some health policy makers continue to advocate for the cultivation of inner resolve and self-control, this position has waned in the wake of behaviorism's mid-century rise, which popularized behavior change as best achieved by altering the environmental contingencies of choice making rather than through inner forces such as "willpower" (Rutherford 2009).

Fitbit
Wearable Technologies and Material Communication Practices

KATE O'RIORDAN

App: Fitbit
Developer: Fitbit Inc.
Release Date: January, 2014
Category: Health and Fitness
Price: Free (pairs with Fitbit device)
Platforms: IOS, Android
Tags: diary entries, letter writing, self-improvement, self-
    discipline, hygienic routines
Tagline: "Fitbit is dedicated to helping people lead healthier,
    more active lives."
Related Apps: Google Fit, Up, pedometer, pedometer and
    weight, diaries, journals, letters

Walking through Time

Fitbit is an app that pairs with a health and fitness tracking device. In this essay I make three key arguments about apps, particularly Fitbit. My first argument is that apps have become the interface environment for proliferating hardware devices. In Fitbit's case, automated data collection is made meaningful through the address of the app to the person using the device. Second, Fitbit constructs an ideal gendered subject for automated fitness tracking. Third, although Fitbit promises a more active lifestyle, the automation of recording directly modulates a more constrained and eviscerated subject than other forms of tracking such as journaling and diaries. I bring in English writer Dorothy Wordsworth's diaries as a case study to offer a historical perspective to the measuring of women's footsteps over time and the changing notion of the fit feminine subject.

Fitbit, as is obvious from the name, operates in the register of fitness. To be fit is to be healthy and suitable. Fitness has resonance with normative discourses of the healthy and appropriate subject; it also chimes with fitness for the purpose of fitting in and raises the specter of the unfit. The flip side of the "taken for grantedness" of the ideal Fitbit subject is the fear of unfit subjects and their challenge to normative gender ideals and technocultural worlds.

App as Interface: Making Automation Lively

The Fitbit app is an interface for the accompanying Fitbit activity tracker, which is first and foremost a pedometer/step counter. The app offers a graphic, interpretative interface for steps, workouts, calories consumed, and hours slept. It assembles this data into graphics and timelines to produce a personal, textual account of data collection and the self. A personalized array of icons representing food, drink, and calorie intake appears alongside icons representing exercise and calories burned.

The app thus operates as an interface that creates an indexical representation of the subject; in other words, the app transforms the tracker's data collection into an enumerated and friendly, graphics-driven visualization. It is through this platform that the liveliness of the user is reinterpreted as calories consumed and calories burned. This representation of the quantified everyday is enlivened by automated direct messages of encouragement—push notifications—and rewards, such as visible acknowledgment of goals and achievements. Fitbit subjects the practices of everyday life to opaque systems of counting and disguises this flattening out through marketing rhetoric focused on activity and achievement. Through a lively, personalized interface, the app delivers an intimate connection to a highly automated platform, which is arguably first and foremost a surveillance tool that collects information on users. The advertising discourse and communicative affordances offered through the app generate a market for Fitbit devices that monitor activity and commodify the subject.

From Apps to Devices and the Internet of Things

Fitbit's interface and practices are part of a long history of recording activity as routinized everyday life. For example, Kate Crawford and her colleagues (2015) contextualize these kinds of everyday practices through a comparative history of weight measuring and wearables. Activity tracking can be thought of as part of new regimes of mediated health promotion, self-tracking, and data ideology (Lupton 2015; van Dijk et al. 2015; Kennedy et al. 2016). However, in the context of the rise of the app as interface, Fitbit also raises questions about what an app is. Fitbit is both a device and an app, and this pairing is significant in terms of accounting for the contemporary social media paradigm in which apps have become both the signature interface and also taken for granted. Looking at the shifting promotional frames for the app bring this to the fore.

From 2011 to 2013 the Fitbit was overwhelmingly framed in media coverage as an app. The medical instrument trade press described Fitbit as part of the app avant-garde (Arnold 2011). In the health and technology sections of newspapers, writers playfully described its appification: "if you're appy and you know it." This aligned with a period of significant app development and currency in the word "app" in trade and popular press, from Apple's launching the first five hundred apps in 2008 to the announcement in 2013 that there were over a million apps in the App Store (Ingraham 2013). In a short period of time, apps shifted from hot new tech industry object to phenomena produced in such numbers that they became a mundane, expected interface.

In 2013, with the launch of wristband-style sensors and their proliferation in the electronics market, there was a slight turn of focus from apps to the language of devices. The launch of the Apple Watch in 2015, for example, contributed to a refocusing of the market on devices and device design. As a device-dependent app, Fitbit is also an instantiation of the "Internet of Things." Framed in terms of big data and apps culture, an app like Fitbit is both about the aggregation of data and devices to diversify its market offering. All transactions and interactions with Fitbit create flows of data, which are stored on the company's servers, creating repositories through which a projected futures market in health data, medical insurance, and drug development is envisioned. However, to date the company has primarily continued to profit from device sales and the emerging sale of subscriptions for "pro" services. The health and activity data that Fitbit sells is not medically valuable, and opportunities to sell the data for drug development look limited.

The promise to link health insurance with activity monitoring has had more traction. Fitbit went public with its stock in 2015, although as one of its detractors commented, "Gadgets that measure fine distinctions in the propensity for heart disease in a population that has a near-zero occurrence of heart disease are not health care reform. It's not serious" (Mullaney 2015). However, the promissory futures of medical data markets are a serious area for investment. For example, Medtronic, one of the world's largest medical instrument companies, announced in 2016 that it was pairing its app with Fitbit devices for diabetes monitoring (Medtronic 2016).

What's New

Since its launch in 2007, Fitbit has been primarily promoted as a pedometer for tracking a user's everyday movement. The company has continued to develop its product line to increase what can be recorded, both through automated collection and user logs. For example, new additions in 2016 were "quick logging widgets and quick water logging" (Fitbit.com). While Fitbit's language of widgets and logging is elaborate and its list of functions keeps growing, the core preoccupation with measuring and communicating steps remains. On one hand, the preoccupation with walking through time reinforces current normative health promotion messages in an era when the couch potato and obesity are conjured as figures of the same media culture as the device. On the other hand, measuring steps connects to a longer history of the personal practice of recording movement. Walking, as a daily practice, and recording and communicating those walks, has featured as an element of daily life, and particularly of female subjectivity, for at least the last two hundred years, through journals, letters and novels.[1]

Gendering the Interface: Upgrading Diaries and Letters

Fitbit is targeted toward middle-class women who are assumed to have time for, and interest in, exercise but whose work or home life is either overscheduled or too sedentary. For example, women working on laptops feature in the advertising material. One advertisement demonstrates the device pushing reminders to a woman presumably at her office to get up and walk around while otherwise sitting in front of a screen. On the UK Fitbit website (as of August 2017) the only images of people using the device are of women. The word "accessory," which is aligned with gendered marketing language, appears on the front page, and the stylishness and slimness of the device is emphasized. Women's fashion designer Tory Burch has a line of designer bracelets to encase the Fitbit device as well as pendants to make the tracker even more stylish, promising to transform your Fitbit Flex into a "super chic accessory" (Fitbit. com). Although Fitbit products are also targeted toward men, the company's address to women is much more dominant in the advertising discourse. References to the device in the press appear much more frequently in lifestyle magazines and health and well-being sections addressing women.

The construction of this kind of female subject as a walking subject has some consistency over time. Counting steps has been a preoccupation for white middle-class women since at least the eighteenth century, as evidenced by diary and journal writing. For example, in Dorothy Wordsworth's journals of 1798, each entry starts with an account of walking: "Walked on the hilltops," "walked through the wood," "walked between half past three and half past five." For certain women, letter and diary writing marked out ritual time in everyday life in the eighteenth and nineteenth centuries. The first known recorded diary by a woman in the UK is that of Lady Margaret Hoby, written at the turn of the sixteenth century (Fraser 1984).

Recording daily walking in diary form was a moral exercise and an expectation of a pious life. This virtuous activity of recorded everyday practices is similarly present in Fitbit's appeal to an audience of disciplinary subjects through its focus on rewards for self-betterment via sustained fitness and health. In other words, the ways in which Fitbit encourages the ritualistic measuring of bodily signals has long been an expected practice of appropriate daily habits and their recording and communication, whether via journal entry or regular data uploads. Like the use of Fitbit, diary writing signaled an appropriately disciplined subject, participating in a normative regime of personal hygiene and good habits, which in turn maintained a particular social world. Fitbit similarly constructs an ideal subject position, engaged both in appropriate activity and in daily rituals of connecting to the device and checking into the online interface or "dashboard." However, unlike older forms of journaling and measuring technologies, as Crawford and her colleagues argue, "the full use of the data is always out of sight to the user" (2015, 494). Unlike the intimacy and accountability of personally recording a day of walking in a diary, Fitbit ports data back to the company and its modes of calculation are opaque.

The aggregate of information Fitbit collects assembles a text that is both a form and an expression of recording, especially when feelings and reflections are added through journaling practices. It is the app, in conjunction with the wearable device, that makes this possible, making an otherwise automated practice of data collection *lively* (Lupton 2015; Ruckenstein 2017; Savage 2013). By offering a more detailed, personalized, and interactive dimension, the app facilitates what the tracker can't on its own, what Minna Ruckenstein refers to as the "work by users that keeps data alive" (2017, 1036). Thus, it is the platform the app provides for the user that bridges Fitbit to older forms of journaling and letter writing through its capacity to provide an extended and mobile space for the representation of activity and subsequent sharing with others. The Fitbit dashboard, for example, allows sharing of all aspects of activity from quantities of water drunk, to hours slept, to steps walked.

Similarly, this mundane logging occurs in nineteenth-century letter writing and journals, which provided lists of daily activity and habits as well as reflection on the habit of writing itself. Letters were often written at a designated time of day and were very similar in content but sent to different people, often shared with an audience of family and friends. Thus, an account of the same activity, thoughts, and health observations would be sent in similar terms to a friend, parent, and sibling so that three or four very similar letters would be written in one morning with only the address and framing changing from one letter to the next. Likewise, diaries were shared around family and friends. To continue with the example of Dorothy Wordsworth, her brother read her journals, and entries were repeated in letters and sent to a wider circle of friends and later on circulated through publication. Fitbit, then, can be thought of as another disciplinary technology constructing a mundane, gendered subjectivity counted and constrained by normative discourses of fitness and activity.

Through features like push notifications, real-time syncing, and competition with friends, Fitbit compels the kind of compulsive attention of other social media forms. It offers gratification to real-time questions such as Has my data updated yet? Has it all been captured? Have I met my targets or walked enough steps? It provides immediate satisfaction if you like graphs and graphics about activity. For people who are already into using structures such as charts and targets, or who aspire to do so, the app provides a sustained pleasure experience. Personal trainers, for example, recommend it as short-term encouragement for people training for specific events or trying to keep up with daily exercise routines. Its ability to capture the mundane and represent it as productive can be helpful in offsetting depression or apathy, establishing even short walks or sleeping as part of a visible regime of effort. Reviews and user comments often resist criticisms of it and express forms of attachment:

> I love the silent alarms that remind me to eat several small meals throughout the day. I love the little motivating statements like, "I missed you," "Great job!"

"Goooal!" It's funny how these little words can entertain you and keep you interested. This little gadget becomes your friend quickly. A good little helper if you let it be. (Fitbit.com)

The language of love, help, and friendship in relation to the alarms, motivation, and help points to the gendered routinization these practices invoke. The help that Fitbit provides differs from nineteenth-century diary writing in the kind of embodiment offered. In the mid-nineteenth century too much exercise was viewed as unfeminine. In Katharina Vester's social history of dieting, she argues that dieting discourses in the eighteenth and nineteenth centuries were directed toward men. Women were conversely thought to be more feminine if they did not undertake much exercise (Vester 2011). In this period the more sedentary practice of diary writing was the more feminine practice. The contemporary fitness device continues journaling potential as an option but records activity in real time. Although it demands screen time attention to the app, this is conceived of as a mobile activity, partially automated, while on the move. The language of Fitbit is all about saving time and fitting in the otherwise laborious process of conducting and recording daily exercise to a busy and active schedule. In an era when the ideal female subject is not sedentary but mobile, energized, young, and productive, the real-time automation of tracking an on-the-go lifestyle appears to offer an ideal form of journaling. Its promise is of liberation, freeing women up, motivating them, and keeping them on track in the pursuit of an idealized productive persona.

Playing with the contemporary language of app marketplaces, we might think of the Fitbit as an updated version of eighteenth- and nineteenth-century (women's) diaries and letters, and the surrounding marketing discourse of self-improvement as similarly imbued with the expected moral expectations of a virtuous, self-disciplined subject. Perhaps Fitbit is Diary Version 4.1, with paper journals (v.1), Filofaxes (v.2), Blackberry (v.3), and apps (v.4) as previous iterations leading up to the app-device pairing as akin to a version 4.1. The labor of the discursive work in letters and journals is transformed through the automated diarizing of Fitbit, freeing up users for more activity.

But whereas Wordsworth's diaries open out onto a world of nature and the mind, as well as everyday observations on nature and social companions and the mundane marking of steps walked, the range of activity Fitbit tracks and documents is much narrower. It is rather more like the productivity of a treadmill or hamster wheel than of personal exploration and fulfilment. Fitbit supports a life that cycles through a repetitive and relentless pattern of home, work, and gym and activities of steps, workouts, eating, and sleeping, indifferent to location and observations of one's natural or cultural environment. Its advertising imagery is confined to homes, offices, cafés, streets, and gyms— places where steps can be acquired without diverting from one's other expected routines of productivity. Its distances are measured in steps and floors, indicating the primarily urban and work/home confinement. It embeds its users as data producers tied to digital platforms, shackled by surveilling bracelets, for which the pleasures are limited, and limiting.

Marvelous Technology

Kylie Jarrett's (2015a) formulation of the digital housewife is relevant in thinking about the gendered work of inscription into digital culture through diarizing and disciplined subjectivity. Fitbit is marketed primarily to young women and mothers as a lifestyle accessory through which health and fitness are achieved and made recordable and manageable within the everyday. A close reading of the media coverage surrounding the device shows that the discussion of Fitbit is usually assigned to women's health and lifestyle sections of the press. As Jarrett (2015a) notes, it is not that all internet users are women but that much digital labor is feminized and aligns with longer histories of consumer marketing. The graphic on one of the first-generation device screens, the Fitbit One, showed a flower growing and shrinking. Another model, the Fitbit Flex, is colorful and signified as small, slim, and "flex" and usually positioned on women in promotional materials.

Fitbit aims to capture repetitive movements in everyday life and abstracts these into an account of fitness. The colorful dashboard and its recursive, purposeful accounts of the mundane produce a digital account of the self. Jarrett notes that "behind even the most sophisticated technological marvel lie the material energies of living human beings" (2015a, 2). The Fitbit dashboard captures the material energies of living human beings and reframes this in the order of sophisticated technology. However, the framing is very narrow, constrained, and flattened out. At the same time, it makes daily life even more mundane through abstraction. In contrast, diaries and letters captured daily life through prose and short entries and, in doing so, articulated a life in relation to a world beyond simply the medium of writing and its capacity to record. For example, Wordsworth's (1798) entry "Walked through the wood to Holford" is an articulation in relation to a specific experience of place. In contrast Fitbit's record of "October 15, 12pm walk, duration 1hr" is an articulation in relation to the mediated body enframed, ready to be converted into activity charts and relations to calories consumed, sleep monitored. Not quite the same, the app works in part to bridge this gap, visualizing the banality of step counting through friendly graphics and personalizing data by aggregating a user's activity and building a narrative of goal achievement through features like earned badges.

To return once more to Dorothy Wordsworth, she lived her life in relation to a relatively constrained circuit of mobility within the West country and north of England in the UK. However, her diaries, even when taken only in terms of accounts of walking, reveal a much more open and relational sense of self and world than the subjectivity offered through Fitbit. Fitbit is attached to a modern subject and enacts a kind of networked self, seeming to offer new freedoms in terms of both cultural imaginary and interface. However, this everyday technology appears to be more empty and more constrained, with a limited circuit of activity (self-device-work-gym-eat-sleep). Contrasted with Wordsworth, these medial forms of walking seem more enclosed, automated, and informational. Rather than the technological marvel obscuring the material energies of humans, in the case of biosensors like Fitbit the technological mundane obscures the marvels of human energies.

Such devices also obscure the marvels of technology. Fitbit is a wearable computer with considerable power compared to, for example, desktop computers of the 1990s. It is part of an array of devices loosely referred to as biosensors, or devices that record biological data. The more radical promise of biosensors is that we might know the world differently, that the incommunicable and the senses beyond perception could be opened up by such augmentation. Biosensors have been understood as opening up the possibility of new forms of communication whether sensory or nonhuman—for example, registering the signals of the body in new ways or taking readings of invisible signals like bacteria, radiation, or chemical balance. However, their use to open up new forms of communication has been limited. Counting steps has become the dominant market model, and Fitbit intensifies a subjectification of digital culture within a highly constrained and normative matrix of ideologies of self, fitness, and productivity.

Fitbit's intersection of software, device, and ideology produces an ideal feminine subject of communications technology. This is a version of feminine virtue as automated, fit, and productive; it is a subject from whom data can be harvested and commodified but wherein the substance and meaning of active and healthy is eviscerated and presented in ways that are more constrained and confining than through previous media forms. Although health and activity could be reconfigured to create a fuller picture of a lived life, with other apps such as maps and images, layers of apps do not have the same kind of subject-centered complexity as Wordsworth's journals, even with the help of the app to bridge subjective experience with enumeration. This stripping out is perhaps reminiscent of Marvin the paranoid android's observation in the Douglas Adams science fiction series: "Here I am, brain the size of a planet, and they ask me to pick up a piece of paper" (Adams 1980).

Notes

1. Consider Dorothy Wordsworth's two-hundred-year-old diaries, which start every entry with a reference to distances walked.

D Lifestyle/Relationships

Tinder
Swiped: A Focal Gesture and Contested App Visions

STEFANIE DUGUAY

App: Tinder
Developer: Tinder Inc. (part of Match Group, within the
    media conglomerate IAC/InterActiveCorp)
Release Date: August 2012 (Current Version: 8.2.0)
Category: Lifestyle
Price: Free (Tinder Plus starting at $9.99 monthly)
Platforms: iOS and Android
Tags: social networking, hooking up, relationships, swiping
Tagline: "Meet interesting people nearby"
Related apps: happn, Bumble, Coffee Meets Bagel, Hinge

At one point, Tinder's webpage displayed the app's download button with the tagline "Any swipe can change your life." When the app launched in 2012, its introduction of the swipe functionality for sorting users into categories of "like" and "nope" was fairly novel. As Tinder gained popularity, reflecting the increased uptake of dating websites and apps among American adults (Smith 2016), swiping proliferated across apps for other purposes, from selecting rental properties to matching rescued pets with owners (Cunningham 2015). The swipe has also become a culturally recognized logic, with popular songs, jingles, and celebrities making reference to "swiping right" on people and products they find favorable. While the swipe is a simple, mundane gesture that has previously featured in touch-screen technologies, its central role in Tinder's design has been highly influential in shaping users' and the press's perception of the app. Despite Tinder developers' assertion that it is a "social discovery app" (Craw 2014) for meeting new people, media commentary has often associated Tinder's rapid swiping with facilitating casual sexual encounters. This association between Tinder and hooking up poses a threat to the company's bottom line because discourses that link casual sex to immorality, promiscuity, and danger affect Tinder's uptake.

Perceptions of apps are not arbitrary; they stem from the sociotechnical, political, cultural, and economic arrangements in which users are embedded (van Dijck 2013). Tinder's iconic design functionality—the swipe—harnesses material features of software and hardware to facilitate rapid, sustained, binary, and visually focused interactions with the app. Users and the media perceive these qualities in relation to social and cultural understandings of gender and sexuality. As the swipe becomes fused with these perceptions, this focal functionality inhibits Tinder from dislodging popular notions of its app through either rhetorical or design interventions. In fact, attempts to realign users to a more favorable and, by extension, profitable vision risk trivializing the app's purpose altogether. This chapter examines the swipe's integration into mobile technology and how its affordances have been perceived within a particular sociocultural context, contributing to the app's reputation. By reflecting on Tinder's swipe, it is possible to understand the relationship between an app's design, corporate framing, and popular reception.

The Swipe as the Main Attraction

The swipe, as an embodied gesture, has historically featured in social, financial, and technological interactions, often bridging the body and technological interfaces. Waving one's hand or finger from side to side is a common motion that can symbolize greeting, dismissal, or communicate that people or things should be sorted *here* or *there*. In combination with various media forms, swiping enables perusing a magazine or flicking a radio dial. With the development of financial technologies, swiping has acquired a transactional quality, as credit cards and loyalty cards now enable the exchange of money or personal data with a quick swipe. Swiping features as a main activity on touch-screen devices, most famously applied in Apple's patented "slide-to-unlock" iPhone feature (Macari 2012). Some form of swiping is now used to accomplish a range of appified tasks, from composing words with SwiftKey to pulling back a slingshot in Angry Birds. While Tinder's adoption of the swipe extends from its role as an interactional, transactional, and interfacing gesture, the app's configuration of users, software, and hardware emphasizes certain qualities of the swipe—mainly the possibility for it to be continuous, binary, and visually focused.

Tinder's app store description identifies the swipe as the app's central activity. The app presents users with a profile "card" containing a potential match's photo, name, and age. While heart and "X" buttons are present, Tinder suggests, "If you're interested in connecting with someone, swipe right! If not, just swipe left to pass" (Apple Inc. 2016). If users swipe right on each other's cards, they create a match, allowing them to message one another. Stefan Werning (2015) describes Tinder's swipe as a sorting function that conceptualizes infinite space on either side of the screen, allowing users to categorize others into bottomless repositories. As Tinder combines the gesture with a predictive algorithm that attunes to users' selection patterns, the swipe is designed to reward ongoing use with more accurate potential matches. Rewarding consistent interaction compliments how smartphones have become constant companions with which users build intimacy (Mowlabocus 2016). Tinder's flow of new profiles, combined with repeated haptic interaction, fosters an intimacy between users and the app that sustains continuous engagement.

The swipe's sorting functionality presents users with binary options that emphasize visually focused information for decision making. Binary sorting according to attractiveness builds on previous rating games, such as *Hot or Not* and the early version of Facebook, *Facemash*, which allowed Harvard students to rank each other's appearance (Kaplan 2003). A user's photo occupies most of the screen, and users must interrupt the flow of swiping by tapping on a profile to see additional information. Emphasizing appearance accords with a broader visual turn in social media design and practice, with photo and video sharing increasing as technology becomes more accommodating to these formats (Duggan 2013). Rapidly swiping through continuous visual information is common to navigating other visually laden social media, such as by flicking upward to scroll through Instagram photos. When deployed among particular gender and sexual discourses, these continuous, binary, and visual qualities of Tinder's swipe are perceived in ways that influence the app's reception.

Interpreting the Swipe

Tinder's framing and users' perception of the swipe's qualities have developed within a particular sociocultural context. Peter Nagy and Gina Neff (2015) propose the concept of "imagined affordances" as possibilities for action that "emerge between users' perceptions, attitudes, and expectations; between the materiality and functionality of technologies; and between the intentions and perceptions of designers" (5). Tinder's material arrangements present the qualities discussed above, which are imagined to afford different applications according to various actors and their relation to the app. The company imagines Tinder's affordances in its attempt to market the app, while users and the media imagine these affordances differently as they adopt and critique Tinder. These multiple imaginings are shaped through existing discourses about sex, gender, and relationships.

Tinder's early promotion framed the app as conducive to sexual encounters. Bikini-clad models featured on Tinder's home page,[1] while launch videos on YouTube (now removed) included sexually charged imagery. Tapping into associations among alcohol intake, college life, and casual sex (Heldman and Wade 2010), Tinder featured alcohol brands in early videos and promoted the app to sororities and fraternities (Summers 2014). Tinder followed on the success of several hookup apps for men seeking men, and media articles often compared it to these apps (e.g., Muston 2013). Similar to how Grindr's focus on images with minimal information exchange frames encounters as "no strings attached" (Licoppe, Riviere, and Morel 2015; Race 2014), Tinder's developers discussed its swipe as a casual, gamelike activity (Stampler 2014). Christian Licoppe, Carole Anne Rivière, and Julien Morel (2015) found that some Grindr users viewed themselves as hunters and other users as sexual prey. Tinder's CEO, Sean Rad, has similarly stated that when people meet, "You're either a hunter—the way we see it—or you're being hunted" (Tinder 2013a) and Tinder mediates this situation. This frames Tinder's swipe as a mechanism for accessing and arranging casual sex, which carries the baggage of gendered stereotypes and expectations.

Tinder user surveys identify entertainment as the top motivation for use (Carpenter and McEwan 2016; Ranzini and Lutz 2016) and the most frequent users as those seeking sexual partners (Carpenter and McEwan 2016). However, men are more likely than women to look for sex on Tinder (Ranzini and Lutz 2016). Examining user commentary beyond these statistics reveals the endurance of a gendered double standard that rewards men's engagement in casual sex, while women are pressured to engage in sexual activity but shamed when they do (Kalish and Kimmel 2011). The Tinder subreddit (r/tinder) hosts discussions among users, mostly men, about optimizing profiles to attain hookups and includes screen shots triumphantly depicting Tinder chats leading to casual sex. While these conquest stories receive the equivalent of textual high fives from other Redditors, female users repost men's vulgar, forceful, and unwanted sexual solicitations to @tindernightmares on Instagram. Their posts highlight gendered discourses about men's entitlement to sex and reflect trends of digital technology being used for gender-based abuse and harassment (Jane 2016). An experimental study has shown that men tend to swipe right in higher volumes while women swipe right less frequently and are more selective (Tyson, Perta, Haddadi, and Seto 2016). Keeping in mind the gendered sexual discourses that shape perceptions of Tinder, one possible explanation of these trends may be that male users are more likely to perceive Tinder's swipe as a way to access sex and, by extension, social status. Concurrently, women might perceive the swipe's double opt-in as a way to deflect male sexual entitlement and regulate the visibility of their own expression of sexual desire.

With Tinder increasing in popularity, media reports and critiques have run with the association between the swipe's qualities and casual sex to sustain moral panics about technology. Some op-eds lament the swipe's rapidly continuous, binary, and visual qualities as inhibiting young people from forging intimate and long-term relationships (e.g., Eler and Peyser 2016); others compare the swipe's transactional qualities to online shopping (e.g., Sales 2015). Campaigns have also critiqued the swipe as commodifying and objectifying individuals, reducing users to pieces of meat (e.g., Maureira 2014). Alarming news articles about Tinder dates being assaulted or murdered associate unscrupulous swiping with tragic consequences (Sutton 2015), often framing the app as an irresponsible way to meet partners. From this publicity, it is evident that Tinder's initial sexual framing and subsequent perception as a hookup app have contributed to a reputation that constrains the app's uptake for other purposes such as serious long-term relationships or even more platonic connections.

Interventions without a Match

User and press interpretations of the swipe as mainly for hookups narrow Tinder's desired scope of use, limiting its appeal to different user groups and potential advertisers. While referred to as an app, Tinder engages in the "politics of 'platforms'" (Gillespie 2010), navigating tensions between commercial interests and fostering user-generated content to form connections and sustain a business enterprise. Within these politics, platform owners attempt to sway the adoption and development of their technology toward the most profitable interpretation of its affordances. A narrow user base and stigmatized sexual environment is hostile to advertisers, especially larger, higher-paying brands that have a stake in maintaining their wholesome image. Therefore, Tinder's attempts to redefine its app must retain utility for current users while increasing broader appeal.

Trevor J. Pinch and Wiebe E. Bijker's (1984) concept of interpretive flexibility describes technological development as a process of variation and selection as social groups vie over the meaning of an object. Companies infuse commercial interests into these iterative development stages, such as through the incremental shifts that transformed Twitter from "primarily a conversational communication tool to being a global, ad-supported followers tool" (van Dijck 2011, 343). Although the notion of closure has been challenged (Clayton 2002), it is useful to note that attempts at stabilizing a technology's interpretation are often made through two approaches (Pinch and Bijker 1984): "rhetorical closure" (426), efforts to produce dominant discourse that allays the controversy encompassing a technology, and "redefinition of the problem" (427), reframing a problem to garner agreement that the technology in question presents the solution. Tinder has taken both of these approaches to instill its reinterpretation.

Tinder's rhetoric has increasingly steered users toward perceiving the app as useful for meeting people and forming romantic relationships. The company's first narrative promotional video, #ItStartsHere, abandoned dark nightclubs to depict young people socializing in well-lit, diverse settings (e.g., parks, beaches).[2] Tinder also combatted perceptions of being for hookups through social media campaigns, such as #SwipedRight, showcasing engagement and wedding stories from couples who met through Tinder.[3] "#SwipedRight" implies that the "right" approach to swiping is one that yields relationships rather than hookups. Tensions between Tinder's vision and media interpretations eventually intensified with *Vanity Fair*'s article "Tinder and the Dawn of the 'Dating Apocalypse'" (Sales 2015), which depicted New Yorkers endlessly swiping for casual sex. Tinder responded with tweets asserting that its app has many purposes aside from sex, including "travel, dating, relationships, friends and a shit ton of marriages" (Tinder 2015a). Despite subtle and fervent attempts to reframe Tinder, media and user rhetoric maintains its perception as a hookup app. CEO Sean Rad's assertions that the app is broadly "about making introductions" (Recode 2016) were still met with incredulity in press interviews following years of relationship-focused branding strategies.

When perceived as a hookup app, Tinder's swipe solves the problem of finding partners for casual sex. Tinder has attempted to redefine the swipe so that it instead solves the problems of (1) initiating relationships between users, (2) initiating relationships between users and advertisers, and (3) monetizing users' relationship with Tinder. Tinder constrained the swipe's rapid and continuous qualities by introducing "an algorithm that intelligently limits the number of likes a user can make in a consecutive 12 hour period" (Tinder 2015b). The swipe limit was designed to thwart users, third-party apps, and spambots from continuous right swiping (Crook 2015) and to encourage users to become "more thoughtful" (Tinder 2015b) through rationed swipes. This update aimed to steer users away from less casual swiping toward conversations with others that could lead to something more. However, this change was imposed concurrently with the unveiling of Tinder Plus: a subscription model that provided perks to paying users, including unlimited swiping. By allowing revenue-producing users to continue swiping indiscriminately, this update did not improve Tinder's reputation but only ignited uproar among users (Acker and Beaton 2016).

Tinder also attempted to alter the swipe's visual focus by presenting more information about users. A major update allowed users to display Instagram photos, made Facebook "likes" more visible, and enabled users to see friends of mutual friends, extending visible connections one step further (Tinder 2015c). However, this information was not visible from the swipe screen and appeared only when users tapped to explore individual profiles further. A later update imported users' job and education information directly to their swipe cards (Lam 2015). While this design change adds personal details to users' decision-making process during swiping, positioning these indicators of socioeconomic status prominently does little to ameliorate Tinder's reputation of commodifying users.

In 2016 a drastic addition to the app's interface combined rhetorical and hard-coded redefinition attempts. Tinder Social allowed users to see other Facebook friends who were using Tinder and to select up to three of them to form a group. Groups could swipe collectively on other groups and match to chat and make arrangements for meeting up. Promotion on the company's blog underscored the feature's social use: to take "an average night out with your friends to the next level" (Tinder 2016). However, reception to Tinder Social indicated just how ingrained the swipe was with perceptions that its main purpose was for hooking up. Media stories immediately circulated interpretations of the feature as being for group sex (Kleeman 2016). Qualities of the swipe were not easily reinterpreted as conducive to forging friendships, since people are thought to carefully select friends based on more than appearance and minimal information. Sorting groups in a binary manner also presented the problem of discarding potentially compatible users along with incompatible ones if one is seeking to form longer-term relationships. Tinder Social was subsequently removed from the app in mid-2017.

These instances of rhetorical and technological intervention into Tinder's imagined affordances demonstrate the difficulty of dislodging the swipe from its initial interpretation. Despite expending resources on rebranding, the app's developers altered multiple aspects of its interface while leaving the swipe's continuous, binary, and visual qualities largely intact. The necessity of retaining Tinder's trademark functionality, which earned the app a place in the everyday vernacular, has constrained Tinder's avenues for shifting users' overall perceptions of it.

Conclusion

Tinder's swipe, as a gesture combined with software features and hardware characteristics, comprises a network of material arrangements that give rise to the qualities of sustained engagement, binary sorting, and visual emphasis. Users and the media perceive these qualities within their sociocultural context, creating associations between the app and casual sex that conflict with the company's vision and commercial interests. In response, Tinder's development has included rhetorical and design interventions aimed at redefining and modifying the swipe's qualities so that they may be perceived as conducive to forming relationships.

Tinder's difficulty in shifting users' perceptions draws attention to how its sociocultural context complicates the management of platform politics. Tinder must balance tensions between users and advertisers while also maneuvering deep-seated sexual stigmas. The swipe's connection to casual sex, which has led to both the app's popularity and infamy, must become distanced from sexual activity to appeal to advertisers who seek to capitalize on the swipe for product selection. Adding functionality deemed "social," along with swipe limits and increased displays of personal information, allows Tinder to demonstrate that its app has a range of uses that are conducive to selling cars, gum, and TV shows while preserving advertisers' reputations. Although the swipe has become entrenched in interpretations linking it to casual sex, Tinder's interventions push against this perception to bolster its bottom line while simultaneously attempting to maintain the app's signature functionality.

This analysis of the swipe, a simple gesture integrated into app technology, demonstrates how such design elements can influence a technology's reception and development trajectory. I have examined how users, the media, and Tinder's developers have imagined and interacted with the swipe in ways leading to the mutual shaping of the app's technological design and its social, cultural, and economic outcomes. This analysis shows how embodied ways of interfacing with technology can accrue greater meaning within particular sociocultural contexts. The interpretation of this meaning can subsequently spark contention among those who have a stake in an app's future.

Notes

I would like to thank past and present members of Microsoft Research's Social Media Collective for contributing to this piece's development, especially Tarleton Gillespie and Dylan Mulvin who provided feedback on early drafts. Thanks also to Jean Burgess and others at the Queensland University of Technology who supported this work.

1. Accessed through the Internet Archive's Wayback Machine: https://web.archive.org/web/20130403040153/http://www.gotinder.com/.

2. Tinder, #ItStartsHere, promotional video, https://www.youtube.com/watch?v=Pi JV1HIHTIY.

3. Tinder, http://swipedright.gotinder.com/.

Hollaback!
Harassment Prevention Apps and Networked Witnessing

CARRIE RENTSCHLER

App: Hollaback!
Developer: Sassafras Tech Collective
Release Date: October 2010 (Current Version: 3.3.0)
Category: Lifestyle
Price: Free
Platforms: iOS, Android
Tags: gender discrimination, reporting, violence against
women
Tagline: "It's not just an app or a map. It's a movement."
Related Apps: Not Your Baby, SafetiPin, HeartMob,
HarassMap, Circle of 6

Hollaback! is an international anti–street harassment group that began as a blog between friends (May and Carter 2016). Their origin story describes how Thao Nguyen, a young woman in New York, inspired them by using her mobile phone to report a fellow subway rider who was masturbating in front of her. Nguyen turned to the web to share her story via *The Gothamist*, posting a photo of her harasser along with a brief, indignant report. The *New York Post* then picked up her story, distributing it to a larger audience of readers. For Hollaback!, Nguyen modeled what anti–street harassment activism is: the act of reporting street harassment using internet-connected mobile phones. Rather than going to the police or other authorities, who often fail to take harassment seriously, Hollaback! models activism as a form of networked publicity, a practice of app-enabled witnessing that connects individuals through their mobile devices. Modeling this practice, Hollaback!'s early publicity images portrayed the quintessential activist as a young woman of color armed with a mobile phone wrapped in hot pink Hollaback! skin.

Since 2005 the organization has developed an approach to activism as networked witnessing, defining activists as mobile phone and app users. As the London chapter website puts it, "Anyone who has ever or will posted [*sic*] here is part of HollabackLDN" (Hollaback! London n.d.). To post is to be a Hollaback! activist. In 2010 the organization designed their first mobile app for reporting and geolocating street harassment, using a hot pink, comic book aesthetic for the interface's look. Categorized as a lifestyle app in iTunes, Hollaback! approaches activism as part of daily life, a way of seeing harassment and seeing others being harassed that is mobilized through users' tools at hand: their cell phones. Activism is something you can carry with you, on your phone.[1] The graphic star-shaped "Pow!" featured as the app icon signifies a fun and female superhero–like vibe, giving Hollaback! a particular activist look and feel that draws from pink empowerment marketing campaigns aimed at girls (see Kearney 2010) crossed with feminist anarchist iconography of girls with bombs.

By 2016 the app was in its third iteration, enabling users to report on street harassment they had either witnessed or directly experienced and then post it directly to the website of the user's local Hollaback! chapter. The first act of witnessing via the app is one of subject positioning, when the initial screen interface asks users to identify as either a victim of harassment or a witness to it. From the very first action with the app, the user is transformed from a victim of harassment into a primary witness to it, or from a bystander into a secondary witness—or at least Hollaback!'s versions thereof.

We might think of the app as a feminist, but certainly feminized, activist bystander technology (see Rentschler 2014, 2015). The app stands by, ready for when a victim of harassment needs to make a report. It is also ready to transform the bystander subject into a witness. On the first screen, the app informs users, "You're not alone," suggesting that the very act of reporting and sharing one's story identifies the user with the larger movement against street harassment. "You're not alone" functions as a refrain repeated across Hollaback!'s materials, suggesting that using the app connects people to an online collective. With its pink star graphic, a "Take Action" icon within the app defines the act of making a report with taking action, revealing how Hollaback! defines reporting harassment to the organization as a way of participating in activism.

Once users identify their position as someone who has experienced or witnessed street violence, the app asks them to classify the kind of harassment they have experienced/witnessed from a list of ten options: verbal abuse, sexual gestures, inappropriate touching, being followed, indecent exposure, assault, homophobia, transphobia, racism, or "other." Posters can select one

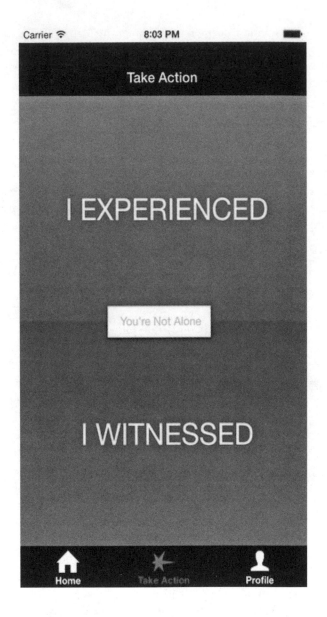

User Identification Options within the Hollaback! App

or multiple options, making the reporting process, in part, a practice of categorizing violence. The categories also function as searchable tags. The app's classification schema represents a particular artifact of the organization's feminist politics. The first six options name harassing behaviors, and the following three name specific interlocking systems of oppression in which harassing acts are produced and experienced. If classification schemes "do work" (Bowker and Star 2000), the category of "other" includes by excluding acts and forms of oppression not initially identified in the schemata. Additionally, while users can name racist verbal abuse in order to specify the particular violence of that speech, the app does not provide the option to name sexism and misogyny.

By not naming them, the Hollaback! app's system of classification reveals a naturalized hierarchy, where acts of gender oppression are not named as such while racist, homophobic, and transphobic acts must be. Naming those experiences of oppression ensures a base-level recognition within the organization, particularly after so many women of color have roundly criticized the broader Hollaback! organization for its white feminism.[2] Perhaps the app's classifications represent a response to these challenges. However, the categories do not, and cannot, adequately address the sources of the problem—for one, a white feminist approach to intersectional analysis that "diversifies 'critically' the content but not the rules of the game" (Bilge 2014, 190)—that are revealed in the app's underlying system of knowledge. Like the tick boxes on census forms, the app's provision of ways to classify harassing violence normalizes and seemingly universalizes gender oppressions as otherwise race-neutral and sexuality-neutral systems. As Sirma Bilge argues, "Intersectionality does not create a shopping list of categories that can be deployed" (2013, 420).

Through its "whitened intersectional" (Bilge 2013) classification schema and its practices of geolocation, Hollaback!'s app models feminist witnessing as a practice of reading Google maps as particular kinds of situated record making, defining the problem and solution to street harassment through a politics of cartographic visibility. It provides ways for those who directly experience street harassment to report it, and ways for bystanders to intervene as witnesses through pins, button clicks, and their own reports. In its earliest iterations, Hollaback! and its partners designed the app with the primary witness in mind: the victim of harassment. "We developed these apps to give folks a real-time response to the harassment that they are experiencing. It enables users to GeoLocate where it happened to them, identify the type of harassment that happened to them, and then once they are home they have the opportunity to really tell the rest of their story" (Emily May, quoted in Feldmar 2011).

Once users categorize the harassment and locate it on the Google map of their local chapter, a screen appears thanking them for "sharing your story," though calling a tagged geolocation a "story" is a stretch. According to the app, a story can consist of no more informational content than "verbal harassment at the corner of University and Main," representing an almost mechanical form of witnessing that closely resembles a police blotter (e.g., noting the date, time, and location of an incident). Through appification, witnessing is reduced to an act of pinning. That act links the classification of a harassing event that likely happened in a particular place and time to a searchable and taggable location on a digital map.[3] The app thus distills witnessing to an act of classification and pinning whose larger purpose includes the expectation of victim-supportive judgment by others—or, as John Durham Peters refers to it, "the discursive act of stating one's experience for an audience that was not present at the event but must make some kind of judgement of it" (2001, 709).

The app deploys Peters's sense of witnessing in the affordances it provides to users to make a more extended account of what happened linked to the pinned location of the harassing incident. To do so is optional but encouraged. The London chapter's website, for instance, encourages posters to talk about how they are feeling and to describe what they want people to know (Hollaback! London n.d.). To witness here means to transmit feeling as well as provide an account. My analysis focuses most specifically on the act of pinning because of the way the Hollaback! app defines its use of pins and classification as a witness "story."

The app enables users to directly upload their stories to the web platform from their phones, as seen in figure 2. I identified Jarry Park as a location around which to submit a post. The location shows up as a pink pin, a space of witnessing harassment as a direct victim of it.[4] Bystander witness reports show up on the map as green pins, following the Green Dot model of mapping bystander intervention, a US-based bystander program. Users can enact another element of witnessing by clicking the "I've Got Your Back" button, an add-on to the app designed by Green Dot to express bystander support (shown in fig. 2). Clicking the button registers as a pink-colored count that accompanies each post. On the right of figure 2, for instance, a count of twenty-two "I've Got Your Back" button clicks accompanies Liz's story of inappropriate touching, signifying that twenty-two users have indicated their support for Liz through this affordance.

Through this button, the app performs witnessing as an expression of technologized affects. Not all that different from the "Like economy" that

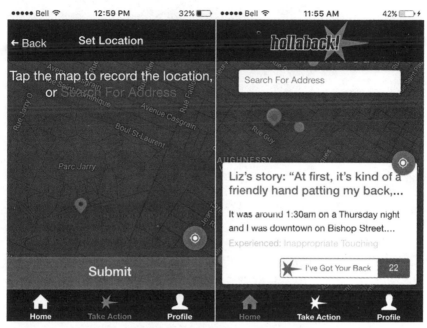

Author screen shots of the app's selection location screen for posters and a Google-Mapped post from downtown Montreal, the city where the author lives

Carolin Gerlitz and Anne Helmond analyze on Facebook's platform, the "I've Got Your Back" button can "instantly metrify *and* intensify user affects—turning them into numbers on the Like counter—while fostering user engagement to multiply and scale up user data" (2013, 1349–50). In their launch of the "I've Got Your Back" campaign with Green Dot, Hollaback! compared the use of their button to clicking the "Like" button on Facebook (May 2011). Through its button click counts, Hollaback! scales up user data to local chapter websites, representing the breadth of a local organization's online footprint in terms of user numbers and then translating those numbers into a register of aggregated feeling. The button, then, can serve as an unofficial means of counting local movement supporters whose actions are understood as a form of other-directed feeling.

Hollaback! explicitly defines activist witnessing as a technological and affective practice, modeled perhaps most clearly in the "I've Got Your Back" button. What the "button count" means as an act of paying witness, however, is less clear. Through the count, the app participates in the social media logic

of popularity (see van Dijck and Poell 2013), but what does that popularity signify on an app for reporting street harassment? Posts with the highest "I've Got
Your Back" counts might signal a qualitative difference among posts, judged
as perhaps more serious, or particularly scary, or especially egregious. Clicks
might also signal an affective response to the ways posters situate themselves
in their report, as someone who asks for help or describes their harassment as
especially injurious. In this way, the button click might serve as an expression
of (admittedly) bare affect in response to the poster's own expression of affect,
a way of supposedly transmitting feeling between clicker and poster.

In addition to accumulating counts and distilling a range of meanings into
a single statement of support, button clicks also link the app witnesses to the
original poster through an email sent to the poster, "letting you know you're
not alone," according to an earlier iteration of the app in 2011 (see Nicola
Briggs, the "subway badass," in May 2011). To pay witness via the app is to be
part of a larger assemblage of social media platforms, other apps, and Google
mapping practices. "This is a big project. It's a map, an app, a film, and a
movement," not just part of a social media assemblage, but constituted by and
constitutive of it (Briggs qtd. in May 2011). As an organization, Hollaback!
draws its vision of activist publics and its sensibility of social justice from the
forms of sociality emerging through technological assemblages of apps, websites, and other forms of media making.

Hollaback! and the Appification of Justice

Witnessing, via Hollaback!, becomes part of a broader Hollaback! assemblage,
a plurality of networked structures that link Hollaback!'s reporting mechanism to other "people, technologies, and materials in order to do something"
(Wiley, Moreno Beccera, and Sutko 2012, 186). Most public discourse about
Hollaback! casts it in terms of upbeat, can-do empowerment talk focused on
the young, usually female-identified users of the app. The app's look and design capture a kind of tough girl, "badass" vibe draped in pink. The organization describes "a culture of badass" as one of its values, embodied by non-
conformists who "embrace bold ideas and encourage risk-taking" (Hollaback!
Halifax n.d.). That tough-girl attitude combines the pink can-do aesthetics
of girl-based marketing with a digilante politics of justice seemingly softened
by the overall aesthetics and affordances of the app's design. Even the act of
sharing posts via the organization's website—a typical action on social me-

dia platforms—enacts forms of nonconsensual use and distribution that run counter to the feminist ethics we might expect of feminist antiviolence organizations (see Kim and Kim 2014; Loza 2014).

Hollaback! mobilizes users as vigilant(e) individuals looking out for one another against harassers (see Ellcessor, this volume). The app's vibrant pink interface represents a vision of girl power enacted by users who dare to report street harassment and turn the camera lens and other tools of documentation onto harassers who seek to control their use of public space. In the process, it embraces a community of vigilance online. Yet vigilance in reporting harassment can easily transform into digital vigilantism, or "digilantism," "a form of do-it-yourself justice online" (Byrne 2013; see also Nakamura 2014) or "crowdsourced justice" (Jane 2017, 3; see also Ankerson, this volume) in which practices of public shaming are brought to bear on those who harass.

In 2013 mayoral candidate, New York city council member, and app proponent Christine Quinn explained in rather ominous language how the app could help in the identification and punishment of harassers:

> We're here today to let New Yorkers—and particularly women and girls—know that people who violate women either by their actions or words won't be able to hide anymore. We will know who they are, what they do, where they do it, and we will put it to an end. (qtd. in Epstein 2013)

A supporter of "stop and frisk" police action, Quinn celebrated the app's capabilities for revealing who harassers are and where they harass with the zeal of a vengeful protector. (Al Jazeera America 2013). The app enables users to take digital photos of harassers while also locating their harassment on the map in order to identify but also presumably to shame, control, and punish them, not unlike the sex offender apps Mowlabocus analyzes (see chapter 14, this volume).[5] The desire to out harassers by posting their photos functions as a kind of feminist digilantism fueled by the faith that doing so will lead to justice for victims of harassment. As Emma Jane explains (2017), rather than offer solutions, feminist digilantism instead diagnoses the problem of street harassment by picturing harassers, describing their actions, and geolocating them.

These very same practices participate in longer histories of vigilante violence directed at communities of color (see Byrne 2013; Nakamura 2014). Based on their representation of street harassment in the 2014 video and in early posts to the New York chapter website that pictured men of color as harassers (see Daniels 2009), Hollaback! has been criticized for its approach to race. Seen in this light, the organization's use of tools for online shaming and other tactics of vigilance easily replicate racist visions of justice, even when the executive director and other Hollaback! members criticize the racist criminalization of harassers and call for more community-based solutions, such as education, bystander intervention, and community safety audits (iHollaback 2015).

Using the app becomes a practice of doing justice informally and digitally (Fileborn 2014). While app users knowingly participate in some informal practices of justice—such as providing support to posters, validating poster's experiences, and warning others of places where they have been harassed—they may not so knowingly participate in the ways others use it. The 2013 iteration of the app, for instance, enabled New York users to report harassment directly to the city council without knowing how the council would use that information (see Epstein 2013). Policy makers can deploy the app's crowdsourced information toward new and revised legislation, and local law enforcement can use the mapped reports to focus their policing efforts on particular neighborhoods, intersections, and transport lines.

In the 2011 US federal "Apps Against Abuse" technology challenge, policy makers and activists saw the development of new apps as key tools of activism against gender violence. Today the Hollaback! app represents one of over two hundred apps aimed at addressing gender violence and sexual assault in particular (see Bivens and Hasinoff 2018). Bloggers and reporters see the app as one of many mobile software devices being used for safety and self-defense purposes (see Adelman 2011). Within a longer genealogy of feminist self-defense, the app offers users particular ways of seeing and reading the city as part of day-to-day safety management, enacting a "street-level politics of mapping" that is meant to "enable . . . others to craft tactical routes through the city" (Lingel and Bishop 2014, 2, 7). For locals, the app re-signifies the city along the lines of a crime map, representing spaces as more and less dangerous based on the density of pink and green pins around certain locations, potentially playing into racist notions of urban neighborhoods as dangerous spaces (Ohikuare 2013).

Within this context, even if users object to the use of their app-provided information in racist systems of oppression, they have little control over what the organization and local chapters do with their information. As one writer explained, the Hollaback! app "brings big data to street harassment storytelling" (Brownstone 2013), extracting data from users who may give it voluntarily but without access to how it will be used. There is no mechanism on the app for users to indicate how they want their information to be used. As

Amy Hasinoff warns, "The key variable missing in the way we think about information flow is consent" (2015, 128). On local chapter websites, people can share others' posts to Facebook, Twitter, and LinkedIn without the original poster's consent. While the current version of the app does not allow users to share others' posts, the app is connected to local chapter's websites that do. By sharing, users can also add comments to a post, reinterpreting the meaning of the experience and drawing additional attention to the story without the original poster's consent. Via sharing, Hollaback! mobilizes a form of social media witnessing without the requisite ethics of care and consent one might otherwise expect from feminist organizations doing anti-oppressive advocacy work on gender violence.

As Hollaback! demonstrates, mobile tools do different kinds of work in the name of feminist justice. The app suggests that notions of what constitutes justice for street harassment are also being appified alongside the applification of reporting of street harassment. This suggests that if harassers can be identified and if harassment can be geolocated, personal and community accountability and redress are sure to follow. Users of the app participate in a form of digilantism that too often pits feminism against antiracism, even when the app enables, and even encourages, users to report racist harassment. This politics of vigilance can make app users complicit in injustices, whether they know it or not. In modeling justice for street harassment, Hollaback! may need to examine its own practices of nonconsensual sharing and its particular faith that mapping and picturing harassment in the city will lead to a less violent world.

Notes

1. A recent study found that women college students relate to their mobile phones as self-defense technologies, as tools that connect them to their larger community networks and create feelings of safety (Cumiskey and Brewster 2012).

2. In 2014 four local chapters in the US and Canada left the organization because of its white feminist approach to street harassment. They left after the Hollaback! New York and Bliss Communications' video "10 Hours of Walking in New York City as a Woman" went viral on YouTube and other social media platforms in October that year, over-representing Black and Latino men as harassers of a white woman.

3. I say "likely happened" because I posted a fake incident of harassment in order to see how the app works. As I discovered, users do not have the option to go back or save a post before submitting. Via the app, submission is the same as publishing.

4. I did not witness harassment in Jarry Park; I simply chose a familiar location to try out the app. I feel compelled to note this, because in order to test out Hollaback!, one must

perform as a witness. This means using the app is necessarily an act of witnessing—whether false in my case or actual in the case of most users.

5. London's chapter disallows posters from uploading photos of harassers due to legal protections against defamation and out of concern for the safety of the poster. Yet their website does not say anything about the ethics, politics, or even the effectivity of publicly shaming harassers (see http://ldn.ihollabck.org/about), a silence that speaks to the troubling digilante politics of witnessing that structure the app's design and purpose.

The Sex Offender Tracker App
Tracker Apps and the Individualization of Risk
SHARIF MOWLABOCUS

App: The Sex Offender Tracker
Developer: BeenVerified Inc.
Release Date: October 2010
Category: Utilities
Price: $0.99
Platforms: Apple iOS; Android
Tags: surveillance, risk, databases, sex offending
Tagline: "The World's First Augmented Reality Sex Offender Locator!"
Related Apps: Offender Locator, Crime Reports, Sex Offender Search, Safe Neighborhood

Before its unexplained removal from app marketplaces in 2016, the Sex Offender Tracker (herein known as Tracker) was the most popular app (by download) within a genre that might best be labeled "crime locator apps." These applications offer users geo-specific information on prior criminal activity and paroled offenders residing in the local area. While some applications offer information on a range of crimes, others, like Tracker, target specific classes of criminality and specific classes of offender. Information about offending is not confined to crime locator apps; real estate and property search services and apps also often make use of this data as they summarize neighborhoods. These days, alongside information on the quality of local schools, details on the crime rates reported on a particular street are increasingly likely to be included in property listings.

However, my focus in this chapter is on applications that specifically address the subject of criminality and offending, because I want to consider what happens when public databases are remediated by private surveillance technologies. Specifically, I want to explore the assumptions this remediation process validates and how the promise of security these apps offer obscures uncomfortable truths about both the accuracy of data and the individualization of risk. I focus mainly on Tracker, which, despite its disappearance, still raises issues and concerns that apply to hundreds of similar apps within the broader genre of tracking apps.

Tracker was launched in 2010 with an advertising campaign featuring internet celebrity Antoine Dodson of "Bed Intruder" fame.[1] Until the app was removed in 2016 by developers, it was available on multiple smartphone platforms for ninety-nine cents. The app's selling point, similar to that of many crime registry apps still available for download, is that it uses augmented reality to provide users with information on registered sex offenders residing in a local area. As the app loads, Tracker displays a universally recognizable icon of a military radar screen. The app's iconography draws heavily upon cinematic and gaming references to the radar/tracker screens commonly seen in such science fiction films as *Aliens* (1986, dir. James Cameron), military action films like *Battleship* (2012, dir. Peter Berg), and first-person shooter video games like *Quake*, but in this case, the first-person positionality of the app tracks interactions between a user and an unreliable and often inaccurate database. Tracker's deployment in this context becomes a metonym for this genre of applications, which serve to remediate public space through the language of militarization.

Scholars have noted the increasingly porous boundary between computer-based game environments and military technologies/environments (Stallabrass 1996; Burston 2003; Huntemann and Payne 2010). Applications such as Tracker further complicate the divisions between the domestic and the military, remediating public space through the language of conflict. With its imagery of radar screens and crosshairs, its promotional rhetoric—these apps are marketed as "trackers" and "locaters," and they position users as "hunters" that "target" offenders—the crime locator genre serves to weaponize the spaces of everyday life.

Once loaded, the app presents users with a digital map of their local environment. Alongside a blue dot marking the user's current location, red pins also appear. Clicking on one of these pins opens a profile similar to that found on a dating or social networking app. This profile features a mug shot of a felon, with details of their height, weight, body markings, and home address. Below this information, details are given of the felon's previous convictions, including the date of conviction and the specific crimes of which they were found guilty. Pins are placed according to the address listed in the profile. As such, a user not only sees how many registered sex offenders reside nearby but also who they are and where they live.

At this point, and given the fact that this software relies on real-time GPS, it is necessary to point out that the rhetoric of Tracker skirts the boundary between truth and fiction. While the radar icon and the map suggest immediacy, *Tracker does not deal in live data*. What the user views on Tracker is the cartographic representation of *historical* data. The only live data captured is the location of the user themselves: the blue dot. This focus on the individual is central to the logic of Tracker, which not only personalizes the risk of sexual violence but also individualizes the responsibility for reducing that risk.

This personalization of risk fits in with broader neoliberal logic, whereby the individual is positioned as an idealized rational actor and, particularly within the context of mobile media, the continual locus of action. This trend can be found in approaches to health and well-being (Peterson and Lupton 2000), genetics (Novas and Rose, 2010), homelessness (Parsell and Marston 2012), crime prevention (Goldstein 2005), and sexual violence (Bumiller 2008). Tracker builds on this rhetoric by inviting the user to see public spaces as inherently threatening and puts the onus on the individual user to take appropriate action to avoid sexual violence. Before I discuss this further, some context on the data these apps use is necessary.

SORNA

In 2006, the US Department of Justice enacted legislation that mandated the creation of a nationwide database of registered sex offenders.[2] Under Title 1 of the Adam Walsh Child Protection and Safety Act (commonly referred to as SORNA, the Sexual Offender Reporting and Notification Act), every state was required to implement and maintain a system for monitoring convicted sex offenders released into the community. This system had to meet specific benchmarks set by the Department of Justice, including a publicly accessible database that detailed the name, address, and prior offenses of each registered offender.

The map screen of the Sex Offender Tracker

Depending on the category of offense, convicted sex offenders in the United States must report to their local police station at regular intervals to update their details for the local database. Some categories of offense require offenders to update their records only when they change address, while other categories require monthly or weekly updating, including details of out-of-state or even out-of-county travel. Failure to adhere to these requirements is a crime, which can result in reincarceration and, in some cases, an automatic life sentence.[3]

Since the outset, SORNA has faced strong criticisms. Civil liberties and offender advocacy groups argue that the legislation provides both "an intermediate sanction" (Button et al. 2009, 417) and an *indeterminate* sanction against convicted offenders. This means that registration measures can be deployed against the individuals for indefinite periods of time (Ho 2008). Meanwhile, a lack of funding has resulted in a patchy and partial national database, with many states failing to meet the requirements of the act (Murphy 2012). One must also acknowledge the challenges that sex offenders (indeed many offenders) face upon release from prison. Securing long-term employment and housing is often challenging, and community notification practices often mean that offenders are intimidated, threatened, and harassed out of communities and jobs on a regular basis. This precarity means that a significant percentage of those who are in the database have no permanent residence to record in the database (Noyes 2010).

The fallibility of the databases—caused by offender transience, parole violations, and a lack of funding—is compounded by the fact that individual states categorize offenders according to their own rubric. Exactly who counts as a sex offender is a matter of geography, and what might count as a minor offense in one jurisdiction (e.g., public urination) might be considered a serious sex offense in another. In one state a sex offender may be required to register in the database for five years, but in another they may be have to remain in the database even beyond death.[4] This explains, in part, the proliferation of sex offender registry apps that are state- or region-specific. A search in app marketplaces for "sex offender tracker" results in many state-specific apps (e.g., Arizona Sex Offender Search or North Carolina Mobile Sex Offender Registry) with iconography and interfaces that are similar to those of Tracker. These disparities are further compounded by SORNA's lack of benchmarks or ceilings for recording, which creates confusion and misunderstanding when it comes to making sense of information provided on sex offender databases.

It is from these unreliable and underfunded registries that Tracker draws its data. Furthermore, while these databases are in theory updated on an ongoing basis, how regularly Tracker refreshes its records while in operation is unknown: the only piece of accurate information the user can be certain of is the blue dot on the app screen, the marker that situates the user in relation to the data. As such, Tracker develops a spatial relationship—not between the user and the registered offender but between the user and the unreliable and often inaccurate database.

Individualizing Risk

Tracker therefore operates within both the digital landscape of poorly maintained databases and the physical landscape that the user inhabits. Contradicting the fallibility of the registers from which it draws, the application offers the assurance that it can reveal where nearby sexual offenders are, which fuels the assumption that to be forewarned is to be forearmed—that having knowledge of where sex offenders reside and what they have been convicted of mitigates risk and helps the individual protect themselves and their dependents.

To repeat my earlier claim, Tracker fits neatly into the current cultural belief in individualism, a belief that finds its roots and energy within contemporary neoliberal ideology:

> Under late-modern conditions individualisation is a necessity, experienced as liberating yet simultaneously an obligation. . . . The person becomes responsible for what they do and, therefore, personally culpable for their own failures. Individualisation is not just rewarding; it is also penalising. (McGuigan,2010, 328)

Tracker is a product of the North American neoliberal zeitgeist, encapsulating a constellation of beliefs, values, and anxieties that characterize this period of late modernity and that coalesce around the individual. First, echoing the free-market fire sale of public institutions that characterized the Reagan and Thatcher administrations, Tracker represents the privatization of public goods. While taxpayers support the running of police databases and websites, Tracker draws information from those resources and offers it for sale on an individual basis. In many respects this illustrates what Jamie Peck and Adam Tickell (2002) term "rollback" neoliberalism, whereby the state retreats from the management and support of public services, providing space for commercial providers to intervene on a for-profit basis. Indeed, Tracker is a metonym for the US criminal justice system. An increasing number of prisons and correctional facilities are run on a for-profit basis, while probation, custody, and other offender management services are increasingly contracted out to commercial organizations (see Schichor 1995). It is within this context of the prison-industrial complex that Tracker finds purchase, for, just as the US criminal justice system increasingly relies on private companies to manage incarceration, so Tracker shifts responsibility for public safety away from the state and onto the shoulders of the private individual. Read through the logic of neoliberalism, Tracker becomes a privatized policing tool that invites the individual to police both the registered sex offenders in the local area and their own movements through public space.

Second, moving from the economic to the social sphere, Tracker reflects Loïc Wacquant's argument that neoliberalism actively punishes the most disenfranchised and marginalized members of society:

> These castaway categories . . . have become salient in public space, their presence
> undesirable and their doings intolerable, because they are the *living and threaten-*
> *ing incarnation of the generalized social insecurity* produced by the erosion of stable
> and homogenous wage-work.(2009, 4; emphasis in original).

Wacquant's reference here to a "generalized social insecurity" echoes Mary
Douglas and Aaron Wildavsky's (1983) critique of risk culture and the methods by which *individual* risk is inscribed onto *public* space and groups of people who are believed to occupy that space. Of course, as Scott Lash identifies,
"There is in fact no increase in risks in contemporary times. There is instead
only an increase in perceived risks" (2000, 49). Tracker (re)constructs the sex
offender as the living embodiment of this perceived risk. It deploys the sensa-
tionalist rhetoric of media and politics (Horowitz 2007) and helps to ensure
that *any* convicted sex offender is liable to be forever marked as both deviant
and threatening.

  This inscription of risk contradicts empirical evidence, which confirms
that sex offenders pose less risk of recidivism than other categories of (nonsex-
ual) offending (see Levenson et al. 2007). Ignoring such evidence, the focus
on (indeed the promotion of) *perceived* risk allows the contemporary criminal
justice system and, more broadly, the neoliberal ideology of the US government to co-opt the sex offender for particular aims. Upon release from prison,
the convicted offender is transformed, via registration and notification practices, into both a symbol of and reason for the growing sense of insecurity
that communities and individuals feel. Along with migrants, the homeless, sex
workers, the poor, the mentally ill, and the long-term unemployed,[5] the sex
offender is positioned as an object of hatred and fear, not just for his crimes,[6]
but also for what he represents and contributes to—namely, the purported
dissolution of society:

> It is this diffuse and multifaceted social and mental insecurity, which . . . haunts
> large sectors of the middle class, that the new martial discourse of politicians and
> the media on delinquency has captured, fixating it onto the sole issue of physical
> or criminal insecurity. (Wacquant 2009, 4)

Tracker becomes another tool for inscribing this belief in increasing criminal
insecurity and a phantasmagorical method for managing that insecurity. Just
as sexually motivated criminality is falsely believed to be on the rise in the US,
Tracker offers a false belief that such criminality can and must be managed by
the individual.

Finally, in mapping alleged risk and insecurity onto physical space, Tracker encapsulates a key response to neoliberal ideology. In his study of public space in Sydney, Australia, Malcolm Voyce identifies the privatization of public space, whereby commercial entities provide "public" services and offer spaces of leisure and consumption to the "right kind" of public. He cites the shopping mall as a prime example:

> The mall is constructed to form a predictable controlled environment which acts like a prison in reverse: to keep deviant behaviour on the outside and to form a consumerist form of citizenship inside. (2006, 273)

Along with gated communities, gourmet market places and waterfront developments, malls represent a new form of "quasi-public" space (273) that looks and feels public but that is in fact regulated, surveilled, and owned by private companies who control access to, and activity within, that space. This global redistribution of space within urban and suburban environments serves to further codify space according to class. While the environs of the city have long been organized according to socioeconomic status (as well as racial and ethnic categories), the corporate takeover of public spaces has served to legitimate such segregation. In doing so, it has turned the ever diminishing spaces of true public life into zones of expulsion, deprivation, and decay.

As public spaces become represented as ever scarcer, anxieties about those spaces become ever more sexualized and imbued with perceptions of violence. The release of the augmented reality game, *Pokémon Go* in 2016, for example, was accompanied by a moral panic suggesting that the game encouraged young people to visit spaces of risk and danger. These dangers and risks were often sexual in nature, such as concerns about sex offenders using the game as bait, sex shops that just happened to harbor the sought-after Pokémon, and users wandering around the spaces of Manchester's gay village, looking for Pokémon on Canal Street (see Cockroft 2016).

Using a similar rhetoric of sexualized fear, Tracker (incorrectly) suggests that registered sex offenders present an imminent danger and that this danger is outside, in public space. In spite of clear evidence that points to the fact that offenders are often known by their victims (e.g., Grubin 1998; Jewkes and Wykes 2012), the public understanding of sex offending in apps like Tracker once again often inaccurately perpetuates such activity as located away from domestic space. In a similar light to Voyce's "malls-as-inverse-prisons" thesis, Tracker invites self-regulation on the part of its users. We are invited to monitor, police, and restrict our movement through public space according to the regime of the app: "Let's get outta here," says Antoine Dodson in the video commercial for Tracker, and that is precisely what Tracker encourages the user to do—retreat from the public into the apparent safety of the private.

Conclusion

For many, the figure of the sex offender is utterly repellent. Crimes of a sexual nature often ruin the lives of victims, and in far too many instances the legal framework surrounding sex offender prosecution—exacerbated by structural misogyny—has failed victims. One might like to believe that judiciaries treat sex offending more seriously today and that perpetrators are dealt with in a responsible and appropriate fashion by the state. Michel Foucault (1975) reminds us that incarceration was (re)cast as a productive power in the eighteenth century, with its aim being to transform and rehabilitate the prisoner. But this is patently not the case. The privatization of the correctional system in the US has coincided with a substantial reduction in (costly) rehabilitation programs (see Wacquant 2009), and it is increasingly unlikely that those who appear on sex offender registers have ever received any form of rehabilitation. In this context, it is all too tempting to see Tracker and apps like it as a solution to the "problem" of paroled sex offenders. If prison isn't making these offenders less dangerous to the general population, then at least Tracker offers a method for protecting oneself. It is my hope that the analysis I have offered in this chapter points to the limitations of such an argument.

Applications such as Tracker are at best ambivalent cultural objects, at worst dangerous tools of misinformation. Buying into Western culture's infatuation with the actuarial (Hannah-Moffat et al. 2009), they promote the idea that offenders can be known and their accompanying risk measured by the app user. Alongside concerns about the misrepresentation of data, the politics that underpin these services erode the very thing that may help to protect potential victims of sex offending. Given that the majority of sex crimes take place in the home, research repeatedly demonstrates that resilience building, community organizing, and an engagement with "the Commons" helps to protect vulnerable people from the inevitable dangers that come with living in the world (see Newman 2002). The privatizing of information held on registered sex offenders works against this sense of communality. It suggests that the sex offender is no longer a community issue but an individual problem. Under the logic of neoliberal individualism, the user is offered (partial and illusory) information and then invited to process this appropriately to manage the threat of sexual violence on their own. This line of thinking is dangerous, not least because the emphasis on personal responsibility that Tracker and other apps articulate tacitly (re)frames victims of such violence as having failed to protect themselves. As such, the promise of security offered by Tracker is in fact a call to immobilize and retreat into the private sphere, presumably the only space where one can be responsible for their own safety.

Notes

1. Antoine Dodson rose to internet fame in 2010 when he was interviewed by WAFF News, following a break-in at his home in Huntsville, Alabama. Dodson's sister was the victim of an attempted sex attack, which was foiled when her brother disturbed the perpetrator. YouTube comedy duo the Gregory Brothers immortalized Dodson's interview in an auto-tuned video that became a viral hit.

2. The Jacob Wetterling Act (1994) previously required the creation of state registries for known sex offenders. SORNA mandated the creation of a national database.

3. This is the case in California, where the "Three strikes and you're out" law has been in place since 1994, albeit in a somewhat modified form. See Law Office of the Los Angeles County Public Defender (n.d.) for further details.

4. As is the case in Florida (see Florida Department of Law Enforcement n.d.).

5. Categories that, of course, often bleed into one another.

6. An overwhelming percentage of the offenders registered on databases are male.

Companion
Personal Safety and Surveillance Apps
ELIZABETH ELLCESSOR

App: Companion: Mobile Personal Safety
Developer: Companion
Release Date: November 2014 (Current Version: 4.1.40)
Category: Lifestyle
Price: Free
Platforms: iOS and Android
Tags: safety, college, blue light, companion, city, women
Tagline: "Wherever you go, get there safely with a
    Companion!"
Related Apps: Circle of 6, LifeAlert, Guardly, Watch Over Me,
    bSafe

Never Walk Alone

Companion, developed by a team of undergraduate students at the University of Michigan in 2014, is a free app for iOS and Android that offers users a virtual companion as they move through what might be unsafe spaces. Users open the app; select their destination, route, and mode of transit; and select a "companion" from their list of contacts. The companion (who need not have the app) receives a text message with a link to a map, where they can watch as the user's geolocative dot moves across a map toward their destination. At any point, the user can push a button that indicates "I feel nervous" and enables their companion to contact them or take other action. This data is saved by the app, with the hope of using aggregate data to identify areas in need of more police attention. Alternately, in cases of immediate danger, the user can push a button in the app to contact local law enforcement (police or, often, university campus security).

In addition to the virtual companion features, the app uses several passive triggers during a trip to monitor the user's safety. These include removal of the user's headphones from the jack, a deviation from the route, a longer duration than expected, a sudden change in pace (such as running) or orientation (as when pushed over). When triggered, the app asks, "Are you OK?" and gives users fifteen seconds to respond; no response prompts the app to then contact the user's companion and the selected authorities. As of August 2017, Companion offered integration with Apple Watch, including the "I feel nervous" and "Call 911" options.

Companion is one of many apps that might be described as personal safety monitors. These include apps such as Circle of 6 and Watch Over Me (anti–sexual assault interventions), apps that offer protection for vulnerable populations such as children or the elderly (Find My Kids, Tracker Assist), apps that adopt a more militarized "security" framing (Guardly, AlarmForce), and wearable devices such as the Safelet. These apps and devices go beyond the disciplinary enforcement of bodily norms common to "everywear" technologies (such as the Fitbit) that track movement, heart rate, or other bodily indicators (Gilmore 2015). While personal safety apps partake in the datafication of bodies, they also construct particular bodies and spaces as at-risk in ways that perpetuate gendered, racist, and ableist norms and complicate relationships between individuals, social networks, public spaces, and societal resources. They have historical precedents; they are neither entirely novel nor entirely reduced to earlier media technologies. Rather, personal safety monitoring devices and apps partake in the normalization of an able-bodied surveilled neoliberal subject, responsible for their own protection and distanced from reliance upon or engagement with civic forms of safety.

"Your Personal Safety Service and Mobile Blue Light"

Personal safety monitors take advantage of mobile, geolocative, motion sensing, and social media technologies, but they are not entirely novel. The original tagline for Companion—"Your personal safety service and mobile blue light"—while unwieldy, points directly to two antecedents. First, the "personal safety service" recalls the personal emergency response systems (PERS) that emerged in the 1980s and 1990s and were primarily used by the elderly and people with disabilities to facilitate medical aid. Second, describing Companion as a "mobile blue light" recalls a history of university security measures in North America.

The leading PERS is LifeAlert, known for its advertisements in which an elderly woman cries, "I've fallen and I can't get up!" This representation, though dramatic, was accurate: PERS were largely used to prevent or respond to falls, and their primary users were women in their seventies and eighties (Robinson 1991, 43). The original LifeAlert device was a wearable panic button; when pressed, it would connect users to a monitoring center, where staff would then contact local emergency services if needed. These monitoring devices were widely promoted as alternatives to assisted living, and as such carried hefty monthly fees to cover the costs of monitoring.

Early on, Andrew Dibner, founder of Lifeline Systems, recognized that such devices could benefit "unsupervised children, younger disabled persons, and anyone in a threatening environment such as employees alone at night, those subject to marital violence or racial harassment" (quoted in Robinson 1991, 46). Personal and mobile, PERS were tethered to an individual and had the potential to offer reliable, personalized monitoring services related not only to health but to generalized bodily safety as well. Nonetheless, the successful deployment of PERS among populations understood to be at risk in terms of their health and personal safety may have made it difficult to reach audiences who did not perceive themselves to be at risk of falling and being unable to get up. Furthermore, the association of PERS devices with assisted living, along with advertising depictions of elderly users wearing devices in private homes, localized these services to domestic spaces. For users engaged in a broader geographical sphere, PERS systems (with their reliance on landline phone systems) were incompatible with ensuring safety in work, public spaces, or during transit.

As PERS devices spread among one audience, different safety innovations emerged that specifically targeted college students, understood as a young, healthy, and mobile demographic. In the 1980s and 1990s, in the wake of increased reports of campus sexual assault, many colleges installed "blue light emergency boxes" (*New York Times* 1989). These were essentially phone booths installed across campus that connected to university security or police dispatch, sometimes accompanied by video surveillance. Blue light phones, like campus escort or bus services, were largely promoted as sexual assault prevention and safety options for women. Yet quickly, blue light phones became a generalized "safety" feature on college campuses; a 1995 *Philadelphia Inquirer* article described a mother as "reassured" by the blue light phones at the University of Pennsylvania, saying that due to their promised safety, "I would want my son to come here" (Goodman 1995). In this example, gendered worries about rape are replaced by racialized fears of crime in North Philadelphia.

Companion invokes the legacy of the blue light phone explicitly. Co-founder Danny Freed recalls that the app's development team "noticed that few people used the emergency blue light stations around campus" (Heath 2015). In fact, campuses were starting to remove blue light systems entirely, as, according to Bakersfield College spokesperson Amber Chiang, "Folks have cell phones. And [blue light phones] are not being used" (Barrientos 2012). One of the advantages to the blue light system was the perception of safety generated by visible emergency services; in their absence, the metaphor of the blue light helps to generate this affective response. A recent *New York Times* article on sexual assault on college campuses refers to several tools that "seem like digital updates of 'blue light' poles" (Singer 2015). Drawing upon 2016 data collected by App Annie (see Vonderau, this collection), it appears that users are using the terminology of the "blue light" to find safety apps. A search using this phrase in the iTunes app store yields a number of safety apps, including Companion, bSafe, and Watch Over Me.

Companion offers one key advantage over these antecedents, as it can generate passive alerts without requiring action from the user. The reactive nature of PERS buttons and blue light phones is evident in that their typical use involves contacting emergency services *after* an incident has occurred, thus relying on the individual to be conscious, alert, and interested in seeking aid. With its multiple "triggers" and "Are You OK?" prompts, Companion treats a lack of response as evidence of an incident, outsourcing the response either to a social contact or to the app itself, which can directly contact selected emergency services.

"Help the community identify sketchy areas"

Beyond the personalized services of Companion, the app aggregates data, which it then may share with local police or security forces. To date, such sharing is in place only in Ann Arbor, Michigan, home of the University of Michigan. However, the development team intends to build relationships elsewhere to share such information and better protect "the community" from "sketchy areas."[1]

Built into this, of course, are assumptions about the nature of both the community (Companion users) and the "sketchy" areas. The app originated with collegiate communities in mind, and a normative campus community is core to Companion. Its promotional video illustrates these dynamics quite clearly. It shows two young women activating the app: First, a woman of color prepares to leave the library, activates the app, and selects her white female roommate, "Clarissa," as her Companion. Second, a white woman exits what looks like a bar or restaurant, selecting a young white man, "Jacob," as her Companion. Clarissa is shown receiving the text message as she brushes her teeth, while Jacob is sitting in a desk chair when his phone alerts him. The video then shows each of the women who requested a Companion being followed, literalizing the app's tagline—"never walk alone." Clarissa, still brushing her teeth and looking at her phone, walks behind her friend, while Jacob rolls along in his desk chair following his friend. Both walks take place at night, through darkened sidewalks of the sort that might surround a college campus, including retail and residential areas.

In this video, young women are figured as potentially at risk as they traverse non-campus spaces at night. Gender, space, and time are linked to create a vision of otherwise valorized bodies as at-risk. Companion is built upon these assumptions about risk, community, gender, and space, which are equally enforced through the app store—which categorizes Companion and similar technologies as "Lifestyle" apps (recalling the "women's pages" and "lifestyle" sections of newspapers)—and through user reviews, which frequently comment on the app's utility for women traversing cities alone.

All of those seen in the video appear young and able-bodied, as the users walk to their destinations, reflecting a normative vision of collegiate bodies. Furthermore, Companion allows users to be accompanied on either walking or driving trips, reflecting modes of transit that may be more common among able-bodied, geographically constrained, and economically secure users. The passive triggers and estimated times that Companion uses to estimate "safe" progress depend on data from Waze, which similarly predict times based upon optimal conditions. The enmeshment of time and location in the app's predictions of a user's movement through space mean that alternate kinds of transit fall outside of its algorithmically constructed expectations for "safe" movement through space. There are no options to track progress by public transportation, bicycle, or other forms of transit, including assistive technologies such as wheelchairs or crutches. Though Jacob is seen following in his desk chair, this metaphorical deployment is very different from what a wheelchair user might experience. Disabled individuals often have a framework for risk shaped by their own bodies and needs and by "being accustomed to everything taking a long time" (Sparf 2016, 1). As an accessible path may diverge from recommended routes, add time to an otherwise safe journey, or entail changes in pace (slow movement through difficult areas, quicker elsewhere), it is incompatible with a safety technology that privileges a path optimized for the able-bodied. Even the unpredictability of public transit can cause difficulties in matching a trip to its predicted features, resulting in glitches, false alarms, or ineffective use conditions. The default features and limitations indicate a specific intended user base and limits the expansion of the service.

In addition to normalizing a young, able, and gendered collegiate community and user base, Companion engages in a racialized construction of space that has real and dangerous implications. Currently, Companion gives the local police aggregate "data about where people are feeling nervous around the surrounding Ann Arbor area" (Heath 2015). This description neatly separates campus from the surrounding town, figuring one as safe and the other as (potentially) "sketchy." As a common slang term, "sketchy" might be understood to mean something like "unsafe," "questionable," "creepy," or otherwise off-putting. Often applied to individuals as well as spaces, "sketchy" is a term with particular resonance among a young and largely white demographic; what makes someone or someplace "sketchy" may easily also be what marks it as poor, ethnic, nonwhite, or otherwise unfamiliar or undesirable. In fact, SketchFactor, an app released in 2014, was criticized for exactly this implicit racism, as the "app depends in large part on the wholly subjective judgments of its entirely anonymous and self-selecting user group" (Capps 2014). Identifying "sketchy" areas and sharing such information with authorities, then, suggests a nearly unavoidable rise in racial profiling, selective police surveillance, and similar discriminatory practices.

In the 1990s, John Fiske wrote that "surveillance is a technology of whiteness that racially zones city space. . . . [It] enables different races to be policed differently" (1998, 69). Though he referred to audiovisual forms of surveillance, appified and social forms of surveilling public spaces can produce similarly chilling effects. Companion's aggregation of users' affective "nervousness" means that data cannot be pulled apart for analysis. User biases—including overrepresentation of some users in using the feature, as well as racial, ethnic, and class biases—are rendered opaque, as the accumulation of this data facilitates the treatment of subjective assessments as objective information. Furthermore, tracking devices bring users into inherently unequal data relations, because "the user never sees how their data are aggregated, analyzed, sold or repurposed, nor do they get to make active decisions about how the data are used," stored, deleted, or shared (Crawford et al. 2015, 487–93). Thus, the data provided by the Companion community to preserve its own safety is not available to that community and is not accountable to revision or oversight by the users who generate it. Companion's use of the data, and the data's use by police or other security forces, is thus difficult to critique in any detail. Anonymized, aggregated, and delivered as a crowdsourced form of knowledge, this data exhibits a "computational objectivity" in which the capacity to digitize and algorithmically parse input about the world becomes a source of truth and basis for action (Gates 2013). As a result, the racial and class biases of users can become those of the technology, and those of the state, all through a naturalization of "data" as information rather than anecdote or affect.

Emergency Contacts

The involvement of police or other security authorities in use of Companion data ought to further prompt consideration of the relationships between personal safety monitors, societal emergency services, and the role of social networks in providing security. Personal safety monitors encourage users to enact a preparedness that accords with neoliberal expectations of personalized risk. Rather than relying upon government-provided, shared, and public resources in case of emergency, users are encouraged to enter into uneven and opaque relationships with technology companies and to outsource their safety to the cloud and to their social contacts. Whom to contact in case of emergency— and who has the agency to make such contact—becomes an increasingly unanswerable question.

The most obvious corollary to the "I feel nervous" and "Call Police" buttons in Companion is a direct call to 911 or police dispatch. The national emergency number 911 is actually a highly localized service. It is a patchwork system, a network of call centers that take reports and dispatch appropriate local emergency services, including police and ambulances. Services connected with 911 are funded on the state level and operated by individual counties, employing a range of legacy and digital technologies. However, a direct call to 911 is culturally ubiquitous—well known, presumed functional, and regularly invoked in a range of media and safety contexts. Furthermore, 911 has ties to telephone systems that have ensured the service must act under the universal service mandates common to telecommunications in the United States (often, 911 is funded out of state universal service funds). In other words, it is mandated to serve all audiences to the best of its ability, including rural, disabled, non-English-speaking, and poor citizens.

Such guarantees are by no means iron-clad; recent police shootings and killings of black men and women and people with disabilities indicate that "individuals located perilously at the interstices of race, class, gender, and disability are constituted as non-citizens and (no)bodies by the very social institutions . . . that are designed to protect, nurture, and empower them" (Erevelles and Minear 2016, 98).[2] There is reason to be suspicious of policing, of police handling data, and of the limits of citizen influence in shaping US policing cultures and policy. The surveilling eye of the state is racialized and invested in maintaining existing unequal power relationships (Fiske 1998; Magnet and Rogers 2012); 911 calls, like security cameras and screening technologies, can be differentially prioritized, interpreted, and acted upon. Black Lives Matter activists and others have proposed ending policing, supplementing it with citizen corps of volunteers, and using live video streams and other social media to maintain public safety without involving police. Yet these solutions also pose risks, as users are asked to proactively take responsibility for their own safety and that of their communities, often in concert with relatively opaque technologies operating in a capitalist system.

By taking on the role of emergency media, apps like Companion act as intermediaries in citizens' right of access to 911 or other emergency services. They are *not* tasked with universal service or subject to government or citizen oversight. Thus, Companion and other apps contribute to the narrowing definition of citizenship by extending discriminatory, normative notions of community, gender, ability, class, race, and space to a market-friendly, neoliberal environment. As emergency services become digitized, personalized, privatized—appified—they also become increasingly removed from the mechanisms by which citizens have had access to intervene in, and advocate for, public services of various types. Instead, these apps foster an environment in which individuals are asked to take responsibility for their own safety by entering into new, data-driven and sometimes paid, relationships with nonprofit or corporate technology companies rather than relying on social safety nets.

Such neoliberal emphasis on managing risk, taking personal responsibility, and exerting preparedness and resilience is evident in many areas of contemporary disaster discourse. These ideas increasingly guide policy and popular responses to extreme weather and natural disasters, have long been deployed in advice given to women looking to prevent sexual assault, are now applied to public safety, and displace discussions of larger issues (such as climate change, rape culture, or systemic racism). Furthermore, discourses of preparedness tend to ignore differences that enable some people to act in accordance with recommended strategies and exclude others; women, the poor, and disabled people, in particular, often exhibit lower levels of recommended "preparedness," which is inherently tied to different material conditions and relationships to institutions of power (see Leyda and Negra 2015; Smith and Notaro 2015). While we might look at an app like Companion as an instance of "public failure" (Vaidhyanathan 2011)—an intervention by the private sphere into services that the public sphere has failed to provide—such technological solutions replicate existing divisions and introduce new fault lines.

Companion and other personal safety monitors also complicate the notion of the "emergency contact," displacing social responsibilities onto individuals and their social networks. The possible dangers of relying on an app to alert your social network to danger, and then relying on that network to take appropriate action, is evident in anecdotes from Companion users. Several reviews of the app in the Apple App Store refer to "trials" of the technology in which text messages were not sent, leading users to opt out before ever using the app in a dangerous situation. A student at Stetson University described a similar test run in which the selected "companion" received the text alert but thought that it was a virus (Guerin 2016). Such misfires, in a truly dangerous situation, could be tragic. They further indicate that communication is a necessary part of marshalling social surveillance in the interest of personal safety. The cultural context of socially surveilled safety monitors is at least as important as the technology; in order for personal safety apps to be instantly understood—and ubiquitously used—society would need to fully embrace security as a constant personal project, a fantasy of ongoing preemptive action (Andrejevic 2017).

Conclusion

Companion and other personal safety monitoring apps and devices intervene directly in the norming of bodies, spaces, and relationships between individuals, communities, and institutions (such as emergency services and policing). They arise from particular histories and particular conceptions of space and safety. Doreen Massey argues that the "gendering of space and place both reflects and has effects back on the ways in which gender is constructed and understood in the societies in which we live" (2013, 186), an argument we could extend to discuss the classed, racialized, and ability-based constructions of space in our universities, towns, and larger contexts. Too often, apps like Companion extend hegemonic notions about space and belonging, citizenship and authority. While apps may counteract such messages, it is important to consider how the very outsourcing of what have been public resources to personalized and privatized contexts may also produce unintended effects. By asking individuals to take app-based preventive responsibility for their own safety and to deputize their social networks as protective forces, Companion and similar apps undermine the possibilities for meaningful oversight, critique, and improvement of public services.

Notes

1. This language comes from a 2016 version of Companion's website. "Companion" (), http://www.companionapp.io. The site has since been updated.

2. In 2015 alone, 965 people were fatally shot by police; of those not exhibiting threatening behavior, 3 in 5 shot were black or Hispanic. Black men comprised 40 percent of those fatally shot (Kindy et al. 2015). Mental illness and race are also strongly tied to police violence, as atypical behavior often escalates official violence. The 2016 shooting of Charles Kinsey, the black male therapist of a Hispanic autistic man in Miami, Florida, illustrates these perilous intersections.

Social Networking/Communication

Yik Yak
From Anonymity to Identification
TAMARA SHEPHERD AND CHRISTOPHER CWYNAR

App: Yik Yak
Developers: Tyler Droll and Brooks Buffington
Release Date: November 2013 (Current and Final Version:
    4.10)W
Category: Social Networking
Price: Free
Platforms: iOS, Android, Web
Tags: anonymity, publics, affect, design
Tagline: "Yik Yak makes the world feel small again by giving
    you a feed of all the casual, relatable, heartfelt and silly
    things people are saying around you."
Related Apps: Secret, Erodr, Whisper, Nearby, Scout, Cloaq,
    Fade, Sneek

Introduction: From Boom to Bust

"What's on your mind?"

When the Yik Yak app launched in 2013, its text box asked users that question, a familiar one within the social networking sphere. Yik Yak was unique, though, in the sense that the bulletin board-style social networking app did not use profiles to identify its users' publicized thoughts. Yik Yak's appeal was based on posts, or "yaks," being anonymous but proximal, arranged according to location. Users were identified by randomly assigned icons that were specific to individual conversation threads. Those threads unfolded through responses that were not from networks of friends or followers but from users within a certain geographic proximity, such as a five-mile radius, thanks to the GPS and locative features of modern smartphones. The company heavily targeted and marketed on North American college campuses, and this is where the app gained its greatest traction among users but also stirred the most controversy for encouraging hate speech and bullying.

As the app evolved, Yik Yak gradually reined in its controversial anonymous features and progressed toward increased identification of users in the name of reducing harassment on the app. In 2016, for example, the app's developers introduced the option of personal profiles, which soon became mandatory. Ostensibly a move to protect users and reduce harassment, we argue in this chapter that the shift in the app's features was instead about protecting the app as commodity, a way to bring Yik Yak in line with social media's demands for hyper-identification and personalization. Yik Yak shut down in April 2017 after the app finally ran out of money. This fate reinforces its status as a paradigmatic case study of the challenges inherent in fashioning apps as commodities through the use of venture capital. This model provides the basis for breakneck growth through the rapid acquisition of users, but it also obliges apps to monetize those users through practices that may ultimately alienate them and drive them away.

Anonymity in Internet Culture

Anonymity has long been one of the most celebrated and reviled affordances of computer-mediated communication. Sherry Turkle (1995) foresaw both the freedoms for identity play that come from online anonymity and its potential threats to the integrity of lived, embodied experience. More recently, the prevalence of cyberbullying, online hating, and "alt-right" discourse online has tipped the balance between optimism and pessimism in favor of a more critical take on expressions of hate in anonymous environments (e.g., Citron 2014; Phillips 2015). The anonymous space that has garnered the most attention in this vein is 4Chan's /b/ random board, a descendant of classic bulletin board systems structured by threaded conversations of posts by users, identified by their chosen screen names or under the default handle "Anonymous." Although this degree of anonymity has proved politically and culturally productive—giving rise to the hacktivist collective Anonymous and the proliferation of internet memes (Coleman 2011; Shifman 2013)—/b/ can also be seen as a central breeding ground for the way trolling shifts into hate speech along lines of entrenched racism, sexism, homophobia, xenophobia, and so on (A. Taylor 2014). And so, as Whitney Phillips and Ryan Milner claim, we arrive at "the ambivalent internet," where "reduced social risk, often spurred on by anonymity, allows these tinkerings to veer into territory that participants might be inclined to avoid in embodied spaces, for better and for worse" (2017, 202).

This inherently ambivalent quality of internet culture is often lost in the moral panics around anonymity online, which were evident across press coverage of Yik Yak. As it rose to popular attention in 2014, Yik Yak was framed as an easy platform for anonymous expressions of racism, homophobia, and misogyny within a suspect youth culture of always-on mobile connectivity. A sit-in protest at Colgate University following persistent racist and threatening messages on the app initiated a more public conversation about Yik Yak and anonymity. Other reports described how the app facilitated threats of mass violence on several college campuses, including a provocation toward gang rape at Kenyon College's women's center (Mahler 2015). A dedicated Tumblr site was created to document "The Absolute Worst of Yik Yak." As a response to such coverage that threatened to alienate Yik Yak's investors, the app began introducing a number of design changes in 2016 intended to mitigate cyberbullying and harassment by moving from anonymity toward identification. We examine these design changes in the context of Yik Yak's culture, conceptualized here as a hybrid public space, and the political economy of the app industry. While the design of the app was altered in order to address moral panics around anonymity and generate increased revenue, the various changes resulted in a loss of activity that ultimately led to the app's demise.

Yik Yak's Emergence and the App Economy

Yik Yak's initial selling point was its unique combination of anonymity with proximity, where mobile phones' location-based awareness associated yaks with others posted from within a five-mile radius. As a designed value, proximity modulated total anonymity by suggesting a closeness to other users not based on individual identification and social capital but, rather, on a close correspondence between the imagined community of users and the local (often campus) community. In this way, Yik Yak didn't strictly adhere to the definition of a social network, typically defined by the presence of visibly networked profiles (boyd and Ellison 2007). This new space for mediated sociality, when combined with an attractive brand based on a distinctive color scheme and the cute yak icon, proved to be a hit with both investors and college students. Yik Yak was ideal for a generation accustomed to inhabiting a hybrid space dominated by ubiquitously connected smartphones and mundane software apps providing sociality and connectivity.

We consider this particular hybrid space through the lens of what Zizi Papacharissi has termed "affective publics": "networked public formations that are mobilized and connected or disconnected through expressions of sentiment" (2015, 126). While Yik Yak's affordances could have provided a space for the mobilization of affective publics based on shared geographical and political concerns, it instead primarily provided a space for users to experience shared affective experiences through jokes and gripes, tapping into the irreverence that tends to mark the structure of feeling emblematic of internet culture (Miltner 2015). Accordingly, Yik Yak's thematics drew upon playful tropes surrounding the yak, inspired by the 1958 song "Yakety Yak" suggested by one of the creators' parents. The cute aesthetic of the app, bolstered in a 2015 redesign that made the eponymous yak on the app tile even cuter, promoted whimsy and irreverence within its discursive culture. As a visual pun, the yak, along with other vaguely Himalayan elements like snowcapped mountains and the "Basecamp" and "My Herd" features, provided a sense of thematic unity consistent with this playful tone.

In April 2015 the app introduced small, brightly colored, circular reply icons to identify anonymous speakers—conveyed in line drawings of a canoe, a pair of paddles, a map, a compass, and a boot, among other exploration or camp-related objects—that served to couch the app's support for potentially controversial expressions in a more irreverent style of signification. For example, the practice of responding to the author of a yak (marked as "OP" for "original poster") often involved conversations where interlocutors referred to one another by icon, as "red canoe" or "yellow paddles." Because identity as marked by icons shifted across different conversation threads, and hashtags couldn't be used to build on collective identification, Yik Yak tempered the affective publicity that Papacharissi finds in Twitter. There, political statements constitute "public displays of affect" tied to personalized performances of cultural identity (Papacharissi 2015, 7, 95). In Yik Yak's case, anonymity limited the performative dimension, and the icons infused even the most serious discussions with a whimsical quality emanating from the app's sign system. Individual users were assigned different icons across threads, meaning that posters couldn't be associated with a consistent identity over time to be held to their views or actions.

Instead, the vetting system on Yik Yak followed an upvoting and downvoting model, where a yak would either receive validation through upvotes, be made to disappear through downvotes, or languish through a lack of engagement. As a quantitative study on Yik Yak posts observed, "The most commonly endorsed yaks were humor and information sharing and seeking, suggesting that Yik Yak served as a source of entertainment for college students as well as a platform for campus community building" (Clark-Gordon et al. 2017, 8). Such humor was often based on local phenomena and events, including "overheard on campus" type posts, reflecting the designed affordance of proximity and its "My Herd" feature; the latter allowed users to choose a particular location that they could then participate in from beyond campus. This innovation was touted as a way for the app to attempt to maintain its user base while students were on break from college. However, the introduction of this feature could also be regarded as an attempt to forestall the app's incipient decline as college students left for the summer and users left for other apps.

Design Changes away from Anonymity

The system of identifying a poster through a cute reply icon, consistent only within a particular thread, was changed in response to negative publicity that pointed out how, despite its cuteness, Yik Yak harbored sexist and racist sentiments (see fig. 1).

In July 2016 Yik Yak thus introduced profiles, claiming that they would render "your experience interacting with other Yakkers in your local feed so much richer" (Droll 2016a). Rather than the reply icons, each user was now identified by a randomly assigned handle and an icon of a standard geometric pattern. The default handle and icon were no longer consistent with the cute, playful tenor of the app; in fact they were made deliberately boring in order to push users to themselves provide a unique handle and upload a profile photo. The architecture of the app thus suggests a path of action for users to follow toward greater identification. Moreover, the cumulative upvote and downvote "Yakarma" measure, which used to be seen only by users themselves, was now visible to others on their profile. By tabulating all of an individual user's upvotes and downvotes, Yakarma now helped to differentiate an authentic user from a spammer or scammer (Bergstrom 2011).

The addition of profiles linked to Yakarma scores was the most visible and dramatic, but not the first, of Yik Yak's design changes in response to criticism of the app's support of harassment and hateful expression (Mahler 2015). Since its inception, Yik Yak had steadily introduced successive modifications to its

Screen shots of sexist and racist yaks

architecture, where code was deliberately altered to encourage less anonymous communication. Handles added in March 2016, private chat added in April 2016, and profiles added in July 2016 were new default settings that eroded the potential for anonymous communication by rendering the platform's preferred path of action as the one of least resistance for users (Kerr 2010). Such developments in Yik Yak's architecture steered users toward increased identity verification, promoted through an appeal to more "real, authentic connections" (Droll 2016b).

Yik Yak also redesigned its proximity features over time to regulate uses of the app. As one of the core original features of the app, proximity, rather than networks of friends of followers, was the dimension that linked user conversations, but it had to be regulated over time in response to criticisms about anonymity. For instance, during Yik Yak's initial burst of popularity with high school students in early 2014, concerns over a spike in cyberbullying facilitated by the app resulted in creators Tyler Droll and Brooks Buffington partnering with Vermont-based company Maponics to apply geofences around elementary, middle, and high schools based on their GPS coordinates (Graber 2014). Somewhat ironically, while proximity was intended to group users into communal relationships not based on identity or networks, it became a way to regulate hate speech through constraints, containment, and exclusion.

As Yik Yak retooled its app to prevent negative social interactions, support for ephemeral expressions also eroded over time due to the tighter controls over anonymity and proximity. In this sense, the app's design changes undermined its particular form of affective public, an alternative kind of discursive space for ambivalent and irreverent internet culture. Ironically, it appears that the app's incremental gravitation toward profile-based social networking that was intended to gain users and quell investors only hastened its demise.

The Political Economy of Identification

From its peak surge in popularity in late 2014 (when it was reportedly valued at between $350 and $400 million) until January 2016, Yik Yak steadily expanded to a sixty-person workforce, supported by $73.5 million in venture capital funding (CrunchBase 2016). But by April 2016 the momentum of this period was lost: Yik Yak fell from its top-ten position in the app store and saw the departure of its Chief Technology Officer. The app was reported to have even used algorithms to systematically downvote any yaks that mentioned competitor apps such as Fade, Sneek, and Unseen (Constine 2015). In this light, Yik Yak's move toward greater identification in the name of community building might be seen as a means of fostering "fantasies of unity and wholeness as the global [that] in turn secure networked transactions as the Real of global capitalism" (Dean 2005, 67). However, Yik Yak's very existence was predicated on an anonymous variant of affective publicity that the app could not uphold if it was to capitalize on user data through identification.

Yik Yak's primary user base was the sought-after millennial demographic attainable on college campuses. The app, itself created by fraternity brothers Droll and Buffington, expressly targeted college-age students not only through design decisions such as geofencing but also through marketing strategies. "Ride the Yak" college tours, the "Peek" function focusing on distant campuses, blog profiles of various North American universities, and app events revolving around college sports all combined to render the affective public space of Yik Yak in the language of college-specific practices. Even in a somewhat anonymous space, such an affiliation works as a mechanism for amassing social capital among users who build affective investments in the app. For instance, the short-lived feature "Yik Yak Famous" granted users with the most upvoted yaks a star-shaped badge and exclusive avatars that could be shared among their herds. Such a feature, while it compromises the values associated with anonymity, reflects the overwhelmingly commercial logic of social media as a space for self-branding via reputation metrics (Hearn 2010). Its facilitation of self-branding through the visibility of Yakarma on profiles further shows how the app sought to ameliorate its public relations problems while retaining active users through features premised on longer-term investments in Yik Yak for building social capital. While Yik Yak served as a space where publics of various sorts could form, it was most often a space for the shared experience of sociality rather than an expressly political platform, especially given the tendency of most of its college-age users to obscure and depoliticize the profit-seeking nature of social media apps (Dean 2005).

As an extension of common app market pressures to hit growth targets by increasing monthly active users, the changes made to Yik Yak's functionality in 2016 were met with attendant revisions to its Terms of Service policies. Yik Yak's final set of policies reflected a move away from anonymity and toward identification for at least two purposes: first, as mentioned above, there was a public relations imperative to assuage moral panics around cyberbullying and harassment on the app. The app's Rules & Info page sternly warned against bullying, harassment, defamation, spamming, or posting other users' personal information (doxxing). Top questions in the FAQs included "I'm being bul-lied/targeted" and "I reported a Yak like 50 times. Why is it still up?" With the roll-out of private chat, Yik Yak established a safety center on its website, containing tips on how to report inappropriate user behavior and a set of guidelines for users, colleges and universities, educators and parents, and law enforcement. While such policies are helpful for users, they also served the interests of Yik Yak in preventing negative press and retaining, if not growing, the number of active users.

Continual growth in the number of active users is the lifeblood of apps built on speculative venture capital funding that projects future profit. Such profit in social networking platforms typically follows the model popularized by Facebook, where user data feeds into an advertising infrastructure built on target marketing. For Yik Yak this presents the second reason for the move to-ward identification, as is apparent in the Terms of Service. The developers' in-centive to follow individual users on the back end in order to "personalize and customize" user experiences entails displaying ads, "including advertising that is targeted to you based on your location, as well as your Yik Yak activities" (Yik Yak 2016). In much the same way as any other commercial space online, users need to be identified and authenticated in order to maintain the integ-rity of targeted advertising as a primary profit model of the web (Lessig 2006, 48). Ironically for Yik Yak, the move toward identification that was necessary in order to monetize users involved vacating its market niche as a distinctive sort of affective public space based on relative anonymity and proximity.

Conclusion: Mundane Hardware and the App Life Cycle

As Yik Yak became more invested in user identification, it also declined significantly in popularity. This correlation presents a chicken-or-egg problem that is endemic to apps as cultural artifacts within a tech start-up world characterized by the drive for constant newness. Most new apps that catch on court controversy, especially those targeted toward youth; in the press coverage of Yik Yak, anonymity equaled cyberbullying. Design changes intended to assuage such fears by mitigating anonymity might have alienated users (Neff and Stark 2002). On the other hand, users were likely already moving away from the app, and design changes were intended to keep them interested (Perez 2016). Behind the scenes, such speculation about user interest determined the amount of venture capital that Yik Yak was able to secure, which had been nil since November 2014 (CrunchBase 2016).

This dynamic embodies the tensions at the heart of social media app culture. As smartphones have become increasingly integrated into the lives of young people, there has been greater interest in capturing their activity and information through the development of social apps. The app industry's funding model facilitates this by making an abundant amount of capital available early on during the novelty phase. In this cycle, the novelty and the funding reinforce each other until the novelty wears off, growth stalls, and the funding dries up. This, in turn, necessitates the adoption of increasingly drastic design measures that often serve only to hasten the app's demise. In the case of Yik Yak, this involved undermining the very features that had set the app apart in the first place. All of this suggests that what is mundane about apps may not so much be the software but, rather, the manner in which the devices themselves have become so deeply ingrained in everyday life as a portal for commercialization spurred on by venture capital. Yik Yak's brief life cycle was a function of the emergence of the smartphone as mundane hardware and constant demand for new spaces and services to animate such devices.

Yik Yak's original features of anonymity and proximity that supported expressions of privilege, in addition to cyberbullying and hate speech, further reveal how app culture is built not only on continual newness but also on a pervasive attachment to individualism and meritocracy. Anonymity on Yik Yak was initially framed through the absence of a profile, which "levels the social playing field so that content is judged on the quality of content, not by who said it" (Droll 2015). The enduring investment in a level playing field across app culture largely reflects the white, male, techno-utopian ideology generative of social media platforms that claim to be democratizing in order to themselves shape public discourse (Gillespie 2010). In this way, even with the possibilities for new publics that might have emerged from the app's unique combination of anonymity and proximity, Yik Yak instead only offered another typical manifestation of a meritocratic fallacy. The addition of profiles amid the app's waning popularity suggests that, for Yik Yak, the holes in this fallacy were becoming larger and more difficult to ignore. Instead of asking "What's on your mind?," the app's opening question changed in 2016 to ask "What's happening?" Perhaps this was an attempt to curb the antisocial elements of Yik Yak. Unfortunately, the main thing happening at that time was the app's inevitable decline due to the inherent contradiction between anonymity and surveillance capital in the fleeting lifespan of an app.

WeChat
Messaging Apps and New Social Currency Transaction Tools

FINN BRUNTON

App: WeChat / Weixin / 微信
Developer: Tencent
Release Date: January 2011 (Current Version 6.3.31)
Category: Instant messaging
Price: Free
Platform: Cross-platform
Tags: social, messaging, payments
Tagline: N/A
Related Apps: WhatsApp, Allo, Facebook Messenger

To say "a brief review of WeChat" is like promising a brief review of the fifteenth century CE, trees in general, or the city of Mumbai. (The fall of Constantinople and the end of the Byzantine Empire? Two stars. The Sengoku period of civil war? One star. Luca Pacioli's development of double-entry bookkeeping? A+++++). Where would we begin? It's an app, yes. It runs on the common smartphone operating systems. It's produced by the company Tencent in China, where it has better market penetration than Facebook does in the United States. It's used by more than 700 million people.

Then the litany begins, sounding like the amphetamine gabble of a desperate salesperson adding features in a failing meeting: It's a messaging app. And a ride-hailing service. A lot of the interaction involves short voice messages, but they're not voicemails. There are games. You can make an appointment with your doctor and pay your rent. There's a Facebook-esque social validation Skinner box called "Moments." There's a video-capture and sharing function ("Sights"), custom stickers, QR code recognition, Yelp-style reviews, video-conferencing, photos with Instagram-style filters and Snapchat-style captions, real-time-location applications, and machine translation. You can clock in, log your hours, and track reimbursement expenses for your job. Renew a visa; shake your phone to message with a random person; get a date; report a traffic violation and get a reward from the police; buy movie tickets; book a hotel room and then set the room to "sleep," closing the blinds and dimming the lights; start a business; shop; pay and get paid; gift and receive holiday money. (More on that later.) In this comprehensiveness, it hearkens back to the undeliverable, all-encompassing software projects of yore, vaporware glaciers like Xanadu and Chandler whose remit kept expanding as their production grew more dysfunctional. Chandler, as chronicled in Scott Rosenberg's *Dreaming in Code* (2008), would turn calendaring, emailing, notes, and an indeterminate number of other things into peer-to-peer networks; Ted Nelson's decades-in-development Xanadu, the original hypertext platform, would transform reading, writing, graphics, music, and intellectual property, creating a universal user interface with a global franchise empire. Such projects end up mired in whiteboard diagrams of universal application, like Lope de Aguirre, stuck on the coast of Venezuela in 1561 with no support and two hundred soldiers, announcing himself king of "Peru, Terra Firme, and Chile": tragic fantasies of conquest. Except that WeChat is real, and it works.

How it works is where we begin to review such a massive chunk of function. In the two sections of this review, I make four arguments about what we can learn from WeChat. First, that we can understand some apps as an emerging form of infrastructure, and, second, that messaging apps are particularly suited to this infrastructural role because they are adaptable technologies within a social "typeform." Third, that the most powerful expression of WeChat's expansion into an infrastructural role can be seen in how it became a payment platform, making money conversational. To make this argument about the "chatification" of money, I look at how the WeChat team transformed *hongbao* holiday money into something specifically suited to the affordances of a smartphone app. Finally, I close with a warning to the reader: the very properties that make WeChat such a fascinating study—a messaging app turning into an infrastructural component for many aspects of everyday life—makes it, and other apps, into a system with a new kind of capacity for emergent coercion.

Iterative Infrastructure

WeChat is, or is becoming, an infrastructural app. "Infrastructure" is a tricky term to precisely define; Susan Leigh Star famously framed it as a "call to study boring things," and when we invoke the word, we often tend to mean some class of essential services and large-scale *stuff*: electrical pylons and gas pipelines, freeways and subways, satellites and logistics—things "big, ubiquitous, and foundational," as Sanford Kwinter puts it (Star 1999, 337; Kwinter 2008, 29). Even that sketchy list mixes wildly disparate items, though, and what's "essential" is the furthest thing from a fixed category. An "emerging essential," Christian Sandvig's phrase for the internet viewed as infrastructure, perfectly captures this (2013).

The internet used to be something that ran on the infrastructure of telephony, all those copper lines and ceramic insulators and creosote-smelling poles and switching systems, socketing a Dreyfus-model Ma Bell handset into the rubber cups of a modem, a "modulator-demodulator" whose very name implies its purpose—to make digital signals transmissible over the (actual) infrastructure. Now the internet has itself become enormous, spatially and financially and physically and socially, something without which large, brittle swaths of contemporary society would simply and entirely collapse. Furthermore, as Sandvig's phrase suggests, it is still emerging globally and in new domains of human activity. Sandvig draws on Paul Edwards's excellent negative definition: "perhaps 'infrastructure' is best defined negatively, as those systems without which contemporary societies cannot function" (2003, 187). While it's too much to say that metropolitan Chinese society would fail if WeChat suddenly disappeared, many components of civic life—from parking tickets, to trains, to businesses and access to money—would be plunged into chaos. It's not quite infrastructure yet, but it's edging toward that condition.

"Bridges are, with skyscrapers and dams and similar monumental structures, the visual representation of our technical mastery over the physical universe," writes Rosenberg in his history of Chandler—a software project bucking for an infrastructural role—nose-diving into the landscape. "In the past half century software has emerged as an invisible yet pervasive counterpart to such world-shaping human artifacts" (2008, 8). There is something more significant than invisibility at play in software, though, that is particularly salient to understanding apps: iteration and agility. If you've never seen it, the WeChat I've described above sounds like a user interface nightmare, recalling the interfaces of 1990s cyberpunk movies and anime—an eyestrain metropolis of tiny related icons, or a fractal abyss of submenus inside submenus. How would you familiarize new users with such a system?

When you launch WeChat—and, more to the point, when WeChat originally launched—you don't see a wall of complexity. You see a stream of messages from different people. You know the format: a bar above, with navigation and utilities (search, return from a particular conversation) and a bar at the bottom: Chats, Contacts, Discover, Me. Chats is . . . chats, a timeline of activity. The rest are equally obvious. It is a typeform.

In the history of industrial design, a "typeform" describes the archetypal expression of a particular manufactured object: it's what you think of when I write "toaster" or "barstool" (Gantz 2014, 89). It's the way a thing "*should* look," in some particular moment, and creates a kind of useful inertia, mingling the conservative choices of manufacturers, who need to work with the established typeform, and the tacit knowledge of customers, who know what they're looking at and where the handle is (Gantz 2012, 163). WeChat epitomizes the *typeform* of a messaging app, as, say, the Brooklyn Bridge does a bridge, or the wall outlet, whatever its plug configuration, says "electricity."

What is different is that WeChat, like other apps-as-infrastructure, can constantly change inside its typeform seams. An app can update in ways that a bridge, a hospital, or a high-tension power line cannot. WeChat didn't emerge as a Platonic solid from the mind of Tencent executive Xiaolong Zhang, president of the WeChat group; it shipped as a messaging app and then incrementally, update by update, filled with implicit complexity. Google Chrome offers another example of this—arriving as a stripped-down typeform of a browser that checked for new updates and, day by day, became a RAM-hungry hulk with hundreds of additional features, as though you'd bought a weed whacker and somehow, four years later, ended up with a combine harvester. Clay Shirky wrote of Lei Jun, founder of Xiaomi—a mobile phone company whose phone software is updated every week—that "if you want to sell a complex object to the general public, then the interface is the product," and behind the interface you can add upgrades as you go (2015). You can build a novel kind of reciprocal infrastructural role on this basis: an "emerging essential" that becomes more ubiquitous and vital by modifying itself in response to what emerges as most essential about it.

The fact that it's a messaging app comes to the fore here. WeChat is protean, iterative, not the same app it was earlier this summer or last year—but so are other apps. Everything running on a smartphone with a data connection can be assumed to be regularly updating and incrementally changing, what Jonathan Zittrain calls a "contingent" connection (2008). But the use of WeChat is primarily not between a user and the app, but between the user and whomever they are in conversation with, and in that lies the possibility for an enormous sleight-of-hand and explains how WeChat's tangle of possible utility can actually be used—with a strategy reminiscent of Alan Turing's "imitation game."

As a mode of interaction, texting produces the perfect format for both machines and humans to interpret. It's far easier for software than spoken voice, where recent breakthroughs are still constrained by acoustic and discursive challenges (and WeChat will, paradoxically, automatically transcribe short voice messages into text to pass along to software as needed). Users are already thoroughly and reliably familiar with the techniques involved in texting and messaging. Every SMS notification, every moment of blocking a stairwell or lurching into traffic while captivated by WhatsApp or Line or Viber or Facebook Messenger, is effectively training for the use of the common interface. New applications of the same interaction style are proliferating. From fitness (Lark), to personal finance (Digit), to personal assistants (Magic), to games (Lifeline), to logistics (Taobao's 阿里小蜜), to news (Quartz), to payment (many) and customer service (Rhombus and more), whole categories of media activity that at one time would have implied custom software platforms, tools, or programs, instead work through messaging. In those environments, the handoffs between humans and machines can be seamless: an exchange that triggers an automated response and one that pops up as a thread on a person's desktop or device, eliciting their response, are indistinguishable.

WeChat, therefore, did not start building custom pages or controls for all of those diverse functions listed above—there is no "Pay my Shenzhen parking ticket" button set by default. Instead, they created a category of "official accounts," which were less a shift of functionality than a shift of context. You added them to your contacts knowing they were not people but businesses, organizations, and government agencies, and opening their channel enables different kinds of possible actions. This blurs seemingly distinct categories of communication: depending on what you follow, you can get notifications from a hospital, a train line, a fellow dieter you've never met, a business promoting to you, and your mother. Yet it also renders them more distinct by keeping them in the main flood of messages and separate from "Moments," which is for intimate social activity alone and has no branded company or media content, auto-posts from games, and similar clutter.

This is a striking arrangement: rather than Facebook or Twitter, which frame themselves as social networks and then expand the scope of "social" to include thoughts from Kanye, brand updates, sponsored content, and viral campaigns, WeChat is a front end of anything that can be texted (or sent as a brief audio message), of which socializing is a *subset*. Rather than social activities adapting to environments developed for business collaboration, everything else—contact with institutions, shopping, traveling, employment— becomes just another chat, another mundane mobile interaction.

Money Talks

Businesses can therefore be run more or less entirely inside WeChat: clients, payments, scheduling, reservations, announcements and advertising, mediating between employers and employees. That's only possible because of the chatification of another conversational technology, one of the oldest and most powerful: money—or, as Nigel Dodd restated Viviana Zelizer's work, "diffuse social media" (Dodd 2014, 8). Zelizer is an economic sociologist, and her work refutes the way many people, including economists, tend to think of money as a neutral, abstract, uniform technology for denominating prices and facilitating exchange. Zelizer documented and studied many ways that money is "earmarked," or valued differently for different purposes and people—think of allowances; money set aside, payments in intimate relationships or as life insurance, gifts and handouts and charity and compensation for wrongs and windfalls. "I arrived at a clearer distinction among three components of money: accounting systems, media representing those accounting systems, and practices that govern people's use of accounting systems and media," she wrote of the conversations following her book *The Social Meaning of Money*, and how all those components "vary with, respond, and inform people's negotiation of interpersonal relations" (Zelizer 1997; Zelizer 2016) That is, with some institutional exceptions, we make sense of money in practice by understanding it in terms of other people, our relationships, and the future and the past.

Electronic forms of money, though they may sometimes purport to be somehow more rational or abstract, amplify this tendency. Zelizer puts it nicely, reflecting on her earlier work: "If writing the book today, I would expand the last chapter's prediction of the paradoxical ways in which new electronic technologies would facilitate rather than blot out the sort of monetary differentiation and personalization the book documents. Consider what is happening with mobile money and multiple payment systems, from Bitcoin to Venmo, Apple Pay, Square, M-Pesa, local currencies, and more" (2016). These mobile technologies and digital currency systems, each with their own also integrate with WeChat and allow trading and transaction by text. Money, managed through the WeChat wallet system, becomes another kind of emoji, another set of stickers—a messaging practice and a way of texting. You can trade Bitcoin, buy tickets, and make gifts with precisely the same back-and-forth used to confirm whether someone is at the restaurant already. This may not seem that surprising to those of us used to electronic payment systems, but to understand its significance, let's consider, for example, the gamification of *hongbao* on WeChat.

Chinese "holiday money" is a tradition running back to the Qin Dynasty, a rich example of Zelizerian earmarking—coins on red strings given by grand-parents to children. It is money understood not just as abstract, depersonalized "market money," perfectly interchangeable (in the language of Karl Polanyi [1957]), but as a gesture of family connection and hierarchy. It goes from old to young, married to unmarried, earners to nonearners, in a ceremony of respect and obedience and the promise of the prosperous, happy lunar new year to come. *Hongbao* places a particular premium, therefore, on money as a specific artifact. It concerns units of money with certain physical properties: crisp new notes, unspent, sometimes with lucky serial numbers and given in a quantity ending with an even number, in decorated red envelopes. *Hongbao* may be the most culturally powerful and widespread expression of physical-cash-as-earmarked-artifact in the world. In 2014 WeChat tried to make it digital.

I've argued above that we can see the application of WeChat to domains as different as employee time cards, hotel reservations, and policing as an app as "emerging essential," a kind of infrastructure still taking shape—at first, by acting as a kind of de facto interface for many other kinds of infrastructure. Money (cash and otherwise), payment networks, and systems of finance are infrastructures, and with the *hongbao* strategy we can see a worrying implica-tion for app-as-infrastructure, one we should consider carefully. Faced with the challenge of inserting their app into a centuries-old tradition built on the physical properties of banknotes, the WeChat team remade the act of exchange as something suited to the particular experience of apps and mobile phones. They reinvented *hongbao* as a game, one played across social networks, with new forms of interaction. Along with sending *hongbao* money to one specific person, you can send it to a group; you can choose how much will be given, and—most important—in how many "envelopes." Choosing a number of envelopes fewer than the number of people means that only the first to check their phones will get the gift. While you have to choose the lump sum (¥0.01 to ¥200), you can let the *hongbao* platform decide how it's divided between envelopes—some recipients will get a token gift, and someone will get something big. (Notice that the first-come game element rewards keeping your phone on and readily available and constantly checking your WeChat notifications.) You can also send automatically generated small sums in culturally significant lucky amounts. You can include text, and your recipients can send thank-you notes.

As probably goes without saying, recipients of WeChat *hongbao* can't send their own gifts—or receive gifts they've been sent—without linking their bank account or credit card to WeChat's wallet. The game structure is a kind of loss leader for getting people to add their account information, rendering it more frictionless the next time they may have an opportunity to make or receive a payment passed through WeChat; it's a clever approach to making WeChat the automatic, reflexive interface for yet another infrastructure—the movement of money. Once the connection has been set up, the WeChat wallet becomes an inclined plane toward other kinds of financial services, from getting paid to investing and managing wealth.

One aspect of infrastructure I've waited until the end to bring up is *neutrality*: that the roads are open to all, the subways cannot decide whom they will or will not carry, public utilities operate by shared standards, freight rail will carry your grain and mine without discriminating between us. This neutrality is far more complex than it may at first seem, of course, both in theory and in practice, especially once the complexity of digital technology is added to the mix. WeChat's project of absorbing payments into its overarching system of chat—of appifying money—is indicative of a larger event: an infrastructural "emerging essential" that can be easily and invisibly changed after the fact. This is the speculative—but, I think, not improbable—warning with which I would like to close my review.

Apps metamorphose, and they don't have even the contingent, embattled version of neutrality that holds for, say, the US Post Office. At the level of design and modification inside the app messaging typeform, WeChat could subtly incline certain kinds of activities over others, pushing particular models of social engagement, interpersonal interaction, and work-life balance. It can try things out, see what the market will bear, make tweaks. Once it has become the default access point and interface for infrastructural elements—from your job to contact with your family or the police to the way you pay for things—it can begin to change their experience and use. In my view, payments are a particularly dangerous set of technologies to be thus incorporated. An app can reward you with discounts for buying more time and data on your phone through their wallet, for instance, make some choices more affordable and attractive than others, and gather and consolidate purchase data for as-yet-unknown forms of surveillance, soft coercion, and price discrimination.

"The payments game, in other words, is being played over databases: who shall collect, fence, own, leverage the commons of transactional data currently locked up in cash purchases?" Thus ask Bill Maurer and Lana Swartz in their study of new payment systems seeking to capture the infrastructural common space of transacting money. "Who will bring purchase histories together with payment information together with locational, credit, social network, or other histories?" (2015, 226) It's a very important question, and the development of an app like WeChat suggests that for payments and many other pieces of app-mediated civic infrastructure, the answer will be subtle and pervasive and taking place at least in part by an over-the-air update—seemingly happening not out in the world, but on the screen in your pocket, which is *everyone's* pocket.

I would like to highlight an MA student at New York University, Xijie Rui, whose excellent research for my spring 2016 digital money class—on digitizing hongbao *gifts—provided the starting point for the latter third of this essay.*

Snapchat
Phatic Communication and Ephemeral Social Media

JILL WALKER RETTBERG

App: Snapchat
Developer: Snap Inc.
Release Date: 2011 (Current Version: 10.15.1.0)
Category: Essentials
Price: Free
Platforms: iOS, Android
Tags: photography, messaging, social media
Tagline: "Life's more fun when you live in the moment :)
    Happy Snapping!"
Related Apps: Instagram, Facebook, WeChat

Snapchat is an app for ephemeral content. On Snapchat you can't browse through friends' old posts, or see who your friends are friends with, or what your friends like. Snapchat is not an archive like Facebook or Twitter; it is a conversation: a constantly moving stream.

In this chapter I argue that Snapchat's ephemerality forces us to rethink our definitions of social media and even to rethink what we take to be the key elements of the internet. The app emphasises phatic communication, as social media have always done, but takes it further than Facebook, Twitter, or Instagram: with Snapchat the phatic connection between people is the key element, often far more important than the content that is shared. This is the case both for snaps sent between individuals, which are often ritualized and playful, and for the kinds of storytelling that emerge in users' individual "Stories" and in features like the collective "Our Stories" and the "SnapMap." These sequences of images and videos emphasize phatic qualities such as immediacy and a sense of shared experience and may even represent novel forms of narrative and intimacy that reconfigure our understanding of social media more broadly.

When Snapchat launched in 2011, it was the antithesis of the web. There were no archives, no links, no back button. The app reminded us that there have always been two poles to the digital: the immediate and the archival. Snapchat's ephemerality seemed strange to a world used to the relative permanence of the web, but ephemerality is not a new trait in digital communication. Long before Facebook, the computer-mediated communication of the 1980s, 1990s, and early 2000s was full of conversations that disappeared as fast as they scrolled off our screens, in MUDs (multi-user dungeons) and MOOs (MUD, object-oriented), on IRC (internet relay chat), in chatrooms, and on webcams. As Wendy Chun writes, "Digital media is degenerative, forgetful, eraseable." Text and images on computers are constantly transmitted and re-generated, she reminds us, because that is how digital memory and networks work (2008a, 160).

But by 2008 or so, when social media had become mainstream (Rettberg 2014a, 62–64), we had almost forgotten how digital communication disappears. We came to think of the archival, encyclopedic nature of the web as being the natural state of all networked media and communication. Chun sees it as ideological blindness: "New-media scholars' blindness to the similarities between new media and TV is ideological; it stems from an overriding belief in digital media as memory" (2008a, 153). It wasn't just scholars. It was all of us. Snapchat lets us see how the ephemeral is reclaiming digital media.

Snapchat as a Platform

When Snapchat initially launched, it was primarily an app for sending private visual messages, or "snaps," to friends. Since then it has changed considerably, from a visual messaging app to a media platform with advertisements, news, collaborations with existing media channels, and sponsored content. "My Stories," which were introduced in 2013, allow users to post snaps to a feed where they are visible for twenty-four hours. "Our Stories" (2014) are shared stories centered on events or places, where Snapchat publishes a selection of snaps contributed by users to a public story about the event. The "SnapMap" launched in 2017, allowing users to share snaps to a map and to browse other users' snaps by navigating this map of the world, showing shared events, such as festivals, as well as individual snaps.

Snapchat quickly became immensely popular among younger users and gained traction with older users as well. On an average day in late 2015, 41 percent of eighteen-to-thirty-four-year olds in the US used the app, according to the company's website. In Norway, 56 percent of the population ages eighteen and older are active Snapchat users, and 70 percent of them use it every day.[1] Many users follow accounts run by media companies or by celebrities and influencers in addition to following their personal friends, and 25 percent of daily users post snaps to their My Story (Snap Inc. 2017). In combination with Snapchat's "Discover" channels and Our Stories, this means that the app is not only a well-established site for personal communication but is also a significant media distribution platform. According to company reports, more than twice as many eighteen-to-twenty-four-year olds watched Snapchat's Our Story about the Republican Party's first debate leading up to the 2016 presidential elections as watched the event on television (Turton 2015). But you can't rewatch that Our Story about the Republican debate. It disappeared after twenty-four hours. It doesn't even seem to have been reposted illicitly on YouTube. If Snapchat is like television, it is like television the way it was before Netflix and YouTube, before Smart TVs and DVRs, even before people plugged home video recorders into televisions in the 1980s. Snapchat is ephemeral the way television was before most Snapchat users were born.

One of the most commonly cited definitions of social media is by danah boyd and Nicole Ellison (2007): "We define social network sites as web-based services that allow individuals to (1) construct a public or semi-public profile within a bounded system, (2) articulate a list of other users with whom they share a connection, and (3) view and traverse their list of connections and those made by others within the system." This is not really a definition of social media, although it is very frequently cited as such. boyd and Ellison are defining *social network sites* and wrote the definition before "social media" was an established term. A social network site is a website, not an app. boyd and Ellison's definition describes Facebook perfectly, but Snapchat counters each of their three points.

A Snapchat user has no profile. All you see are snaps a friend may have sent or posted to their feed in the last twenty-four hours, and if they haven't been active, they are invisible to you. Users can see a list of their own friends but not of other peoples' friends. This is a huge shift in how we understand what social media can be and what social media can do. Snapchat returns us to more ephemeral media; media that are more "degenerative, forgetful, erasable" (Chun 2008a, 160). Through apps like Snapchat, we are returning to an internet that is a conversation more than it is an archive. That ephemerality may lead us to take the app less seriously. It appears frivolous, mundane, boring. But the ephemerality we see in apps like Snapchat is not only reemphasizing social connections through phatic communication; it is also opening up new forms of storytelling and narration.

Conversational Intimacy

In the beginning, Snapchat's ephemerality was seen as key to the privacy of the app. You could be honest on Snapchat. Users knew the only person who would see their Snap would be the person to whom they sent it, and the snap would disappear once viewed. If the recipient took a screen shot, the sender would be notified. Before the Stories feature was released in 2013, Snapchat was primarily used for bonding with close friends, and this is still a foundation of the app. A survey of first-year university students in the UK found that half the snaps they sent were selfies, usually with text added. Seventy-five percent of respondents were at home when sending their last snap, most sent it to a single close friend, and most were in a good mood when they sent it. Eighty percent of the respondents used Snapchat to communicate with no more than twelve people on a regular basis, making it a far more personal form of communication than Facebook and other major social media services (Piwek and Joinson 2016).

Snaps are about intimate connections where the very "unimportance of the content stresses the meaningfulness and importance of the friendship as exactly intimate," as Jette Kofoed and Malene Charlotte Larsen (2016) found in their study of twelve-to-seventeen-year olds' use of Snapchat. The mainstream media perception, before many adults started using Snapchat, was that the app was mostly used for sexting, a view that was strengthened in the popular imagination by the "Snappening" hack in 2014, where tens of thousands of private snaps were released by hackers (Johnston 2014). Although there were many reports of young men meticulously going through the snaps to make them searchable by country and level of nudity, there were also reports of people being disappointed to find that most of the snaps were boring photos of everyday life. As Kofoed and Larsen note, "Intimacy, in the case of *Snapchat* and youth life, seems to be carried by the mundane sharing of ordinariness" (2016.).

Earlier analogue and digital photo sharing practices have emphasized documentation, archiving, and curation, whether in the family photo album or online services like Flickr or Google Photos (Rettberg 2014b, 24; Bourdieu 1990). Nathan Jurgenson writes, "The logic of the profile is that life should be captured, preserved, and put behind glass. It asks us to be collectors of our lives, to create a museum of our self" (2013). Snapchat refuses permanence.[2] In one ethnography of Snapchat use, the college students interviewed did not think of the app primarily as a photo-sharing service. Instead, they saw it as "a lightweight channel for sharing spontaneous experiences with trusted ties" (Bayer et al. 2016, 956). Snapchat is a conversation, not an archive.

Photos shared on Snapchat are intentionally ephemeral. Katharina Lobinger proposes that we think about three modes of photo sharing: "(1) sharing photographs to talk about images, (2) sharing photographs to communicate visually and (3) phatic photo sharing" (2016, 475). Lobinger writes that Snapchat is a site where you often see the third form of sharing: "People engaging in phatic photo sharing describe the photographs they typically exchange as gratuitous, useless photographs because the motifs of the photographs do not matter at all. . . . The photographs rather serve for keeping up a continuous flow of visual contact with friends" (482).

Phatic photo sharing is often ritualized. For instance, groups of friends may regularly share "good-night" photos (482). Such ritualized, phatic photo sharing can be a kind of gift giving: "Sending a photo gift to a friend (giving) creates the obligation to accept it and to return the gift in an appropriate manner (reciprocating), which can be either an adequate visual or a verbal re-action to the gift" (482). On Snapchat one of the rituals is sending daily snaps to maintain "streaks," where the app shows how many consecutive days you have exchanged snaps with a friend. Here, as in other phatic communication (Miller 2008; Rettberg 2017b), the connection, not the content, is important.

Phatic playfulness is further encouraged by features like the selfie lenses, geofilters, and doodling. The selfie lenses, introduced in September 2015, are one very specific way Snapchat encourages this playful use of images. Lenses apply effects to faces: adding a dog's nose and tongue; making you look like a unicorn with a rainbow flowing out of your mouth when you open it; distort-ing your face or smoothing your skin. These lenses have become part of the aesthetic vocabulary of Snapchat, used by everyday users and celebrities alike. By 2017, similar filters had spread to other social media, such as Facebook and Instagram. They are used to express emotions, to make friends laugh, to look more beautiful, and to mock politicians. As Snapchat has become increasingly monetized, there are more sponsored lenses—for instance, allowing users to see themselves looking like the lead character in a movie, or adding facepaint and confetti in appropriate colors for a football team before a major game. Geofilters, another fun and informational element, are texts and images that can be overlaid on any image taken in a certain place. They are often created by users who have uploaded their designs for specific places, but they can also be sponsored by companies or event organizers, or they may be created by Snapchat for a festival or event.

Stories

It is not surprising that phatic communication is important in video messaging. The emphasis on phatic connections in narratives is less explored and is evidence of how aspects of oral storytelling (Ong 1982) are returning in digital narratives. The description Snapchat published in the iTunes store with the 2013 update that introduced Stories emphasized the interminable nature of narrating one's life: "Your story never ends and it's always changing. The end of your Story today is the beginning of your Story tomorrow" (Etherington 2013). This is quite different from the description in the support section of Snapchat's website in April 2016: "Stories are compilations of Snaps that create a narrative. Stories honor the true nature of storytelling—Snaps appear in chronological order with a beginning, middle and end" (Snapchat 2016).

While most users simply post images and videos taken throughout the day, with little connection between them, the latter definition fits the way Stories have come to be used by influencers and brands on Snapchat. For instance, when Paris Boy, a popular American Snapchatter based in Paris (username, trevortuck), was traveling home from Kosovo, he began his story with a video of the view out his window and the words "so long kosovo" on the screen as we heard the sound of Simon and Garfunkel's "Homeward Bound" playing. Next we saw a series of very short video clips from the journey to the airport, all in Paris Boy's signature black-and-white: checking out at the hotel desk, walking to the taxi, getting into the taxi, looking out the front window of the taxi, and so on. There are twenty-six of these brief video snaps from taxis, airports, and planes before we finally see the door to his apartment opening and the first words in the story: "Made it!" The next snap is of Paris Boy's face; he speaks directly to the camera and starts talking about something different—in effect, starting to tell a different story.

After the fact, this video could have been posted to YouTube, and it is far from endless: it has a clear beginning, middle, and end. But watching it unfold on Snapchat, we see each snap at the moment it is posted and must wait for the next snap to appear. This is the video version of a diary or blog where each entry tells of the immediate past (Rettberg, forthcoming), but here it is taken to the extreme. *This is the moment I am experiencing now. And now. And now.* And yet, by sequencing them together, moments are threaded into something we may interpret as a story. The ephemerality of Snapchat gives us an episodic narrative that is always in motion, always changing.

Collective Narratives

"Our Stories" are an explicitly collective form of storytelling, using what Scott Rettberg calls "contributory participation": contributors choose to contribute but have no control over the final product (2011, 198). Sharing a snap to Our Story or the SnapMap is thus placing fragments of self-narration in a larger context, both among snaps and stories from other people and left to curation decisions by production managers at Snapchat (Luckerson 2015). Curators select events and locations, design filters that work as prompts to encourage users' contribution to certain kinds of snaps, and then select and order snaps to include in the Our Story.

The selection process and brevity of Our Stories (usually only two to three minutes) means they tend to present a topic or event in a cohesive manner that can never show the totality of what people are actually experiencing or even snapping in that time and place. The Our Story for Australia Day 2017 is a striking example. On special days and holidays, such as Independence Day in the US or Constitution Day in Norway, there are filters in these countries that prompt users to share certain sorts of content: Uncle Sam with fireworks or diving into a pool, or more explicit prompts for Norway, like "What I like to eat on 17 May," or "All dressed up."

Snapchat generally caters to its main demographic by curating Our Stories that feature people in their teens and twenties having fun, but this seemed particularly egregious in the Australia Day Our Story, which was heavily dominated by young white people drinking and partying by the pool, completely ignoring the tens of thousands of people who were in the streets of every major city in Australia for #changethedate protests, demanding that the date of Australia Day be changed so that it celebrates Australia, not the anniversary of the European invasion of Australia. Coming just days after the app's thorough curation of Our Stories from the protests around Donald Trump's inauguration, this was a striking example of how partial and unrepresentative Our Stories can be. Despite this, on Twitter viewers were discussing the story excitedly, and many described imagining themselves at the parties, celebrating with the Australians. They felt connected to the people in the snaps. The phatic aspect of the story worked, at least for viewers who felt they were similar to the people represented in the story.

Perhaps the apparently less curated snaps available on the SnapMap will allow for a more diverse view of the world. Browsing Paris on the SnapMap in July 2017, users could see tourists smiling and laughing at the Eiffel Tower and other popular sites, but they could also see protesters burning cars on the other side of town. Zooming out to view the global SnapMap, the Kaaba in Mecca is consistently one of the busiest spots. The seemingly uncurated stream of videos of white-clad masses visiting the Kaaba is strikingly different from the carefully curated Our Stories of Ramadan published on Snapchat in 2015 and 2016, before the SnapMap feature was launched. These Our Stories featured English speech and writing and were very clearly aimed at explaining an "exotic" phenomenon to a Western audience.[3] The videos on the SnapMap are far more mundane, with little or no attempt at explanation. There is a sense of immediacy and authenticity to the snaps on the SnapMap. We don't know who posted them, but we know that somebody was there, using the same mundane app as we are using to watch the videos. Not every snap shared to Our Story appears on the map, however. Snapchat presents a view of the world that appears to be immediate, collectively shared, and open to all, but it is in fact selective, based on criteria that users are not made aware of: perhaps the politics of the company's founders, the interests of its sponsors, or the whims of algorithms.

Snapchat is becoming a platform for media companies, advertisers, influencers, and news companies. Older audiences are arriving. That may well change the way the app works. It may kill the intimacy and drive young people away. But Snapchat has already changed the way social media work and the way we think about what social media and networked communication can be. We are learning new ways of telling stories and communicating with each other that are closer to the phatic communication of conversations and oral cultures. As I noted earlier, this has been the case for decades in text-based digital communication, but Snapchat is the app that made conversational, ephemeral video central in online communication. Established social media like Facebook, Instagram, and Twitter are following Snapchat by increasingly emphasizing video sharing and video messaging, as newer apps like musical.ly are expanding how video communication is used (Rettberg 2017a). YouTube and earlier video platforms are archival and content-centric. Snapchat's ephemerality and the playfulness of filters, lenses, and doodles bring video into an oral paradigm that highlights the phatic and the conversational.

Notes

1. US numbers are from Snapchat's website, April 2016, and Norwegian numbers are from Ipsos Social Media Tracker Q1 '17 (http://ipsos-mmi.no/some-tracker).

2. The company servers may act as more permanent archives, but these are not accessible to users.

3. Our Stories were more commonly referred to as "Live Stories" at this time.

Foursquare
Checking In and Checking Out of Location-Aware Social Media
GERMAINE HALEGOUA

App: Foursquare
Developer: Foursquare Labs, Inc.
Release Date: March 2009 (Current Version: 11.8)
Category: Food & Drink
Price: Free
Platform: iOS, Android, watchOS
Tags: location-based social media, industrial strategies
Tagline: "Discover places that your friends and experts love."
Related Apps: Swarm, Marsbot, Gowalla, Loopt, Yelp

In 2009 an app called Foursquare ushered in a location-based social media explosion with a "Check In" button—a button that would allow users to announce their physical location to a social network with a single click. This button appeared as a bright, horizontal bar under the name of a given location and spread across the expanse of the screen. It coordinated with the bright green or blue stickers that decorated storefronts entreating users to "Check-In

Here on Foursquare." Based on the interface design, there wasn't much else to do with the app. Participants could search for names and locations of venues, read or write tips about nearby restaurants or bars, and "shout" messages to friends, but in order to fully utilize the app, you and your friends had to check in. The check-in unlocked game mechanics, virtual rewards, specials and discounts at participating businesses and was a way to connect with people in your social network and to record and remember the places you visited.

For the millions of users who downloaded the app and activated Foursquare accounts, the Check In button *was* Foursquare. However, this was not the case for all users. While front-end users (those who downloaded Foursquare from the app store) cultivated personal and community behaviors around checking in, back-end users (API [application programming interface] developers, data clients, advertisers, and investors) became more interested in the data the check-ins produced. Five years after the app's launch, the company made a choice to cater to certain uses and users over others, ultimately privileging check-ins as geospatial data to inform corporate business decisions (i.e., location intelligence) rather than check-ins as social practice (i.e., location announcement).

A study of Foursquare over time reveals something about the not so secret life of apps: that the back end—in terms of both computer programming and industrial strategies—is far more complicated than front-end user experience. The user interface, platform, and monetization strategies evolve and change with help from front-end users but often without their knowledge or consent and sometimes to their displeasure. On the surface, apps are designed for usability in that they appear simple, streamlined, and user friendly, but the social decisions that make them so are messy and contradictory behind the scenes. The "usability" of the interface and discrete packaging of software and services in the form of an app obfuscates the back end and promotes more than algorithmic opacity but an industrial opacity as well. The user-friendly face of the app makes the work of the application, and of the industrial strategies that make it work, invisible and unexpected to certain types of users.

In this chapter, I employ Foursquare as a case study to analyze the complex relationships between different types of app users and the companies they support. I highlight how the form and discursive construction of apps both distance front-end users from the function of the app and invite them to participate in its development and success. I focus on a variety of tensions to make this case. The tension between the flexibility of Foursquare as "perpetual beta" versus the reification of dominant uses and users, the contradictory practices of gamification of location announcement and gamification of iterative design in app development, and the industrial evolution of apps that depend on data generated by devoted yet delimited users.

The Rise and Fall of the Check-In

Within the first two years of Foursquare's operation, the location-based social media (LBSM) app reached 10 million users worldwide. This number doubled within the next two years and continued to grow. Foursquare soon became an app that was emblematic of mobile social media (Humphreys 2007) and outlasted several of its competitors, such as Gowalla and Brightkite, to continue to attract investors. Technology analysts noted the success of Foursquare as a test case for the marriage of location announcement and social networking and as an incentive for companies like Facebook and Twitter to offer location-announcement functionality (e.g., "Places") on their platforms.

From the outset, a confluence of perspectives about the significance of mobile social networking and location announcement circulated among different types of Foursquare users. Dennis Crowley, the cofounder, CEO, and public face of Foursquare, marketed the app as a tool for place discovery and meeting up with friends. Local businesses experimented with Foursquare as a way to reward loyal customers and attract new customers through friends' recommendations. Investors and advertisers saw an opportunity to accrue "location intelligence" and generate profiles or targeted ads based on the types of locations people tended to visit. Meanwhile, scholars continually noted creative and disruptive ways in which "checking in" became meaningful in the everyday lives of local populations around the world, such as performing identity or playing against the house rules by "jumping" for badges (Cramer, Rost, and Holmquist 2011; Halegoua, Leavitt, and Gray 2016).

By 2014 Crowley publicly proclaimed the death of the check-in and announced a major shift in the composition and purpose of the company. In May of that year, Crowley told *The Verge* plainly, "Listen, the point of the company, this whole thing, was never to build an awesome check-in button. That's not the thing we got out of bed and said, that we wanted to build the most awesome check-in button in the world!" (Popper 2014). Instead, they wanted to build a "location intelligence company" to power other geolocation and location-based services through their API and location-aware tech and were using the check-in to do so (Heath 2016). While encouraging users to check in, Foursquare was using participants' data to build services like Pilgrim, Pinpoint, Attribution, and Place Insights that measure the effectiveness of ad campaigns based on foot traffic and audience behavior.

In 2014 the company split Foursquare into two entities: Foursquare, a location-based recommendation service, and Swarm, a location-announcement service with functionality that resembled the original system. To front-end users' dismay, as expressed through crying emojis and irate hashtags such as #killswarm, the check-in was dead (Hu 2014). The revamped Foursquare resembled a mobile travel guide that encouraged users to discover "the best food, nightlife, and entertainment in your area" and eliminated the social networking and communication features that had led to its success. In 2016 Foursquare claimed to have fifty thousand active users, over five billion global check-ins, and customers like Apple, Microsoft, Twitter, Pinterest, Garmin, AT&T, and "100,000 more developers who rely on our location data" (Glueck 2016; Larson 2014).

Designed Invisibility

Mobile media infrastructures are continually disguised as nondescript structures or natural objects meant to blend into the landscape and render their materiality and politics invisible (Parks 2010). App interfaces are designed to be similarly mundane. The interface presents software as handy, discrete packages that distance front-end users from the industrial and technical workings of the app. The invisibility of the apps' code and politics are often shrouded in design choices and discourse that construct app software as powerful yet mysterious or as indistinguishable from magic. Foursquare's interface, not unlike other apps, obscured any messiness or tensions in industrial strategy by offering a consistent tone and visual style that presented a sense of coherence and universal use. The company's original logo emphasized the simplicity and playfulness of the check-in with a visual reference to the eponymous school-yard game and exhibited large, categorical buttons that indexed functionality such as "Friends," "Tips," and "Check-In Here." Minimal, uncluttered interfaces in the service of user-friendliness presented a limited number of modalities for engagement, circumscribing the additions and adjustments that front-end users could contribute to the platform and prioritizing certain actions over others.

While the original Foursquare icon, an image of a looming checkmark, reminded users that the app was driven by their individual check-ins and participation, the updated app logo was designed to look like a superhero emblem (Brown 2014). The closer the company came to retiring the check-in, a symbol of user participation, the more "magical" the inner workings of the app were purported to be. Executives and designers began to refer to the app as having superpowers that endowed front-end users with an "almost magical power . . . to find the best things wherever they are" (Butler 2014). As Chun (2008b) describes, the "sourcery" ascribed to software fetishizes source code as magical and erases the structures and ongoing vicissitudes of a program's execution. In this case, obfuscating source code behind the rhetorical construction of location intelligence as superpower not only obscured the code but also the industrial decisions that repositioned user agency and engagement with the app.

The "magic" of Foursquare worked through the gamification of a disjointed form of iterative design that harnessed front-end user activities to meet back-end user desires. Within the process of iterative design, Foursquare privileged location announcement as data rather than cultural practice but encouraged the opposite imagination among front-end users. Front-end users became invested in the check-in; they incorporated the check-in, the virtual loot, and social connections that it enabled into their everyday lives in meaningful ways (Lindqvist et al. 2011). Because the design and discourse of the app placed the Check In button front and center, it was difficult for users to imagine a function, purpose, or business model that would accrue value in any other way. The emphasis on the act of checking in overshadowed the company's efforts to leverage the data produced to power a different sort of system—one that would no longer incorporate its early adopters and growing user base in the same way. As the app users' social media traces, complaints in online forums, and dwindling numbers imply, the majority of participants most likely never expected that by checking in they were contributing to the reinvention and dissolution of a platform they enjoyed using. For front-end users the check-in became habitual (Chun 2016), but once it did, the company used data produced through these routinized, rhythmic actions to change the app. Ultimately, what elicited irate comments and solemn emojis from front-end users was not only that the app was changing but also that the imagined value and imagined relationships to monetization ascribed to their own habitual practices did not coincide with the future directions for the company.

Just as the graphical user interface (GUI) provides an imaginary relationship that both promises and distances the user from direct manipulation of code (Chun 2011), the app interface simultaneously distances the user from exposure to the back end while "unmask[ing] the reality that it's the back end that matters" (Anderson 2010). Where GUI belies code, the app interface belies data. The production of locations and personal mobility as data and metrics were foundational to Foursquare. However, the meaning of these data points differed among front-end and back-end users. For the front-end app user, the seamless design, game mechanics, and social engagement around the production of data framed location announcement as self-quantification or social surveillance rather than "dataveillance"; for the back-end user and company executives this imagined relationship was inverted.

Conspicuous Mobility and Permanently Beta

Although Foursquare actively recruited small businesses, regional franchises, and national brands to cultivate a presence on the app, individual Foursquare participants were more interested in the social connections and personal archives they could forge rather than aspects that connected them to vendors and their products. However, both entities assigned value to their participation on Foursquare, and the app, with its purposefully designed flexibility, encouraged various uses and users to coexist.

Foursquare aimed to remain adaptable in terms of who the company *could* include in their potential revenue stream and user base, even if Foursquare could not incorporate these users immediately. For example, the fact that Foursquare founders insisted on a flexible GPS verification system and did not eliminate points or mayorships for inaccurate check-ins can be read as an effort to cater to different and evolving types of front-end users and activities (Halegoua, Leavitt, and Gray 2016). The strategy of keeping the design and software of a digital platform in an ever evolving "permanently beta" state in an effort to keep a system potentially open to different types of use is not unique to Foursquare (Neff and Stark 2002). As a study of Foursquare indicates, although maintaining flexibility in app design may allow for the incorporation of new or different types of users or a company's ability to pivot or innovate, an app that remains in beta might conflict with front-end user expectations and imaginations of how their data will be utilized and monetized.

A pioneer in the LBSM market, Foursquare continuously asked users and investors to imagine and reimagine its potential for profit. Over a six-month span, Foursquare promoted itself as an app that could generate value as a promotion and marketing program for venues and advertisers, as a location intelligence company with a robust API and lots of data, and as a recommendation system for local places of interest. During Foursquare's initial release, the potential for monetization focused on providing advertising, sponsored messages from retailers and brands such as Bravo and American Express, as well as metrics and check-in data to brick-and-mortar businesses. In addition to offering specials or discounts to people who checked in, businesses could pay to advertise or access more detailed data about users and their locations. Eventually, Foursquare partnered with the online, mobile food-ordering companies Grubhub and Seamless to offer customers the opportunity to order directly from the restaurants and stores that appeared on the Foursquare app (Etherington 2014). Another revenue stream flowed from Foursquare as a trust and recommendation network, where friends played the role of advertiser and marketer within their social networks through their check-ins. This effort culminated in Foursquare's "Explore" tab, which transitioned the app into a mobile, location-aware, Yelp-like recommendation service based on user profiles, check-ins, and tips.

All of these efforts further positioned check-ins as central to the social and cultural practices of front-end users as well as financially lucrative for Foursquare. The rhetorical construction of the check-in solidified a particular imagined relationship between the practice of checking in and the monetization of personal mobility. According to Foursquare's founders, one element that differentiated the app from other LBSM services was "turning life into a game" in an effort to use software to change behavior (O'Reilly 2010). Conspicuous mobility on Foursquare (represented in the form of a check-in) restructured urban experiences as transactions or potential transactions while promoting the ability of social media to intervene in patterns of mobility in order to change habits (Wilson 2012). The premise behind this idea was that if a member of a social network frequented a particular venue, then other members of the network might alter their routine in order to frequent that locale as well.

While Foursquare's monetization strategy was presented to front-end users in terms of leveraging mobility patterns as social indicators and influencers of consumption practices (location announcement), back-end users plotted ways to turn geospatial data into business insight (location intelligence). Foursquare gamified physical presence and broadcasted personal mobility through the form of the check-in, and in doing so, Foursquare gamified front-end user contributions to iterative design. By checking-in, front-end users were building a new system of location intelligence that benefited back-end users and would ultimately undermine the functionality of the check-in that had initially allowed for creativity and flexibility of use. The app gamified the digital labor of producing locations and mobility as data. However, this labor was intended to improve software not for front-end users invested in location announcement and play but for data clients as well as the next round of users, who would (hopefully) be more interested in a recommendation engine for "the best places" rather than checking in.

Conclusion

Treatises on the "new economy" continually inform digital media users and designers that software development involves participation in an "ongoing dialectical process" of innovation, testing, and revision along with versioning, expansion, and obsolescence (Neff and Stark 2002). Gina Neff and David Stark (2002) argue that this process can be exploitative or inclusive and that the interpretation of the practice of participating in the design and evolution of products depends on "the participants level of organization in order to counter the instability's social cost." Within this structure of planned instability, those who download and use an app can be understood as the labor pool that supports iterative design as well as the population that temporarily benefits but bears a fair amount of social cost. The case of Foursquare's evolution highlights the ways front-end app users often participate in iterative design processes under conditions of industrial opacity, and sometimes at their own expense.

The appification of the check-in as an interface between location as experience and location as data relegated users to the realm of "produsers" creating earned media for the company rather than as the beneficiaries of iterative design. Through their check-ins and social interactions with place, platform, and each other, front-end users fueled and maintained the dialectical cycle of iterative design in app development even if they didn't always realize they were participating in a perpetual proof of concept. Users were organized as sustainers of the app in the form of Superusers (select Foursquare users who helped keep information about check-in locations organized) as well as those who contributed check-ins and tips on a regular basis. However, these users lacked agency within the industrial shifts and social decisions around the app they

helped to produce. While front-end users contributed to processes of innovation, testing, and revision, they lacked input and insight into the procedural
trajectory of versioning, expansion, and eventual obsolescence.

The opacity of the layered, volatile processes at work behind user-friendly
app interfaces distance front-end users from what might be happening behind
the scenes while simultaneously encouraging them to imagine their own participation and monetization in particular ways. What location means, how it
is monetized, and how "location intelligence" is made understandable to different types of users (front-end users, data clients, advertisers, app developers,
etc.) is contradictory and copresent. Distinguishing the app as a discrete program with magical powers masks these differences and industrial relationships.
While app store users might imagine apps as permanently beta in particular
ways, they don't always recognize their contributions to mundane software
cycles as temporary or under development—that their social practices, which
are so carefully and concisely constructed through the app interface, might
power the creation or reinvention of other software or business strategies that
could lead to the demise of the app they were originally invested in sustaining.

WhatsApp
WhatsAppified Diasporas and Transnational Circuits of Affect and Relationality

RADHIKA GAJJALA AND TARISHI VERMA

App: WhatsApp Messenger
Developer: WhatsApp/Facebook
Release Date: January 2009 (Current Version: 2.17.80)
Category: Communication
Price: Free
Platform: iOS
Tags: call home, chat, family, affect public/private,
   community
Tagline: "Simple. Secure. Reliable Messaging."
Related Apps: WeChat, Facebook, Snapchat, Viber, Line

In 2011, when the Telecom Regulatory Authority of India (TRAI) put a cap of one hundred text messages per day in India, avid Short Message Service (SMS) users experienced quite a setback. Although this was done to prevent overloading mobile users' inboxes with spam messages, it also created a roadblock for younger generations using SMS to communicate. This paved the way for the instant messenger mobile application WhatsApp to enter the Indian market with full force. Yahoo engineers Brian Acton and Jan Koum developed the smartphone app in 2009. Facebook purchased WhatsApp in 2014, earning Acton a net worth of $4.9 billion (Forbes 2016). Interestingly, the app's user base and the two cofounders' monetary success continue in spite of (or perhaps because of) their slogan of "No ads! No games! No gimmicks!" (Yarrow 2014).

As smartphones became cheaper and more available, SMS popularity waned and WhatsApp began to replace texting. While Indian youth were often the early adopters, the app soon entered the lives of parents and grandparents, especially as their children moved out to study in India or abroad. When parents and grandparents started using the app, it also turned into a form of surveillance (e.g., the "Last Seen" feature provides information about when someone logged on to WhatsApp, and this resulted, as noted by several interviewees, in various kinds of relational conflict). But the app also allowed Indian users to conduct conversations they couldn't otherwise in offline spaces. WhatsApp provided a virtual seemingly safe space that was more convenient than any other social media messaging tools. WhatsApp was also cost effective—the app runs through Wi-Fi or data plans with competitive and relatively low-cost pricing. Not having to spend money on SMS and yet being able to send innumerable messages opened up mobile connectivity to many more users. It connected communities without the need to "log on" to an account with IDs and passwords, as was required by Facebook or Snapchat. After introducing even more features, WhatsApp became, for many Indians, the most convenient way to connect with their relatives and friends abroad.

Our main focus in this chapter is the pivotal role WhatsApp plays in the lives of Indian women—whether in diaspora outside of India or living in India—for whom this app has become a digital space to communicate more freely. We draw on more than fifty interviews, participant observation, and our own use of WhatsApp. We investigate the use of WhatsApp by Indian women within and across India and the Indian diaspora in the United States. For comparative purposes, this study also includes interviews with a convenience sample of students living in the United States, Norway, China, and Spain, as well as class discussions and exercises with students from an online undergraduate class in international communication. The interviews were unstructured, and the main goal was to understand women's agency, labor, and affective, relational circuits of care and connectivity to family within and through digital diasporas as connected migrants.

Through these interviews and our personal use of WhatsApp (in Tarishi's case as a twenty-something Indian woman living and working in India, in Radhika's case as a fifty-something US citizen of Indian origin living in northwestern Ohio but connecting to family and friends in India and elsewhere) we gained insight not only into the role of WhatsApp in contemporary Indian digital diasporas but also into the app's role in the construction of Indian women's everyday life. While not all of our interviews focused exclusively on WhatsApp, the app featured prominently as a relational tool for connection. While WhatsApp has not historically been as popular in the US, largely because of the accessibility of unlimited mobile texting plans, in the second decade of the twenty-first century, this may be changing. US users have begun to discover the utility of WhatsApp, just as some non-Chinese users are beginning to incorporate WeChat (see Brunton, this volume). This seems to be the case especially when US users are somehow internationalized through jobs or through interactions with people of diverse international backgrounds.[1]

Apps like WhatsApp and WeChat play key roles in shaping intergenerational connectivity in home and work life, with the app itself becoming a relational space for storing affective moments and sustaining transnational intimacies. WhatsApp has also become a popular app for international travelers or migrants from Global South regions or some regions in Europe. When these travelers move to the United States, the app becomes a crucial technology for connectivity with "back home" (again, like the function WeChat serves for Chinese travelers or migrants to the United States; see also Li's chapter on Tencent MyApp, this volume). The intergenerational nature of the coauthoring of this chapter by women of Indian/South Asian origin currently in the United States (one having lived here for over thirty years and the other having recently arrived as an international student) plays a reflective role in how we unpack the data we collected through interviews and participant observation. Consistent with Radhika Gajjala and Melissa Altman's (2006) methodological interventions into research on digital/material global contexts, we employ multiple qualitative methods under the umbrella approach of "epistemologies of doing" where the researcher is also immersed in the use and inhabitance of the digitally mediated spaces she is researching.

We use a specific sample of Indian women to explore why WhatsApp appeals to migrant and diasporic communities, and the role the app plays in the formation of networks of family and friends across geographies. Our research shows that the app's range of users is wide and complex. In the case of the specific sample of South Asian women interviewed, they note that WhatsApp has become a practical everyday go-to app for those who use their smartphones in free Wi-Fi zones while on the move in a way that, arguably, other messaging apps such as Facebook Messenger, Snapchat, and Instagram have not. This is primarily because WhatsApp is familiar to most of their family and friends in South Asia and in the South Asian diaspora. They also prefer WhatsApp's focus on encrypted message delivery compared to the more public, more heavily marketed and self-promotion-oriented uses of social media such as Twitter, Instagram, and Facebook. Notably, many of our interviewees praised WhatsApp's ability to create private groups of close circles of family and friends.

We find the formation of what we call *WhatsAppified diasporas* of Indian women. In Gajjala's earlier work that articulates and extends definitions of "digital diaspora," she notes:

[Digital diaspora] occurs through affective transformations and acquisition of multiple, layered, technical, cultural and other literacies. At the same time users are also under the illusion of being in global social space, with a perceived choice in where and how they engage this space. In addition, while the commonly invisible labor and structure of hardware manufacturing and complex software coding makes this platform for access to globality possible, users get to feel as though they are in fact "empowered" to produce their particularities within this global space. This allows the formations of various niche local places online within global places. (2011, 396)

A WhatsAppified diaspora, then, refers to the mobile phone–based digital diasporas that form through, in this case, Indian women's use of WhatsApp. As previous research has demonstrated (Gajjala 2011), digital networks become places that are lived in—through affective immersion—as relationships are built as people work, dance, and build in that virtual world. People do not just connect; they live with one another. They do not simply transfer information and transmit affect; they store memories of being with other people, and they have physical and spatial experience of the three-dimensional environment. A mobile phone text messaging application does not have the kind of three-dimensional immersive potential of computer games or of virtual worlds like *Second Life*. However, it nonetheless becomes a place for emotional archives and affect circulation as people carry their relationships in their pocket as they travel. The idea of inhabiting relational digital place through the formation of WhatsAppified diasporas follows the idea of digital diaspora that transforms the subjectivity of even the physically non-traveling subject into that of a global diasporic subject. A WhatsApp ping in the middle of the night for someone in the United States might be the result of a traumatic relationship breakup. A small segment of a Bollywood song or a religious chant sent by a relative from India might serve to comfort a homesick graduate student. Time zones are intermingled, and so are various social contexts.

Interfaces

Interviewees made comparisons among WhatsApp and other contemporary social media platforms like Snapchat and Instagram. A couple of them noted the similarity of WhatsApp interaction to early text-based tools such as "Talk" via a DOS (disk operating system) machine where the screen would be split in two as users texted/chatted synchronously, as well as to MSN (Microsoft Service Network) Messenger and Google Chat. However, while some of the women interviewed are old enough to have potentially been immersed in early internet chat spaces like Usenet or MOOs and MUDs, the conversations rarely prompted comparisons to pre-2005 modes of digital sociality, despite known continuities between earlier modes of internet-based communication and mobile app–based communication (Banaji 2015; Leurs 2015).

There are also notable differences. Shakuntala Banaji, in examining caste, class, gender, and rural-urban nuances in how children in India between nine and seventeen years old access smartphones and computers, notes how WhatsApp is part of the topology of everyday discourse and life for these young people, whether they directly accessed the app or not. She writes, for instance, that these technologies are "familiar and cherished" (2015, 14), and they compete for time that might be used for housework, schoolwork, and social activities outdoors and offline. Mobile phones are often shared, so using WhatsApp also means a negotiation of agency and individual expression, a space of contestation that leads to connectivity to the global through very local, interpersonal, and relational contexts that are mediated by socioeconomic factors. Writes Banaji:

> While mobiles were generally shared between family members, particularly mothers and children, comments about media amongst these lower- and middle-class children in urban areas were diverse and passionate, ranging from "I want to become a film actor, comedy is so cool" and "I sleep with my smartphone under my pillow," to "every one of us has to comment on a picture or post something funny to feel part of the group" and "I can do without sleep but I can't let go of WhatsApp." There was a lively understanding of marketing ("You know about ad blockers?"), the possibility of parental surveillance ("My older brother checks my posts and informs parents") and the dangers of spiraling identity expectations ("I worry if I look fat on Facebook"; "Boys are always saving sexy pics from Snapchat." (14)

These statements indicate personal relationships that individuals are able to build through the mobile gadget itself, even if that gadget is being shared among many family members. For the person above who was posting in order to feel like a part of the group, for example, it might be difficult to vocalize comments in person due to various cultural, social, or economic barriers. The app allows them to erase these boundaries. Nothing is public here—they are safe within one-on-one chats. With the ability to delete these messages or archive them, they are less worried about privacy as well.

Public versus Domestic/Relational/Communicative

Like other text messaging apps, WhatsApp allows domestic/relational/familial communicative spaces to be carried around in a mobile device. However, given there is no extra cost to using WhatsApp (unlike other text messaging tools in India), our interviews revealed that the app also allows additional circuits of connection through family relational—specifically, domestic communicative—space. We found that Indian women negotiated agency, autonomy, and self-expression in relational space through the use of time for leisure and work while also making choices on what becomes placed in various digital publics. Although younger and single women and men often described the use of WhatsApp with their friends and talked about messages from parents and extended family members, their mention of how their mothers, aunts, grandmothers, or married sisters used WhatsApp allowed us to understand this weaving in and out of domestic affective (and feminized) space. Even when their uncles, fathers, or grandfathers were mentioned, their use was largely domestic—passing on religious messages, food recipes, or other such domestic-oriented and affect-based affairs.

Yet the use of the app by women in domestic space also orients them outward—into an assertion of social agency away from family contexts. As coauthor Tarishi notes of her mother's WhatsApp usage, technological challenges often forced her mother to borrow Tarishi's phone to troubleshoot or send messages. However, once her mother was involved in a messaging group with friends, it was difficult to take the phone away from her. Tarishi was not allowed to participate in that group or read its messages—there could be jokes her mom would never allow her children to read. The name of the group was something her mother would otherwise never use in public—it included the word "sexy." This example demonstrates an interesting turn to agency away from domesticity even as Tarishi's mother was connecting through the domestic. As Tarishi notes of her mother's use of the app, this side of her personality is not one she would present on Facebook. WhatsApp, it seems, provided her a veiled space of agency.

Another respondent, "SM," has a range of emotions for the app but also attests to being more expressive in WhatsApp. She says:

There are typos sometimes that might create misunderstandings. It also is an addiction and I end up wasting a lot of my time; but there are somethings that I can talk about very freely. My sister-in-law is having a baby—it's easier to talk to her on WhatsApp rather than in person because sometimes family members don't like when the *nanad* tells her *Bhabhi* things. Sometimes, my friends share more freely about their strained relationships on WhatsApp than in person.

These contact zones and their contradictions require us to examine, yet again, redefinitions of domestic and public spaces. In our everyday we are simultaneously in public space and domestic space thanks to apps that allow for interpersonal connectivity as well as mass broadcasting. Apps leave users a choice for how communication is conducted—whether for private, intimate, one-to-one, and small group communication or for mass broadcast to a larger public while maintaining varying degrees of privacy.

In our interviews, users described Facebook, Twitter, and Instagram as public spaces for following and reading information and posting announcements that were more public than WhatsApp. One informant noted, for instance, that Facebook is "just a platform where people show off" (conversation between "M" and "S," 2015). For interviewees, Facebook was a social network where one either shares baby pictures with family or shares social activities in public. In most cases, Facebook was not perceived as a place for private interaction among close family and friends or a space for relationship building. Even using Facebook chat (now its own stand-alone app, Facebook Messenger) was tedious for some, as it required an account on the social networking site. This is interesting in light of longer histories of social networks online: when Facebook and MySpace were new, youth and teenagers perceived these as their hangout space—private for them and their friends and not intergenerationally accessible. When Twitter arrived and Facebook became more global, users grappled with increasing "context collapse" as family, friends, and coworkers all created accounts and existed in the same semipublic expressive space online (Marwick and boyd, 2011, 9).

In contrast, WhatsApp acts as a private space where users "can chat to only the people [they] want to" (conversation between "M" and "S," 2015). Interviewees noted a preference for WhatsApp when chatting with friends and siblings but used Skype video calls when speaking with parents (conversation between "B" and "M," 2015). WhatsApp was, after all, a replacement for SMS. Even with a public broadcast message—a message that goes out to a number of people on your contact list—users can have one-on-one conversations without interacting with all of those who received the broadcast. Apart from a status message, one is not obliged to publicly share details about their lives. Although WhatsApp has now introduced "stories"—similar to Snapchat and Instagram—the primary aim of the app remains conversation and mundane communication.

Interviewees' preferences for WhatsApp, rather than Facebook, Instagram, or Twitter, reveals a negotiation of digital/networked publics and privates, but it also reinforces a tension "between rich affective interaction and cheapened commodified exchange" (Jarrett 2015b, 204). Thus, in moving away from Facebook and toward WhatsApp for personal interaction, we see that these users are implicitly turning away from public commodification of affective exchanges and alienated exchange values. Since WhatsApp does not require logging in or account creation in ways that Snapchat and Instagram do, the communication seems less commodifiable. Additionally, the text-based nature of the app—unlike the photo-based interfaces of the other two—makes it easier and convenient.

The move toward performing intense affective work as well as sharing and benefiting from affective work from others through close networks contains the affect flows within private familial space or in the space of close friendship and does not flow out into public and global commercial and monetized exchange. The app allows emotional/relational labor to be privatized by providing a domestic space of nonmonetized work—where the interaction is named as both pleasurable and as obligation/duty to particular close networks (most often family in the case of the Indian women interviewed).

Radhika's own experience, for instance, feeds into the understanding of this personal use of WhatsApp and the transmission of affect through relational sharing and negotiation. She started using WhatsApp intensely in 2015 soon after an emergency visit to India when her then ninety-three-year old mother was hospitalized. Her sister, who was her mother's primary caregiver (along with her spouse and son), requested that they all stay connected via a WhatsApp group for the six siblings so that she could keep them informed about their mother's health. After realizing WhatsApp's convenience while traveling across continents, Radhika also put together several other groups with nieces, nephews, sisters-in-law, graduate student advisees, and so on. She was perpetually connected to these select groups of people in yet another mode, even though she was already connected to many of them through Facebook, email, Instagram, and Snapchat. Originally from India and simultaneously part of the Indian, larger South Asian, and United States diasporas while living in Norway, now Radhika too became immersed in the WhatsAppified Indian digital diaspora—a text-based chat app on a mobile phone turned into an appified community in her pocket. Along with this "gadgetized" community came an intensity in relational exchanges with family members and friends that contributed to the production and circulation of affect across digital and mobile interactive spaces.

Similarly, as Tarishi moved to the United States in 2017, WhatsApp became her primary point of contact with her parents. Her mother did not need to set a time to call her, as many previous travelers with relatives in different time zones had. A "good morning" message may be met with a "good night," but immediate responses were possible. A family group also keeps things exactly as private as her family needs them to be: the four members of her family interact on the app several times a day with little regard for time. For exchanges with the extended family there are more groups, but only selected information needs be shared with them. It is this nuanced specificity and multivalent affective connectivity—through the gut and intense but also efficient and digitized—that helps build and maintain communities through WhatsApp.

As was the case for Radhika, Tarishi, and many of our interviewees, we see how WhatsApp produces, transmits, mobilizes, and circulates affect, sustaining networks of family and friends across physical and generational distances. These WhatsAppified diasporas offer forms of self-expression and opportunities for (re)claimed agency that reshape notions of private and public communication. For Indian women, this also materializes as a form of communication with close friends as they connect to social spaces of home (rather than work or social spaces of the global "outside") and manage new modes of self-expression and agency around control and disclosure. These include expressions of delight, discovery, amused annoyance, and even anxiety about *lack* of response—all emotions that affect the long-term development and negotiation of familial and intimate social relationships. Moreover, communication is strategically chosen and miscommunication is managed internally within the relational spaces through various features such as emojis, shared jokes, memes, and YouTube clips.

From Social Networks to Appified Community

Even as we continue to explore how hierarchies of power play out through social media apps (Gajjala 2011), the present review of WhatsApp draws mostly from interviews with middle- to upper-class women. Even if interviewees refer to usage by women care workers who do not share their class privileges, the vantage point is still classed. Access to smartphones is predicated on independence in decision making and cash flow, so the use of WhatsApp by women and men outside of middle- and upper-class urban centers is varied and depends on different types of literacies beyond just ownership of smartphones.

The Indian women we interviewed use the app as a portal for sociality and community building. Compared to a platform like Skype or picking up the telephone and making a call, WhatsApp provides an asynchronicity that gives them a choice in how and when they make themselves available. At the same time, WhatsApp offers instant and synchronous communication, which is reminiscent of using SMS or MSN text messaging, where asynchronous and synchronous are implicit social options for using the interactive tool. As Koen Leurs notes of adolescent migrant communities in Europe, even as smartphone messaging apps take over the role played by older platforms like MSN, "the underlying communicative and symbolic principles of instant one-on-one exchanges are largely comparable" (2015, 143). The affordances of asynchronicity and synchronicity are equally valuable to WhatsApp users.

Further, some of those interviewed also noted that WhatsApp liberates them from having to be at the computer or from having to rely on a data plan. Although tools like Facebook Messenger are also mobile, our interviewees still linked Facebook with computer usage and with needing to negotiate a larger, more public network. Some respondents even implied that Facebook status messages seem like mass communication rather than communication within a friend circle, which influenced their preference for WhatsApp. Additionally, users of Facebook don't always like to add family members as friends, since they perceive that Facebook Messenger requires users to be Facebook friends with the person with whom they message. WhatsApp use, on the other hand, does not come with any associated requirements of creating an account and allows users to negotiate away from having to include family members in the larger networked social spaces that interweave both professional and peer groups, spaces where intergenerational communication within close and extended family networks may not feel comfortable.

In addition, those who prefer WhatsApp noted a distaste for what seemed to be unreliable privacy settings on Facebook. In WhatsApp there is repeated reassurance of the encrypted nature of exchanges and content ostensibly remains private to their chosen chats. Thus the context collapse that Alice Marwick and danah boyd (2011) note in relation to social networking platforms occurs differently in WhatsApp groups. The same young people who may never add an older relative as their friend on Facebook have no issues with connecting with them through WhatsApp.

Conclusion

WhatsApp is a convenient tool for intercontinental and intergenerational conversations, easy group interactions, and media exchange. However, the messaging app has a lot more interesting nontechnical functions that it serves, especially in the circulation of affect and as a space for free expression. For Indian women it is an important gateway, because it enables them to explore different personas that are freer and more expressive. Where the digital gap is massive, with 71 percent of the male population using the internet but only 29 percent women constituting the space (India Exclusion Report 2017), access to WhatsApp is essential. The app allows private conversations and allows them to perform the persona they like. Through our interviews, conversations, and personal experiences, we also found that among the Indian digital diasporas, WhatsApp has also become a tool for relational negotiations within close friend circles as well as in immediate and extended family networks. Even though the distance allows for breaks in relationships, WhatsApp provides more choices for who we share our private lives with when we want them to be social but also limited, affective, and meaningful.

Notes

1. My own use (Radhika), for example, of WeChat to communicate with Chinese graduate students has increased, and we note many similarities in the kinds of intergenerational conversations and affordances that come from the two apps: WhatsApp used among Indian family and friends across generations and time zones, and WeChat used among Chinese family and friends.

F News/Entertainment

This. Reader
Trending Topics and the Curation of Information

DEVON POWERS

App: This. Reader
Developer: This. Media Company
Release Date: May 2015 (Current and Final Version: 1.3.1)
Category: News, Social Networking
Price: Free
Platform: iOS
Tags: content management, community, failure
Tagline: "This. Is a Great Place to Be a Link."
Related Apps: Flipboard, Longform, Nuzzel, Pocket, Feedly

You Can Get with This.

There's a reason why This. Reader had a period in its name. Launched in 2014 as a "digital shelf" for managing content, This. limited sharing to a single link per day, in hopes of letting users "find just the best from those curators they trust" (Oswaks 2015). The name "This." was borrowed from a practice that by that time had become standard on social media—namely, drawing awareness to a link, image, or video one shared through a bold declaration: This. As if to say, Not that, but This. Nothing really matters but This. That was then, this is . . . This.

That bit of punctuation, the period, was a symbol: a tiny command to pause and pay attention. From early on, the app's CEO and founder, Andrew Golis, imagined that This. would capitalize on our considered moments, with the aim of being "the first place people go when they sit down on the couch at night and think: What's the best thing for me on the web right now?" (Oswaks 2015). Originally created under the auspices of Atlantic Media before forging ahead as an independent company, This. peddled an undisputed, elegant, and seemingly welcome definitiveness. Because in a world of unlimited choice (of what to read, to click, to watch, to gaze upon), didn't we all really just want to be told?

The This. era became one in which the tyranny of choice increasingly balanced itself against plain old tyranny, contingent upon an emerging belief that it is sometimes better to be unequivocal. The serendipities of the early web were carefully hemmed in by knowing curation. Gut feelings and mic drops won over those who were skeptical of relativism and subtlety. The world that invented affirmative consent still salivated for Christian Grey. One way apps began to mirror the drift toward certainty is through offering "personalization" or "intelligence" to anticipate and meet needs that have not yet been expressed. Take, for example, Google Now or Waze's ability to warn that traffic might slow down the morning commute, or a notification from Amazon that your latest shipment of laundry detergent is on its way, just in time. This. joined a host of apps attempting to do the same for news and information, some by providing stories only from sources a user selects (Flipboard, Newsify), others by supplying only a digestible serving of must-reads (theSkimm, Nuzzle). But This. set its sights even higher than its peers, harnessing the renewed allure of definitiveness in order to cast doubt upon the entire idea of abundance itself. Compare: In a 2011 sketch on the satirical IFC television series *Portlandia*, some hipster friends quiz each other on how well they are keeping up with the ever lengthening list of essential periodicals (have you read that thing in *Mother Jones*? The *New Yorker*? *McSweeney's*? *Dwell*?). It's a competition no one can win, but in This.'s world, why even try? This. was the answer to cultural one-upmanship.

But as it turns out, the metaphorical full stop of This. was, and is, a grand paradox. Forced to suspend operations in midsummer 2016, Golis, his heart heavy, declared in a July 15 email to users that This. "never got big enough to raise long-term capital or begin to build a sustainable business." For sure, the demise of This. is reminiscent of what Douglas Rushkoff (2016) calls the "growth trap": the pathological drive toward expansion that determines the fate of so many digital companies. Yet the imperative to grow quickly posed a glaring irony for an app that promised specialness and distinction rather than abundance and miscellany. Even more inconsistently, the singularity assigned to what users shared via its network in no way lessened the hard work of filtering. To distill content down to its essence—and condense social media participation down to a lone link—This. wholly depended upon the very excess it feigned to dismiss. The compound effect of these troubles was that despite its noble if questionable mission of operating around the logic of taste, This. ultimately could not reject the economics of the web, where information trends toward the logic of attention and sharing necessarily takes on a promotional cast. The rise and fall of This. Reader should provide a cautionary tale for any outlet seeking to control and direct the distribution of information within our contemporary media environment (Goldhaber 1997; Davenport and Beck 2001; Powers 2017).

This. Is How We Do It

In May 2015, after an invite-only beta period for its website, This.cm, developers launched This. Reader app, geared toward its mobile user base. Branding itself as a "social magazine," Golis explained the objective of This. in an open letter to the ten thousand people who had already joined the platform:

> When you sit down on the couch at 8 o'clock at night, with a beverage of your choice, and fire up your phone or tablet or computer, where do you start? The day's almost over, things are quieting down. You don't want noise. You don't want the latest headlines. You want a well-told story. A provocative argument. Something beautiful. You want the web at its weirdest, its most ambitious. Its least "virally optimized." This. connects to that web with an extreme curation mechanic (each user sharing just 1 link a day) and an awesome seed community of curators (you!). (Golis 2015c)

Golis proposed a scenario that contrasted sharply with what we've come to assume about social media engagement. The imagined This. reader arrived at This. Reader seeking not the frenzy of a newsfeed or scroll, but a deeper experience—reflective rather than distracted, exceptional rather than mundane, relaxing and homebound rather than harried and on-the-go. This. Reader, in kind, privileged a stationary mobility—the convenience of that phone or tablet, yet kept in place (namely, on the couch, with a beer). Also implied in Golis's comments was that users of This. Reader would spend more time with the app, with absorbing reads rather than quickly scanned "headlines." In a later interview, Golis reiterated as much, explaining:

> We see the rest of social media in a crazy arms race of volume and optimization, where attention goes to whoever is the loudest or most calculating, not the person with the best taste. By limiting everyone to sharing one link a day, we're letting our

members show off their rad taste without them worrying it will be drowned out. Second, that one link a day limit tends to take "news" off the table. You almost never see someone share a breaking news story on the site . . . Instead, the limit on This. encourages people to share things that are more evergreen: essays and narratives and bits of entertainment that will be interesting for at least 24 hours, not just the next five minutes. (qtd. in Wang 2016)

Executives at This. regularly referred to its members as belonging to a selective club, in all senses of that word. If the ethos of This. was to eliminate the need to choose through placing severe constraints on sharing, it banked heavily on the idea of "the chosen"—pertaining to content as much as it does to the people and publishers that populated the service. Touting both a different kind of reading as well as a different kind of readership allowed This. Reader to appear indifferent to, and maybe even hostile toward, growth. Additionally, This. could position itself as part of a backlash against the undifferentiated teeming of many of the most popular social networks, whose early adopters continue to loudly lament the lost sense of kinship and care such platforms once enjoyed. For example, journalist Jon Ronson wrote in his bestselling book *So You've Been Publicly Shamed*, "In the early days, Twitter was a place of curiosity and empathy," whereas these days, it was ornery and unkind, at risk of becoming a place where users "lose our capacity for empathy" (2016, 290–91). While the change in Twitter's culture can't entirely be attributed to its growth—nor have other maturing platforms fallen prey to exactly the same set of issues—Ronson's observations parallel those of José van Dijck in her analysis of Flickr's history. In tracing the failures of that service, she argued that social media platforms often struggle when the "connectedness" beloved by its initial community clashes with the "connectivity" that is of greater benefit to corporate interests (2013, 91–98). In other words, the affective bonds that bind platform users to one another fester when they are exploited for financial gain—when the goal becomes efficacious, associable, minable, or simply more connections.

That is one reason why it is noteworthy that Golis declared the desire to "deepen our community's relationship to the open web by focusing on the value of a link" (qtd. in Sterne 2015). This was not the talk of someone attempting to build the social media behemoth, but a strategy of using trusted content to lure, then maintain, a dedicated niche of devotees. On March 13, 2016, he echoed that sentiment, saying in an email to users, "When you share on This., you aren't just marking something as special on a website. You're wrapping it up in a package and sending it to people who've given you their trust by following you." There curiously appeared to be an overt attempt to bind "connectivity" and "connectedness": to create a trustworthy and authentic social media community that is valued (and valuable) precisely because of its norms regarding what is allowable in the sphere. Quality became a stand-in for quantity; there may not be much, but it's all good.

It is bizarre, then, that typical quantitative metrics abounded through the app. Account pages noted follower numbers. Instead of "liking" or "favoriting" what others posted, This. Reader used the term "thanks"—how nice!—but those, too, were counted. By the end of its run, none of these numbers was very high. A popular post might have less than a dozen thanks. "Featured" accounts—ones This. leadership had deemed notable—hovered around a thousand followers or so. Name-brand publications such as the *New York Times* boasted fewer than five thousand. Golis and the other staff at This. did not regularly publicize the number of people who used their service beyond affirmations that the total was steadily increasing—perhaps a capitulation to the fact that they couldn't completely disregard the need to appear as though they were growing. It was also perhaps a confession that despite their professed commitment to something other than size, the aim of This. was to be less like Tumblr or Twitter in their humble, homey early days and more like their later iterations. This. wanted to be special, but it also wanted to be popular, essential, and expanding.

Is This. All There Is?

An emphasis on quantification was not the only way that This. Reader resembled other options among the social media pack. Like Pinterest, Twitter, or Tumblr, a user joined the service by inputting an email address and selecting a user name. Next, the app encouraged the user to "follow some folks" in order to "populate your feed with great daily articles." One might use the search engine to find specific accounts or could opt to browse through a list of "featured" users. The featured list included many traditional publishers, such as *Fast Company*, *Texas Monthly*, and *Esquire*, as well as individual users, some of whom were already high-profile social media celebrities. That said, it was not clear how decisions about whom to feature were made, since they were not ranked by any clear measure, such as number of followers. (In a statement in late 2015, Golis explained that featured members existed "to make sure it's super easy to find great links, even if you're new and haven't built a network" (2015b).

Anyone who signed up to This. could also receive This. 5, a nightly email comprised of five "editors' picks." Honing in on what they called the "retro cool" of newsletters (Golis 2014), This. 5 drew from that day's most notable contributions, highlighting the crème de la crème. In keeping with their pledge to provide substantive, intriguing content, This. 5 stories tended toward the longer and the cultural, such as the offerings on July 8, 2016: "Why One Woman Pretended to Be a High School Cheerleader," "Marie Kondo and the Ruthless War on Stuff," and "The (Really, Really) Racist History of Gun Control in America." Other factors also influenced what made the list, such as how often a piece was shared (or, conversely, a great piece that risked going unnoticed); diversity in viewpoints and voices; and a mix of different types of stories, which at times included multimedia (Boghossian 2016). Further, This.'s editorial policy noted that the company sought to amplify independent publishers or voices from the margins—goals that were reflected in the newsletter's overall makeup. In a further show of their editorial vision, in March 2016 This. 5 was joined by a "social newsletter," an addendum to the nightly email that included another smattering of links personalized to the recipient's interests.

To share on This. Reader, a mobile user navigated to the intended page via the browser app Safari, pressed the icon that opened a menu of added features, and chose to share via This. It was the same way someone might share something on Facebook, Twitter, or Pinterest, the difference being that this act is good only once per day. That This. had an app-based reader was less about the possibilities of mobility—for instance, location-based curation or augmented reality enhancements—than the desire to capture users' curative activities wherever they might be and then seamlessly integrate their own filtered content back into the This. network. This. Reader also used an app-based notification to call attention to its daily newsletter of curated links—a further attempt to assimilate itself into a user's content consumption habits. As unique as This. strove to be, its look and functionality were standard.

According to Nicholas John, sharing online has evolved into "a concept that incorporates a wide range of distributive and communicative practices, while also carrying a set of positive connotations to do with our relations with others and a more just allocation of resources" (2013, 176). When This. users shared, they operated within these well-established, positively accented ideologies, even as sharing is well known to have become "algorithmically mediated interactions in the corporate sphere" (van Dijck 2013, 65). At the same time, This. Reader, in certain ways, attempted to reinterpret sharing: ditching algorithms in favor of human curation, making sharing scarce rather than abundant, and rendering it exclusive rather than open. The creators argued that these revisions were necessary in order to develop a quality service and an invested community. But these practices, carried to their logical extension, suggested the creation of a gated community that purported to operate independently yet would wither in the absence of the abundance beyond its walls.

Moreover, a significant contradiction that surrounded the This. Reader was that one link per day was never just one link. Instead, users arrived at This. with the expectation that they were voracious, discerning content consumers; once installed, they acted as arbiters who, in theory, admitted only the most excellent things they came across. That carefully groomed selection created a flow—the This. ecosystem—from which those judicious users made even more choices about whom to follow and what to read. Finally, editors cherry-picked what to consecrate and circulate back into the network, taking ultimate authority of what "the best" content looked like. Acting as publishers as much as they were concerned stewards of an ardent community, one best link was never enough, especially for a company among whose goals was also the effect of a particular editorial politics. That said, perhaps the most central conundrum of This. and its reader was how, exactly, in a world of information excess, meaningful content can truly rise above the din.

This. Is What You Came For

In thinking about This., we therefore confront questions about the value of curation and its relationship to promotion. It's a quandary that besets most media platforms that manage large content catalogs, though for This. the issue presented itself uniquely. For example, while Netflix, Spotify, and the like have worked hard on developing "smart" recommendation algorithms, This. relied on its users, whose entry into the network was predicated on their desire to participate in filtering. (One could participate in This. purely as a lurker, but the system could not sustain itself with only those sorts of members.) In another difference from those other media services, what appeared on This. was freely available elsewhere on the internet; other content providers have in recent years been loading their catalogs with original content as a way to ensure subscribers. One therefore could ask sincerely what users (and would-be advertisers) found worthwhile in a service that redistributed content that had been delivered dozens of times over via other, more popular and more lucrative social networks.

The answer arguably lay as much in curation as in promotion—the capacity for This. to highlight something that might otherwise disappear, for a group of users who would care and pay attention. Golis repeatedly invoked the reader as the beneficiary of its tasteful refinement. Yet Golis's desire to make This. "a great place to be a link" was a promise that was especially compelling for producers of content, who benefited both when users shared their content and when they used the platform themselves (2015a). The hope of finding an attentive audience also could shape editorial policy; as This.'s editorial director, Mayukh Sen, acknowledged, sometimes "editors and writers to reach out to us personally, flagging a certain piece" for inclusion in the nightly newsletter (Boghossian 2016). Even as This. tried to contrast itself to the bounty that existed beyond it, it could not on its own rewrite the dynamics that created that bounty in the first place.

Many others continue to work toward developing apps to manage the onslaught of content online; vague gestures toward taste and quality are always the rallying cry of those fed up with the status quo of surplus, clickbait, and dissonance. One example: Medium's Evan Williams, in a 2017 interview, announced his desire to reform "the architecture of content creation, distribution and monetization on the internet," and vowed to use his platform to distribute better, richer content (Streitfeld 2017). Perhaps he and anyone who agrees with him should learn a lesson from the failure of This. and recognize the steep road ahead. To create an app that privileges content sharing means to play into the promotional tendencies that dictate how, what, and when we share. To provide less choice requires selection at some point along the way, and those selections are rarely benign (and may themselves be a kind of promotion). To create a user base guided by norms that are less vulgar than attention requires more than just attracting a small, caring community. With the economic model for news content squeezed on all sides and the pressures of growth overwhelming developers of all kinds, the story of This. should give pause to anyone who thinks apps can solve the problem of drawing audiences to quality content. Ultimately, there may never be an app for This.

Periscope
The Periscopic Regime of Live-Streaming
MEGAN SAPNAR ANKERSON

App: Periscope
Developers: Kayvon Beykpour and Joe Bernstein
Release Date: March 2015 (Current Version: 1.7.5 for iOS)
Category: Social Networking / Live Video Streaming
Price: Free
Platform: iOS, Android, Web
Tags: live streaming, liveness, media witnessing
Tagline: "See the world through someone else's eyes."
Related Apps: Facebook Live, Meerkat, YouTube Live, UStream

When Diamond Reynolds started broadcasting on Facebook Live seconds after her boyfriend, Philando Castile, was mortally shot by police during a traffic stop in Minnesota in the summer of 2016, a new perspective of a police shooting was made publicly visible. Turning the camera on herself, Reynolds places viewers with her in the front seat as she narrates the scene: Castile is slumped next to her, his white T-shirt pooling with blood; her four-year-old daughter is in the backseat. Through the window we see only the officer's hands gripping the pistol he keeps aimed inside the vehicle. "I told him not to reach for it!" he shouts in a wavering voice. "You told him to get his ID sir, his driver's license," Reynolds replies in a measured tone. "You shot four bullets into him, sir. He was just getting his license and registration, sir." The video continues for ten more minutes and concludes with Reynolds tearfully addressing her audience while handcuffed in the back seat of a squad car:

> Ya'll please pray for us, Jesus, please ya'll. I ask everybody on Facebook, everybody that's watching, everybody that's tuned in, please pray for us. Sister, I know I just dropped you off, but I need to be picked up. I need Alizay to call my phone.

The video is a powerful document chronicling the immediate aftermath of police violence. And it came the day after a cell phone video capturing the shooting death of Alton Sterling by police was made public, sparking outrage and protest in Baton Rouge, Louisiana. Reynolds's video, available on Facebook after the live broadcast ended, was quickly picked up by the activist group Black Lives Matter and circulated as further testament of the violence inflicted on black people by American law enforcement.

Both the Sterling and Castile videos were shared on social media, and both fueled protests over racial injustice and police brutality. Yet, as mainstream media couldn't help observing, there was something remarkable about the live-streaming aspect of Reynolds's recording: "It provided an even more intimate, distressing view of the shooting," a rawness achieved precisely by virtue of being broadcast live (Stelter 2016). Cell phone videos are shared with knowledge of the recording's outcome, but live video fulfills an additional service. As one commentator put it: "Unable to call the authorities as she watched her loved one slip away, Reynolds instead called on the public" (Lapowsky 2016). This video serves as both a testimony of police violence and a real-time emergency broadcast alert that directly calls on bystanders for help. It offers a vivid demonstration of the power of mobile live-streaming apps to bypass traditional institutional structures and provide citizens with direct, immediate access to global audiences and viewpoints that escape mainstream media attention. And in doing so, it exposes some of the fundamental tensions accompanying the reorganization of media systems in the digital age, between broadcast and the so-called post-broadcast eras, between top-down centralized institutions and bottom-up participatory production cultures, between mass media and interpersonal communication. For these reasons, live-streaming apps are material sites from which to stage critical questions about the intersections of "old" and "new" media imaginaries. If early television promised a new expansive global view—"your window on the world"—from the family living room television set (Spigel 1992, 102), what view of the world does live-streaming promise? What do live-streaming and social broadcast platforms mean within the historical context of television? In short, is it premature to characterize today's media climate as a post-broadcast era?

Perhaps no app is better poised to address these questions than Apple's 2015 "App of the Year," Periscope, which first inaugurated social media's live broadcasting craze with the tagline: "See the world through someone else's eyes." An app that essentially transforms a smartphone into a personal mobile broadcast studio, Periscope allows users to stream live video directly from their camera phone to an audience who responds in real time by typing comments, asking questions, or tapping the screen to send a stream of animated hearts as a form of appreciation. Although not the first live-streaming app, Periscope's prelaunch acquisition by Twitter in March 2015 helped boost its visibility. The app earned praise for enabling citizen journalists to broadcast live from the streets, and it ignited antipiracy debates when users broadcast pay-per-view sports, music concerts, and cable programming. Part of Periscope's popularity comes from the sense of immediacy and "rawness" that live video affords: stars share behind-the-scenes experiences, aspiring musicians take song requests, stoners smoke a blunt, bored teenagers seek conversation. When giants like Facebook and YouTube rolled out live video as well, media pundits declared "social broadcast" the next big game-changer for mass media and social media alike. "Forget television," bloggers announced. "The revolution will be live-streamed" (McGarry 2015). The battle over live-streaming may well be social media's next new thing, but the persistence of discourses of liveness, real time, and broadcast offer important reminders that apps like Periscope need to be understood within the historical context of both tele-vision—mediated seeing at a distance—and broadcast, a powerful social institution that arose in the twentieth century. Must we "forget television" in order to contend with its re-configuration, its appification? This chapter considers what is new about "so-cial broadcasting" by examining how live-streaming apps expand the broad-cast imaginary through a reconfiguration of media witnessing.

The concept of "media witness" has inspired much debate since television scholar John Ellis used the term in *Seeing Things* (2000) to describe new ways of experiencing the world that emerged with the rise of electronic media in the twentieth century. For Ellis, broadcast in particular constituted a "profound shift . . . in the way we perceive the world that exists beyond our immediate experience" (9) The dominant viewing position to emerge, Ellis contends, was that of *witness*, a distinct mode of perception intensified by the key character-istic of liveness. Through constant media exposure to images of distant suffer-ing related by people we don't personally know, modern media systems have transformed what it means to witness (Frosh 2006). Reynolds's video offers a poignant example of how live-streaming apps radically alter the nature of these encounters. In quick succession, she calls out to multiple audiences simulta-neously: "Y'all," "everybody on Facebook," and "everybody that's tuned in" coexist alongside direct communication to "Sister" and "Alizay." This hybrid address, which combines imagined publics with specific knowledge of actual viewers, is one outcome of the entanglement of *mass media* and *social media logics* (van Dijck and Poell 2013; Ankerson 2015). Some of the social protocols and aesthetic conventions emerging around live-streaming remediate practices developed in relation to prior media, including broadcast television and social networking sites. Other practices arise from the material-semiotic dimensions of the app itself—that is, its platform, design, and interface. To examine how the culture and politics of live broadcast are being reconfigured through Periscope, I focus on the stories told *about* the app by key stakeholders, *by* the app itself through its user interface (UI) metaphors and affordances, and *through* the app by Periscope's users. I refer to this cluster of activity as Periscope's "design narrative," and suggest it is here that we might locate particular shifts in the meaning of liveness and the experience of media witness in a social media–driven, "appified" era.

Scopic Regimes of Tele-Vision

How are practices of looking, knowing, and experiencing the world mediated through an app like Periscope? One approach is through the concept of "scopic regimes." Film theorist Christian Metz introduced the term in the 1960s while identifying a paradox: despite theater's stronger claim on reality—real bodies perform before us rather than projected prerecorded scenes—films appeal more intensely to our sense of presence, proximity, and belief: "one 'believes' in the film's fiction much more than in the play's" (1974, 12). Metz defines the "specifically cinematic scopic regime" as those ways of staging and seeing the world that are unique to cinema: editing, lighting, camera work, as well as the spatial positioning of spectators, projector, and screen in a darkened auditorium (1982, 61). By illuminating the institutional and medium-specific ways that visuality and perception are organized, the concept of "scopic regimes" insists there is nothing "natural" about vision.

Martin Jay broadened this concept to encompass the larger social milieu in which visual conventions are organized, arguing that the history of Western modernity is a persistent (re)alignment of vision with certain types of knowledge. Jay's work helped extend the notion of scopic regimes to include "the contested terrain" of socially and politically constructed ways of seeing and being seen (1988, 4). Combining the medium-specificity of Metz with Jay's larger historical connections between representation, knowledge, and power, I examine the "Periscopic Regime" through the lens of media witnessing as a way to historicize modes of perception that characterize the broadcast imaginary. How is the scopic regime of television reimagined in the age of "social broadcast"?

"Television" literally means, "tele-" (at a distance) plus "vision," or seeing from afar. The extension of global media systems over the twentieth century has transformed how we see and know the world well beyond our immediate environment. Separated in space but united in time, those tuned to a live broadcast occupy a unique viewing position of *media witness*, which emerged as a result of experiencing mediated liveness-at-a-distance (Ellis 2000; Peters 2001). Viewers of live television see faraway action "almost directly," but the experience is clearly not the same as "being there" (Peters 2001, 718). As John Durham Peters points out, "witness" is an intricate term that forms a communication triangle: an eyewitness like Diamond Reynolds who occupies the position of "being there"; the media-witnessing audience who follows along in real time from afar; and the "witnessing text" itself, the utterance produced by an agent bearing witness. These positions, however, are not equivalent, and each point in the triangle faces different levels of personal risk and moral responsibility. For the media witness, the shared "immediate now" of concern is offset by the safety that distance affords.

Positioned within the scopic regime of broadcast television, the media witness encounters a "world out there" that is constructed by news media organizations covering, interpreting, conferring selective attention, and packaging accounts for the nightly news (Rentschler 2004; Ellis 2009). The audience can assess what is shown, but, importantly, for the media witness *action is not possible*: "It is impossible to offer help or console with a hug" (Ellis 2009, 76). One cannot tap the TV remote during a live broadcast of Syrian peace talks to send little graphical hearts to the delegates negotiating a fragile cease-fire. Viewers can offer no solace to the refugees being interviewed from their temporary shelter. Viewing the world through broadcast television, media witnesses encounter suffering from a position released from an obligation to act.

Yet there is a second, more ordinary component of media witness that Ellis argues must also be considered. Alongside the monstrous, we regularly encounter stories of people dealing with the everyday frustrations of life—heat waves and mosquito infestations, traffic jams and airline delays—a process Ellis calls "mundane witness":

> We witness children arguing with parents; couples in the throes of divorce; strangers thrown together and deprived of outside stimuli; people challenged to change their behavior. Such is the nature of contemporary media witness: the monstrous and the mundane occupy the same space, and the mundane predominates. (2009, 74)

This intertwining of the banal and the extraordinary, the impersonal and personal has become a defining feature of broadcast, found in programming patterns seeking to balance "hard-hitting" news with human interest segments. Although television broadcasters don't know who exactly is tuned in at any moment, they look directly into the camera (teleprompter) and greet a mass of strangers, familiarly rather than formally. Home viewers understand that the broadcast is directed toward them but is simultaneously addressed to indefinite others out there who are also tuned in.

Paddy Scannell calls this a "for-anyone-as-someone" communicative structure, a style of talk that addresses audiences simultaneously as part of a mass and personally as individuals (2000, 12). This collectively imagined "we" is one of the key characteristics of modern public discourse. But as audiences become numerable and fragmented as well as imagined, this style of personal/impersonal address no longer works. We see this in the way Reynolds addresses her audience on Facebook Live, using a hybrid style that merges the for-anyone-as-someone character of broadcast talk with direct interpersonal communication associated with Skype or FaceTime. In light of the devastating events that had just transpired, it may seem trivial to focus on a style of communication. But perhaps it is through something as slight as shifting modes of address that we may find a vantage point for grappling with the changing stakes of media witnessing in relation to new experiences of liveness.

Is the liveness experienced through Periscope the same liveness found on broadcast television? To assume so is to risk conceptualizing liveness in purely technical terms, as an inherent quality of media like radio or theater, while losing sight of what Scannell terms the "management of liveness." Like looking, liveness is not a natural fact but a historical and cultural condition. Examining the discourse of "liveness" and "simultaneity" in the nineteenth- century media imagination, for example, William Uricchio finds no real distinction between "liveness" of simultaneity (telephone) and "liveness" of a storage medium (film) (Uricchio 2004, 125). Motion animates pictures and brings them to life; it was this sense of liveness celebrated in early descriptions of film. Today a similar sensibility informs Apple's "Live Photos" feature, which "brings your images to life" by capturing moments just before and after a picture is snapped. Projected on a movie screen, tuned to a broadcast program, or touched on an iPhone display, liveness is not a technical achievement but a social one. To examine how the Periscopic regime transforms media witnessing, I turn to the confluence of myth, marketing, and materiality that make up Periscope's design narrative.

Media Witnessing in the Platform Era

According to Periscope developers Kayvon Beykpour and Joe Bernstein, the idea for a live-streaming app began in the summer of 2013 when Beykpour was traveling in Istanbul. Protests broke out in Taksim Square and the situation turned violent. As Beykpour scoured social media trying to determine if it was safe to travel, he found himself wishing for a real-time feed from the streets: "Why isn't there a way for me to see through their eyes what's happening right now?" He and Bernstein decided to build a live-streaming app that would "let you broadcast what you were seeing to anyone in the world, in real time" (Pierce 2015). To convey the sense of immediacy, presence, and connection Beykpour was imagining, he referred to the app not as a "live broadcast station" but as a "mobile teleportation service," a term that conjures images of sci-fi fantasy machines capable of defying the limits of space and time. Television's debut as "your window on the world" promised to bring the outside world into the domestic space of the family home (Spigel 1992). Periscope flips these conditions; rather than bringing the world into the home, your vision is transported outside to explore the world at large. By urging us to see from countless situated positions, Periscope's tagline suggests the better view of the world is a partial one, obtained by "seeing the world through someone else's eyes."

Design narratives often begin with a creation myth, the story that appears in investor briefs, press releases, and marketing materials. These anecdotes— riots in Taksim Square, live video as a form of teleportation—get repeated whenever founders are interviewed. It is the story of the app and its core guiding values, as told by its makers and designers. Periscope's website prominently features "Our Story," which explains how the company was "founded on the belief that live video is a powerful source of truth and connects us in an authentic way with the world around us" (Periscope.tv). This statement helps qualify Periscope's tagline by locating the "source of truth" not in the partial, situated perspectives of all the live "someone else's" out there broadcasting, but in the state of liveness itself.

Design narratives don't just tell a story about a product; they inform the big picture, encompassing creation myths, product design, user experiences, and production practices. Stories may emanate from founders and advertisers but can also be found in the built world and design of material things. Analyzing design narratives requires listening to how things talk (Daston 2004). Apps tell stories through their UI metaphors, design patterns, color schemes, default settings, dialog boxes, push notifications, user prompts, and so on. When beginning a live broadcast, for example, Periscope prompts users: "What are you seeing right now?" The answer becomes the title of the broadcast, but it also orients users to approach broadcasting in a particular way: as eyewitnesses who lend their vision to the public.

Even subtle changes in design might have profound implications for media witnessing in the live-streaming era. Consider a minor UI alteration made in the fall of 2016. Since Periscope's debut, users initiated a live recording by tapping a big red button labeled "Start Broadcast." But competition from Facebook Live spurred tighter integration between Periscope and Twitter, and one outcome involved the development of a shared vocabulary for live-streaming across both platforms. "Start Broadcast" was renamed "Go Live." On one hand, this update changed little, since the function of the button—to begin a live recording—remained unchanged. But the instruction to "Go Live" signals something quite different from "Start Broadcast." Starting a broadcast can be a daunting undertaking; it suggests preparation first, setting the stage before the broadcast officially begins. For many would-be live-streamers, "Start Broadcast" takes nerve. Going live, on the other hand, is like checking in: it's a status update. A new name for the big red button that does exactly what it did before might seem unremarkable, but it's a good example of how communicative affordances are tied up with the material-semiotics of design narratives and not just the material features of a technology that enables and constrains action in physical ways (Norman 1988; Hutchby 2001; Nagy and Neff 2015).

Other design decisions, however, can fundamentally alter the ways technologies work, how they are used, and what they mean. Reimagining "broadcast" from a one-way model that distributes signals outward to a two-way mode of communication between broadcaster and viewers affords entirely new forms of social interaction. Broadcasters and viewers can converse, greet one another by name, comment, and ask questions. The audience is not strictly "imagined" as with mass media but also numerable. Everyone tuned in knows how many others are tuned in too. Periscope users receive alerts when someone they "know" has joined the broadcast. But this introduces a new problem: one of the first significant challenges Periscope faced was would-be broadcasters' stage fright. Because social interactions, technological affordances, and cultural expectations are deeply entangled, "friendliness" had to be a communicative affordance built into the system. Periscope's lead designer, Tyler Hansen, explains:

If you are actually on stage, you can read the audience and how well they like things. That's when we started thinking of things like hearts—again ephemeral time-based feedback that was very simple to do for a viewer so that the broadcaster knew you were there. All you had to do was tap on your screen and the broadcaster saw you were tapping on your screen. Knowing someone was there helped a lot. (Truong 2016)

The hearts feature, designed with random paths to give the hearts "human flair," became one of Periscope's most important design solutions (Truong 2016). In making action possible through only the smallest gesture—tapping a screen—these new features profoundly alter the position of media witness. Not only can viewers say something; they can do something too.

Design narratives encourage users to take up particular subject positions and imagine technology in particular ways. PowerPoint, for example, address-es a "presenter," even though individual users are free to occupy or reject the assignment. But the communicative affordances of media systems can literally build in certain possibilities for interaction while preventing others forms of interaction from taking place. Until recently, Twitter afforded the communi-cation of messages no longer than 140 characters. Periscope speaks of adding "features," not fiddling with affordances, but updates constantly reshape the communicative environment. Building in mechanisms to flag objectionable content, block users, follow favorite broadcasters, save and replay broadcasts after they have ended—these technical solutions are important components of the design narrative that profoundly impact the broadcast experience and the meaning of "going live."

Finally, design narratives include the textual conditions of production, the actual content created. In Peters's communication triangle, this is the utter-ance of the witnessing agent, the audiovisual text created and streamed live with Periscope and the Periscope audience. Before the introduction of "Peri-scope Producer" expanded the range of production controls, Periscope users had relatively few options for editing or managing the audiovisual flow of the live-stream. After tapping "Go Live," broadcasters had access to two basic types of edits: "Swap Camera" and "End Broadcast." "Swap Camera" toggles between the front-facing and rear-facing cameras, allowing broadcasters to choose whether to show the world what they see or to show themselves to the world. But it is through the second function, "End Broadcast," that we en-counter a radical departure from broadcast conventions. When I found myself the only viewer tuned in to a Michigan middle-schooler's live-stream, I was unable to hide out on the other side of the screen. "Ask me a question," the broadcaster instructed. I asked her about sports and television before I finally got some inkling of what the show was really about. "You're boring," she said, and promptly ended the broadcast. It was as if the broadcaster turned the channel on me, the audience, because I wasn't entertaining enough. In that instance, liveness was not a technical property but a condition of interaction that we created together (or failed to).

Conclusion

Unlike broadcast, which is bound to scheduling as the primary mode of organizing time, live-streaming is sporadic and unscheduled. The unscheduled nature of time is what allows anyone to "Go Live" whenever, for as long as they want. This makes it possible to witness mundane everyday events like Periscope's first big "hit," the January 2016 #DrummondPuddleWatch (a six-hour live feed of pedestrians navigating a treacherously large rain puddle in Newcastle, England), but also allows witnesses of atrocity, like Diamond Reynolds, to immediately say and show what is happening in ways that modern media institutions are unable to. Platforms like Twitter, Facebook, and Periscope have become crucial tools for activist groups like Black Lives Matter, which coordinates activity through the hashtag #BLM. Yet there is another hashtag that escapes mainstream media attention but, I would argue, is just as crucial for understanding the politics of visibility and the unscheduled nature of time: #BORED. In fact, expressions of boredom are one of the most frequent answers to the question "What are you seeing right now?" By attending to not only the monstrous but also the banal conditions of "mundane witness," we find new ways of expressing and perceiving liveness that transform the scopic regime of television.

Tubi TV
Free, Mobile, Moving Images: TV Everywhere
GERARD GOGGIN

App: Tubi TV
Developer: adRise Inc.
Release Date: June 14, 2014 (Current Version: 4.1.3)
Category: Entertainment
Price: Free
Platforms: iOS, Android, Xbox, Roku, Chromecast, Apple TV,
    Amazon Fire TV, Samsung Smart TV & Blu-Ray
Tags: TV, streaming, failure
Tagline: Movies. TV. Free.
Related Apps: Sparkle; Fullscreen, Netflix

Free Mobile TV?

Who doesn't want to watch whatever TV and video they want anywhere, anytime, and on any media device? Yet this enduring dream of media audiences has proven surprisingly tricky to realize. So why not appify TV? Surely there'd be an app for that—to finally cut the cord and make truly mobile TV a reality.

Cue the Tubi TV app. Billed as the "World's Largest Free Streaming TV Library," Tubi TV was launched in April 2014 by adRise Inc., a privately owned San Francisco–based company. The app opens with a colorful series of tiles, each with an image of a TV show or movie. Hover over the image to read a short description of the show, then tap to watch. Tubi TV's offerings are organized by a wide range of categories. Some are reasonably familiar, carried over from video and DVD store organization into familiar categories of subscription and on-demand television: TV Dramas, Comedy Movies, Action and Suspense, Movie Night, and Fright Night. But Tubi TV (hereafter, both "Tubi TV" and just "Tubi") livens things up with some rebadged categories like Extreme Adventures, Wild Things (Nature), Reality, Killer Kung Fu, and Grindhouse (a schlock collection). A big selling point is the app's efforts to ex-

Tubi TV menu and start screen

pand programming to cater to more diverse audiences, hence categories such
as Anime, Todo en Español, Black Cinema, and Chinese TV. Suffice to say, the
default content is mostly English-speaking, Western, North American shows.

Tubi TV takes slightly different forms across different platforms—
televisions, computer screens, tablets, phones, and gaming consoles—and operating systems. The basic drivers of usability and user simplicity as well as the
effort to capture consumer attention and financial yield remain broadly the
same. However, the experience of watching Tubi TV differs from a number of
competitors because advertisements interrupt the shows that are playing. As
one journalist put it, "Tubi TV is essentially bringing the traditional, terrestrial free-to-air model to internet TV" (Pressberg 2017).

In what follows, I discuss Tubi TV and its place in the world of apps as
well as contemporary understandings of television. I argue that Tubi is a rather
unremarkable app and, like many apps, it may or may not survive. However, it
is worth our attention because it stands in for a kind of perfect fantasy of appified mobile television. Tubi TV links together the key elements of free access,
abundance of choice, and the offer of unprecedented platform mobility. Combined, these affordances would make an ideal mobile television app, with real
scope for audience use and appropriation, along with market disruption. As

I suggest, though, the failure of Tubi TV to play this leading, transformative role sets up key questions about the future of television, especially as mobile media insert into the emergent ecologies of TV. Ultimately it's not a lack of digital affordances nor cultural interest that inhibit the dream of mobile free TV but a more complicated patchwork of historical trajectories and regional specificities in broadcasting, cable, state-run media, licensing, and the social nuances of transnational audiences (Athique 2016) that undermine the actualization of appified TV.

Becoming a TV App

Upon downloading the app, the new consumer of Tubi TV is likely struck by the similarity between its "look and feel" and that of Netflix, the well-known provider in many parts of the world. An obvious difference between the two is that Tubi promotes itself as a provider of third-party content, especially non-mainstream "alternative" content. By contrast, Netflix prominently displays and leads with its own content—"signature" series, like *Narcos* and *House of Cards*, as well as other desirable and well-known television programs and movies. Tubi TV has only a limited amount of high-budget, popular or niche content and in fact makes a virtue of necessity with its cheeky "Not on Netflix" category, showcasing documentaries and movies that the viewer would not expect to find on competing platforms as well as content from Asian television and cinemas that may not be well known to North American audiences.

The app further provides a digital counter that calculates the total of "Movie Hours Not on Netflix." As one review notes, "[Tubi TV] seems to be aware of the fact that most users will be turning to it as a second option after major services like Netflix, and rather than running from that, they've embraced it" (Lovely 2016). However, there is a revealing tension in this strategy of Tubi's reflexivity in "not being Netflix." Rather than offering high-profile original TV series and other content, Tubi is making an unlikely pitch to establish itself as a massive, accessible library of content—something that Netflix and other providers have not been able to deliver and that may be attractive to viewers amid the confusing proliferation of streaming TV options.

As noted, a key tenet of Tubi TV is that its free content is advertiser supported. Tubi claims that it routinely shows "four to five minutes of commercials per 30 minutes of content" (Pressberg 2017), which it presents as favorable to traditional TV, or at least as favorable to the amount of advertising on subscription TV, a point made by one viewer on Reddit's "Cordcutter" chat forum: "in the case of *Tubi*, yes they have commercials but at least it's free, unlike cable which was costing me an arm and a leg to watch programs that also had commercials" (Totodee 2017). However, as with the long history of advertising-supported television, there are a range of other viewer responses and reactions.

Consider, for instance, the publicity Tubi TV generated in May 2017 when it announced it had raised $20 million in capital to expend its "premium ad offering" (Tubi 2017). In doing so, the app cited Nielsen in support of its claim that "75% of the highly desirable target audience—which is Tubi's core audience—report that ads don't bother them as long as the content is free" (Tubi 2017). The Nielsen research to which Tubi refers was the 2017 "Millennials on Millennials" report, which presented audience data on "ad avoidance" (Nielsen 2017a). As it turns out, millennial attitudes toward ads are more complicated than Tubi suggests—Nielsen's report somewhat confusingly states that millennials (i.e., those under thirty-five in their study) are open to ads if the content is free but that a significant percentage would prefer "no ads under any circumstances" (Nielsen 2017b).

What is missing from this research is the kind of advertising that such consumers have in mind and are prepared to encounter. Much of YouTube and other mobile video and television content available through apps routinely serves up ads as prefatory to viewers watching the programs they desire. Similarly, much other social media is underpinned by advertising, although social platforms use a different, targeted model that relies on the massive amount of fine-grained data that sites like Facebook or Google hold and harvest. Tubi is simply not in this league. The advertising model within the app could easily lead to viewer rejection; the frequency of the ads, for example, risks disrupting the viewing experience and subsequently constrains Tubi's ability to build, scale, and blend into the mobile entertainment experience in the way other, more successful digital content and social media have managed to do.

In reflecting on my own experience with Tubi TV, there are clear benefits in easy access to more content, especially lively, diverse, alternative content that is free to the viewer (apart from the labor of watching ads). It is not, however, offering much content that other platforms are not already offering. Nor is the app doing anything particularly "new" or better in how it stores, retrieves, plays, and presents content. As yet, the shared cultural experience of Tubi TV is very minor, with little take-up so far, and its social function as an app, let alone media, is unclear. Mildly diverting, you might say, but otherwise pretty dull in a sea of mobile, online, social, digital, and other TV and video entertainment varieties. So why bother? Well, from a cultural and social stand-point, Tubi's at best mixed success tells us a fair bit about the place, stakes, and predicament of apps in the complex transformations of entertainment media. In particular, Tubi TV highlights the struggle that mobile TV apps face in finding their place in the emergent landscapes of digital media.

Hard to Break Free

The unique claim of Tubi TV lies neither in any innovative app design, tech-nology, database, algorithm, personalization, or customization, nor in the way that it marshals content to harvest user information and data. Nor is there something particular about mobility or location that Tubi inaugurates. The characteristics of the app and its underpinning technology are taken for grant-ed, in a mobile communicative, post-YouTube way. "Taken for granted" is the term Rich Ling uses to conceptualize mobile communication, meeting the requirements of what Émile Durkheim proposed as the essential nature of a "social fact" (Ling 2012). As an app, Tubi TV emerges from a moment where contemporary network television combines with features of mobile devices and communications (especially smartphones, tablets, and laptops) to forge a set of affordances that are increasingly mundane. Against this scene, Tubi makes two entwined claims. It promises to serve audiences abundant, free TV and, in doing so, to provide the convenience of a consolidated "one-stop shopping" experience and liberate users from excessive device and platform switching:

> Escape the claws of subscription fees. We are Tubi TV, the free Internet TV net-work, that's working on your behalf to unlock Hollywood so entertainment is free, without the burden of credit cards. We have the largest collection of premium and unique movies and TV shows. For free, forever. (www.facebook.com/tubitv/about)

Easier said than done, because the reality is that "free TV," especially "free Internet TV," is tough to deliver. Tubi TV's immediate competitors include the many providers who also wish to secure audiences interested in free con-tent. These are not just the people formerly known as television audiences, hearkening back to the pre-broadband internet, mobile, and digital TV world

(Goggin 2011b). Across many markets and through its evolution, much television has been advertiser supported rather than paid for by license fees, taxpayer subsidies, or subscription fees. With the overlapping changes in television in recent years, viewers are increasingly attracted to and demand, or at the very least perceive the following: (1) they can view any content they want, especially new shows; (2) they can view across devices, operating systems, platforms, and digital formats (Tryon 2013); (3) they can share access to viewing content with other members of their household, if not family and friends in other locations; and, (4) they can view content without having to directly pay for it themselves, whether via their internet or mobile bills, television subscription fees, or password sharing (Harper 2012).

The problem for viewers is that free TV via the internet has proven difficult to obtain at the level of quality or availability they desire. This is a vexed, widely shared problem globally, whether audiences are consuming television via YouTube or other video-sharing sites, via catch-up sites or free-to-air terrestrial TV with internet and mobile extensions, via downloading and informal media economies, via new-generation broadband networks, or via the TV apps of manufacturers such as Samsung. Alongside these fraught conduits of free TV, there additionally exists a plethora of reasonably reliable and affordable (for some) subscription or digital content services for rental or purchase (Landau 2016).

A key response to these new ecologies of online and video-on-demand services lies in the redoubled effort of the formerly regnant broadcast TV networks, cable TV, and subscription services to offer their content via online and mobile technologies as well. Much of this content is supported by advertising. Across many markets, a viewer can watch catch-up television punctuated by advertisements, just as its programs viewed on television receivers would be. Intermingled with these online television types are two especially popular platforms: YouTube and Facebook, both leaders in video and widely accessed via mobile and tablet devices (Goggin 2014). Along with Google and Apple, these "global media giants" (Birkinbine, Gomez, and Wasko 2016) have increasingly titan-like counterparts in other regions, such as the Chinese companies Baidu, Tudou, and Youku (Keane 2015; Yu 2017).

In sum, each market will have their own distinctive structures, audiences, and online and mobile TV ecologies (Goggin and McLelland 2017; Pertierra and Turner 2013). Globally, however, it is fair to say that no television app rules all entertainment, except for YouTube (Burgess and Green 2013), Facebook, and their various counterparts in Asian countries, which offer new kinds of platforms for collecting, distributing, sharing, and viewing video entertainment. Tubi TV exemplifies an app's effort to step into this long, complicated global evolution of mobile television (Goggin 2012), and as such, it is emblematic of why the appification of television, at least in this case, is something of an epic fail.

The Instructive "Failure" of TV Apps: What Tubi Tells Us

Tubi TV is a case in point of why the appification of television has been slow to materialize. Tubi taps into a powerful cultural imaginary of television that struggles to be realized, due to the varied state, regulatory, economic, and cultural contexts of established television markets and audiences. Since its inception, a recurring theme in digital and internet TV is the notion that a cornucopia will be finally available to all, an infinite realm of historical programs, as well as the best new material. This was the promise of TiVo, the digital video recording technology that attracted great excitement in the US in the early oughts. The company thrilled users by allowing them to gather all available content across channels and services and letting them choose what to view in a customized, personalized way (Carlson 2006). Yet TiVo use dissipated in the US and even more so in other markets, such as Australia (Meese et al. 2015). What displaced TiVo in the cultural imaginary of television plenitude was internet TV. First YouTube then Netflix and many others envisioned a service where viewers could watch nearly anything. Yet as it rolled out internationally, Netflix proved to have its definite limits, just as predecessor YouTube also did.

In the case of Tubi TV, it's not the technological affordances nor the cultural interest in circulating free, readily accessible television (the taken for grantedness of an unremarkable app) that are in question here so much as the historically and regionally specific trajectories of broadcasting, cable, state-run media, and transnational licensing headaches, not to mention the specificities of transnational audiences (Athique 2016) that essentially stall the effective, profitable, "friction-less," appification of TV. This scenario is evident in the notable differences in how Tubi is accessed across different markets outside its North American base. As an Australian viewer, I can access Tubi TV (at the time of writing) via my laptop computer in different browsers but not via the mobile app. So there are obvious and discouraging platform-specific restrictions and affordances in Tubi's model, which contradicts their TV-anywhere promise. A major obstacle is that mobile apps, in the realm of audiovisual content, are not as global as they wish. Differences in developer and operating system specifications, app marketplaces, government oversight and restriction (e.g., China), and even Wi-Fi connections and download speed affect the accessibility and quality of moving-image media on apps. Further, Tubi does not enjoy the momentum that keen anticipation from international audiences has provided for Netflix, Amazon Prime, or other platforms, which have also been relatively slow to extend their services to markets outside the United States despite the interest.

Another difficulty app developers face in streamlining television into a global, mobile free-access medium is the challenge of intellectual property. In particular, Tubi TV and its confreres confront the difficulties and costs of licensing as well as the many challenges of creating, adapting, curating, and programming content and formats, especially for cross-media and multiple screens (Mayer 2011; Bury and Johnson 2015). To offer programming, Netflix and other streaming providers need to own the rights or negotiate with the rights' holders to be able to gain a license for the content. As streaming online video and television services have developed, it has become clear that no one provider rules them all. Each viewer, or group of viewers (i.e., a household), needs to decide which and how many streaming and other television services to subscribe to, pay for, or purchase, or otherwise how much effort, time, and risk they will invest in finding alternative ways to access content through informal downloads or sharing log-in credentials (D'heer and Courtois 2016; Fortunati, Deuze, and de Luca 2014).

Conclusion

If we were to demystify the marketing claims of Tubi TV, we might see them simply as an aggregator of online video and television content. They are in the business of making as many deals as possible with leading providers of video and television material in order to be as expansive as possible. Their potential "unique" selling point, or value proposition, lies in three things.

First, and foremost, they need as much content as possible. Or at least as much exciting and compelling, desirable and essential content as possible. There is currently a necessity for players in digital television like Tubi to create the perception of the "abundance of choice," as Netflix has successfully done——even if it is crystal clear to consumers that their choices of TV via online and mobile providers are, in fact, highly constrained. On this count, Tubi does not entirely succeed. They are a minor player competing for programming and rights in a tough (primarily) US market, and their efforts to expand internationally have not yielded much fruit, largely due to the rather unclear nature of their expertise and their product offering, calling into question their ability to achieve a global audience.

Second, Tubi TV premised its service on gaining advertising support to pay for the content and to generate revenue and profit. Given the many headwinds, power plays, and economic uncertainties in the three relevant and converging advertising markets—television advertising, online advertising, and mobile advertising (Andrews et al. 2016; Grewal et al. 2016; Quintas-Froufe and González-Neira 2015)—it seems unlikely that Tubi will ever be in a strong position to attract and build audiences, let alone commodify them sufficiently. Here Patrick Vonderau's thoughtful assessment of the politics of content aggregation is helpful: it looks like Tubi TV is courting the fate of other online aggregators, helping to devalue the potential new services and industry segments its model seeks to create (Vonderau 2015).

Third, in a somewhat halfhearted way, Tubi TV is seeking to establish itself as a one-stop shop for customers across many platforms, devices, and operating systems. Here the promise is that Tubi can be watched via its app or in-browser software, "over the top" of other systems. This is not a novel idea, having been achieved successfully by others, especially YouTube, but also by various subscription TV and internet TV providers. Yet if the documentation and user chat rooms noted earlier are any guide, it is more seamful than seamless at this time.

All of this sounds rather pessimistic, but there may still be time for Tubi TV to establish itself in the firmament of new television ecologics. While audiences, revenue, and scale elude it, Tubi is a resonant sign of the times. When mobile media finally arrived as a mass, popular cultural form, it did so via the app. The mobile phone carriers, the handset manufacturers, the mobile cinema, television, and video producers and industries were not able to create the "fourth screen" in any extensive, ubiquitous way—though the years of experimentation, appropriation, investment, and design were critical to its evolution. Rather, television and video entertainment apps have emerged at the intersection of 3G and 4G mobile networks, broadband internet, smartphones and tablets, new ecologies of television, new kinds of audiences, and the social (Holt and Sanson 2014; Lotz 2014). Lots of apps allow all kinds of television providers—advertising supported, public service media, subscription, new streaming services, aggregators, and so on—to make television available via apps. No one app in particular rules, and this is the situation that Tubi TV confronts. Rather, all the apps are interlaced and cross-referenced into a media ecology of television and internet (Cabral and Natividad 2016), with no business model, social form, and audience culture yet trumping them all.

Hillary 2016
Appified Politics in Campaign Apps
FENWICK MCKELVEY

App: Hillary 2016
Developer: Hillary for America
Release Date: July 2016
Category: Social Networking
Price: Free
Platform: iOS
Tags: politics, political engagement platforms, real-time
   control, mobility, affective labor
Tagline: "The only thing standing between Donald Trump
   and the presidency is us."
Related Apps: MiniVAN, Hustle, eCanvasser

Politics Appified

Searching for political apps was a window into a divided American electorate during the 2016 presidential campaign. Results for then-favored candidate, Hillary Clinton, included all sorts of unofficial apps curiously ranked higher than her own official app (though personalized results complicate any claims of objectivity). It was a bawdy mix mostly of arcade games and emoji makers. The carnivalesque results speak to a political reality of "fake news" that attracted international attention after Donald Trump's surprise win (at least for some observers).

Some apps were clear fakes. At the top of my list was an app from a company called OpenDNA. It "borrowed" the Hillary 2016 campaign logo for its unofficial Hillary Elect app that promised in all caps to be "THE ONLY APP YOU NEED TO KEEP UP TO DATE WITH THE HILLARY ELECTION CAMPAIGN!" The app appeared to be more interested in beta-testing OpenDNA's artificial intelligence than Clinton's campaign.

Other apps parodied, supported, and opposed her campaign. Some apps pitted candidates against each other in a virtual race or poker match, taking known formulas and swapping in Hillary or Donald. Others were openly hostile. A description for the *Never Hillary Clinton* game began: "My name is Walker Haber. I'm 13 years old and from Sioux Falls, South Dakota. Ever since my mom ran for Senate, I realized that everybody loves to pick on the candidate . . . . Similar to the game wack-a-mole, you tap on Hillary's face whenever you see her pop up in the window" (Haber 2016, n.p.). For the low price of ninety-nine cents, you can try to beat his high score (which was 96 points as of August 2016).

Donald Trump had apps too. You could "Punch the Trump" or receive inspirational Trumpian quotes all from the same iOS device. Trump had an official app also, America First. The app encouraged users to complete actions for the campaign, like sharing key messages, to win points and move up a public leaderboard. The app sent updates and let its users donate to the campaign as well. Like many political apps, America First was the product of parties, partisans, and political failures that led to its stable set of features (Kreiss 2016). Its developer, uCampaign, built the app "in 2014 in reaction to Mitt Romney's loss in 2012," which CEO Thomas Peters saw as a "missed opportunity to utilize [the Romney campaign's] smartphone app" (Bhaduri 2016).

In this chapter, I discuss another official campaign app, seen as innovative at the time but now tainted by a failed presidential run: Hillary Clinton's official app, Hillary 2016. Launched a week before Trump's app, critics considered Hillary 2016 more sophisticated than her rival's (McCormick 2016; Hinchliffe 2016). Like Trump's app, Hillary 2016 was designed to make participating in the campaign easier for its supporters by creating a friendly gamified way to plug into the campaign. Behind its cartoonish interface, though, the app connected users to a serious political machine bent on winning the campaign.

As scholars on digital campaigning look to engage ethnography and cultural studies to compensate for biases in the field that led to Trump's victory being unexpected for many (Kreiss, Barker, and Zenner 2017; Karpf 2017), my focus on Hillary 2016 continues an emerging interest in political technology as logistical media. As Ned Rossiter explains: "Logistical media—as technologies, infrastructures and software—coordinate, capture, and control the movement of people, finance, and things. Infrastructure makes worlds. Logistics govern them" (2016, 4–5). What is a *political* logistics? How does it coordinate the infrastructures—the smartphones and phone lines, databases and websites— that connect voters to parties and politicians? How did the Hillary 2016 app function as logistical media? What forms of control do logistical media enable for a campaign? Over whom? Though Clinton's campaign failed, her apps endure as important artifacts of political logistical media appified.

Your Own Little Campaign Office

Hillary 2016 opened to what Gizmodo called the "appropriately grim surroundings of [a player's] own 'campaign HQ,' a second-floor office with eight chairs, one sofa, and no staff" (Menegus 2016), or what the campaign described as a real Iowa office! Upgrading your grim HQ was a big part of the app. By its own admission, "Your office starts out pretty bare, but as you continue to play, you can spruce up your space" (Hillary for America 2016). Luckily, the app provided its users with an easy way to spruce up their virtual offices in exchange for their attention, data, and labor.

Users received and competed for challenges deemed important to the campaign through the app. Challenges ranged from checking into campaign events, sharing some campaign content, or completing a quiz. Users who completed the challenges received "stars," which bought items from the app's own store. In-store items such as posters, plants, and furniture could be placed in the app's bare Iowa office. Every action in the app also awarded points and, with enough of them, the status of being ranked in statewide and national leaderboards. Leaderboards rebooted every Sunday, so everyone had a chance to win. The app also provided real-world rewards, including the promise of an autograph from Clinton. Jenna Lowenstein, digital director of Hillary for America described its "endgame" for Hillary 2016 as "online, offline integration so if you take a lot of actions in the app you can win offline prizes like trips to events" (Politico 2016).

Launched in the summer of *Pokémon Go*, Hillary 2016 emerged as the latest example of gamification in politics. Gamification is when "gaming elements, like scoring points and competition, are extended into nongaming contexts like education, business, and on-line marketing" (Peters 2016, 143).

For their app, the Clinton campaign borrowed tactics from the "most addictive mobile apps and games," according to Lowenstein, and applied them to "things that are a little drier," like politics (Politico 2016). Just as users spruced up their drab virtual offices, the app spruced up mundane volunteering tasks with digital badges and leaderboards.

Making politics more like a game has its critics. Game designer Andrea Phillips worries that gamification further trivializes the vital sphere of politics, treating it as *just* a game (Phillips 2010). Game scholar Ian Bogost calls gamification "bullshit," marketing speak to sell new services (Bogost 2011a). These critics argue that gamification changes the meaning of a political campaign from deliberation to play, but that was precisely the point of the app. Hillary 2016's austere virtual office put a friendly face on the reality that the app was a remote sensor of the campaign (Andrejevic and Burdon 2015).

When asked to describe the motivations behind the app, Lowenstein suggested it was about widening the campaigning: "We set out at the beginning of the election to think about what parts of our support base have been under-served by traditional campaigning" (Politico 2016). Lowenstein echoes a trend in American politics to use mobile phones and apps to reach new or disconnected voters (Baldwin-Philippi 2015). Instead of connecting to voters through a webpage or a phone line, Hillary 2016 moved in on the smartphone, a device that most Americans check hourly (Newport 2015). Once installed, the app was one way for the campaign to coordinate supporters' preoccupations and desires, converting them into simple tasks, donations, and moments of attention. In doing so, the app continued the tighter integration of social media sharing and citizen campaigning into the logistical media of campaigning (Baldwin-Philippi 2015; Gibson 2015; Kreiss 2016).

Political Logistical Media: From Clipboards to Smartphones

Clinton's app was just another iteration of a long-standing desire for a political logistics operating on the ground, alongside the voter's everyday. In 1919 Samuel Peter Orth described voter profiling as a key part of Tammany Hall, the political machine created by the Democratic Party in New York state that lasted until the 1960s. Orth explained that precinct captains are "acquainted with every voter in his precinct and keeps track, as far as possible, of his affairs" and that "Tammany's machinery enables a house-to-house canvass to be made in one day" (Orth 1919, 90). Arguably, the canvasser's clipboard was the first mobile political app. Their clipboards were mobile media to record and report from the field, part of a well-oiled political machine of an earlier era.

Since Tammany Hall this logistical desire has led to many new political "applications." Beginning in the 1960s, direct mail allowed campaigns to reach voters directly. Computer databases let campaigns better focus on the most sympathetic voters, connecting on common, categorized issues (Johnson 2016). Improvements to canvassing sped up reports from the field. By the early 2000s, Palm Pilots, the smartphone's predecessor, digitized the entire ground game as canvassers pecked responses with a stylus and later uploaded results to the campaign's database (rVotes 2016). Hillary 2016 must be understood within this (incomplete) history of political logistical media. Gamification is the latest in a long series of innovations designed to improve how a campaign communicates with a voter and coordinates their actions and intentions.

Hillary 2016 functioned as logistical media, in part, through its coordination of data to and from voters. The app included a newsfeed in case you wanted some newspapers to fill your virtual office. Information flowed from the campaign to the app. Users who opened the app received tailored news without any traditional gatekeepers. The app was another example of what W. Lance Bennett and Jarol B. Manheim call a one-way flow. Rejecting the classic model of a two-step flow, where opinion leaders mediate political messages to the general public, "the communication process is aimed at the individual or at the direct messaging of assembled networks of like demographics" (2006, 215). Hillary 2016, for all its poster buying and promise of autographs, was an attempt to use emotion and play to distinguish a campaign channel from its competitors. This one-step flow typically suffers from fragmentation, with voters able to quickly change the channel (or in today's parlance, multitask).

However, gamification further tries to address this challenge of a one-step flow by encouraging social sharing. Rather than hearing about the app directly from the campaign, Hillary 2016 encouraged players to invite their friends to the campaign or send them a bumper sticker. Users became their own medium, personalizing campaign messages (Nielsen 2012). The app rewarded users who shared their phone's contact list with the campaign, a common ask of political apps made friendlier through its gamelike interface (Carson 2016).

Data also flowed directly back into Clinton's campaign as well. Hillary for America had a broad, campaign-wide unified privacy policy that collected information: "when you fill out a form, send us an email . . . or otherwise communicate with [the campaign]." The app collected any contact or identifying information and used that information to "carry out any other purpose for which the information was collected" and shared information with vendors, consultants, service providers and volunteers (Hillary for America 2016). In short, the app collected as much as it could, whenever it could, and shared it with anyone involved in the campaign.

Though there is little reporting on how the app's data helped the campaign (if it even did), it is not hard to imagine the leaderboards being used to identify high-energy supporters. It is certainly plausible that data from the app may have helped the campaign assess the effectiveness of its key messages.[1] Gamification became a way for supporters to self-rank their commitment, or what surveillance scholar David Lyon (2003) calls "social sorting," the techniques employed to identify, rank, and manage populations. Where Lyon investigates the governmental and corporate techniques to sort people, the Clinton campaign elected a more user-generated system. The very act of winning—being at the top of the list—marked an alignment between the game and political logistics of weighting voters' probable support.

Closer to the Skin: Affect and Political Apps

In addition to the potential for new data flows, Hillary 2016 exemplifies the changing emotional relations between voters and parties. The app exemplifies how campaigns harness supporters' affective labor in the heat of the campaign, what Zizi Papacharissi categorizes, using the example of advertising, as "affective labor." It "engages potential consumers through the suggestion of a possible affective attachment they might develop for a product" (2015, 21). Campaigns do the same for politicians and voters. They are trying to tell a story, develop a relationship, and emote with the voter.

Hillary 2016 succeeded as a tool for political logistics according to Dan Carson, a former design lead for Obama's 2012 campaign, by making "new users feel welcome" and for "silly stuff like watering plants, and feeding the dog" that kept users coming back to the app (2016). These subtle emotional touches established a routine, making the app a friendly place. Users could wake up and check their phones for updates, with the app there to harness their anxieties and hopes into tangible campaign tasks. Just as Amazon has launched their own physical mobile buttons to facilitate the quick, almost instinctual, reordering of products such as laundry soap or diapers, Hillary 2016 tried to give supporters a button to push to feel involved. Ideally, voters would feel they were in a good place when using the app. The campaign often depended on their emotional labor. Tasks required a user to share a key message or reach out to a friend to talk about the election. The app's friendly design, in other words, enabled its users to project their emotions into the campaign's messages.

The app could also translate voters' emotional states into campaign donations (though critics found the "Donate" button hard to find). Voters could reach for their Hillary 2016 app and donate in the exuberance of one of Clinton's campaign speeches or in the despair of an unfavorable poll. Drawing on lessons from the 2012 Barack Obama campaign, the Clinton campaign knew mobile phones presented both a problem and solution for modern fundraising. The Obama campaign, for example, found that a quarter of its emails were opened on mobile devices but resulted in a completed donation less frequently. Working with another technology company, the campaign created the "Quick Donate" tool that allowed voters to donate in "one-click" or via text message. Users, without thinking about it, could translate their emotions into donations. These "one-click" donors gave three times as much money, four times as often when compared to conventional donation requests: "We had people giving $1,000 via text" (Teddy Goff, Obama 2012 Campaign digital director, qtd. in Green 2012). Without the hassle of completing a form or the distraction of the internet, "Quick Donate" feature created a new circuit from smartphones to bank statements to campaign finances. Hillary 2016 offered a similar phone-based donation solution, one that had the same potential to connect with the voter affectively, priming them to donate as an expression of their emotional state.

Campaigns Having Always Been Trying to Reach into Your Pockets

These features of the Hillary 2016 app point toward the widening reach of the campaign, or, rather, the campaign assemblage that brings together a heterogeneous mix of professionals, amateurs and volunteers (Nielsen 2012). Increasingly, campaigns depend on what I've called "political engagement platforms" to coordinate their parts in a common medium of communication (McKelvey and Piebiak 2016).

The Democratic Party uses NGP VAN as its principal political engagement platform. It is the result of the merger of the Voter Activation Network, a political database, and NGP software in 2010. Since then, NGP VAN has become a major part of the infrastructure running through the party. As the company explained in a recent product launch, "A lot of our products were initially designed as complex back-end [customer relations management systems], but what's happened within our products is that over the past couple years especially the user numbers have exploded and that's because campaigns are giving more access to users that they would never have given access to before." In reaction, NGP VAN explained, "We're focused on building VAN for more people, and we're focused on building VAN in a way where you can provide interactions through more channels and making it the primary engagement platform for the campaign" (NGP VAN Next 2014). Cutting through the hype, NGP VAN is telling its customers—the people who run Hillary Clinton's campaign, other Democrats, Canada's Liberal Party, and so on—that NGP VAN is trying to be the central media and information technology embedded throughout the campaign and necessary for as much of the routine tasks as possible. The app becomes the ingress of this participation.

In the future, Hillary 2016 might be a prototype for new, public-facing apps that are closely connected to a party's political engagement platform. Where NGP VAN presently includes a campaign-centric app called Mini-VAN, mostly used for mobile canvassing, future campaigns may use the gamification rhetoric seen in Hillary 2016 as a way to better enlist disconnected voters in its political logistics. Badges and prizes may provide a more accessible and intuitive means to align a voter's idle time with the campaign's strategic objectives.

In that case, new iterations of Hillary 2016 will be part of the changing function of the app as a logistical medium for politics. Where NGP VAN provides its own custom apps, another platform called NationBuilder has opened its application programming interface, or API, to developers so that they can program their own unique apps to connect to the platform. Some Nation-Builder apps duplicate the digital canvassing features of MiniVAN. Other, similar apps alter a campaign's communication flows. Hustle, which integrates into NationBuilder, replaces phone calls or emails with SMS text messages. Hustle's website advertises that campaigns can send messages faster than traditional phone calls and with a better response rate. It removes all the cultural baggage of phone calls and embraces the native language of the mobile phone: the text. Its website showcases rapid organizing via chat windows, all connected to a known list of supporters and a system for tracking responses (Miller 2016). In doing so, Hustle, like Hillary 2016, suggests that the campaign assemblage and its logistical media orientated around the smartphone.

Zuckerberg versus Kanye 2020

For all the prior discussion of political logistics, the 2016 election seems to have been won by an altogether different political machine: a mass media machine obsessed with Trump and his tweets. Trump did not need to coordinate a win through logistics alone (though he certainly benefitted from the support of digital advertisers optimizing campaign ad spending (Kreiss & McGregor, 2017)). He instead persuaded enough people to believe in his platform, broadcast online and on cable. Supporters could believe that Trump wanted the best for them and for America, even though it was not always clear what Trump himself cared about during the campaign. A skilled reality TV star, Trump's kayfabe—the code wrestlers obey when they pretend staged events are real—amplified a political double-speak where political extremism becomes indistinguishable from inside jokes (Klein 2017). Winks and innuendos circumvented those things you cannot say on TV to speak directly without being direct, playing to the misogynist and racist desires in some of his voters. The spectacle engulfed the attention and capacities of outsiders and opponents. No logistics, not even a gamified app that invited affective participation, could mute it.

The Trump and Clinton campaigns could be said to represent two sides of political machinery. Trump emphasized a political imagination; Clinton focused on a political reality. Politics aside, neither machine is right or wrong. A political imagination could be very different, perhaps progressive and inclusive. Sophisticated voter targeting, by contrast, can be creepy, an antecedent to the authoritarian state that Trump codes in terms of bans or winning. Nor are these machines mutually exclusive. Did the Obama campaign integrate political imagination and logistics? These questions must occupy studies of politics, aware of both the symbolic and technical expressions of political antagonism, and the app—as an evolution of mobile canvassing and voter participation— helps us address them. Now, what will make the next machine effective? That may be decided in the next US election when Kanye West's dark and twisted fantasy runs against Mark Zuckerberg's meaningful social interactions.

Notes

1. The Liberal Party of Canada used a similar technique when they created a mobile app for their election "platform," a document that displayed their policy statements. As users read the app and lingered over certain sections, they sent feedback to the campaign (Ryan 2016).

G Music/Sound

Shazam
The Blind Spots of Algorithmic Music Recognition and Recommendation
ELENA RAZLOGOVA

App: Shazam
Developer: Shazam Entertainment Ltd.[1]
Release date: 2002
Category: Music
Price: Free
Platform: iOS, Android, Blackberry OS, Windows Mobile,
    Windows, MacOS, watchOS
Tags: music discovery, serendipity, surveillance
Tagline: "Discover music, artists, videos & lyrics."
Related Apps: SoundHound, Musixmatch

Name That Tune

You're in a coffee shop or a pub, or next to your radio, and love the music coming through loudspeakers, but you can't recognize the track. You could ask a person at the bar or wait for the DJ to announce the song. These were some of the ways—along with cool friends, independent record stores, dingy concert venues, and (in the nineties) xeroxed fanzines and obscure email lists—a music junkie would discover new bands and records. Then came Shazam.

Shazam was the original music recognition app. When the software was first released in 2002, smartphones did not exist. Users dialed the number 2580, brought their cell phones as close as possible to the loudspeaker, then waited for a text message with the song title and artist to arrive. The iPhone version came out with Apple's App Store launch in 2008, to rave reviews. In late 2017 Shazam passed one billion downloads to iPhones, iPads, and Androids, and it has been incorporated into the software for Siri, the Apple iOS personal assistant (Thompson 2014). Shazam's nearest competitor, SoundHound, was nowhere near as popular nor as well-integrated into mobile systems.

Shazam worked, its name suggested, like magic. Listeners delighted in the way the app brought human and algorithmic ways of listening together. "I remember 10 years ago," user Mary H. noted in 2015, "holding my phone up in a loud restaurant and impressing my friends with the way Shazam could identify a song" (2015). As an app that aligns with what Jeremy Wade Morris and Evan Elkins call "mundane software," Shazam ventured "beyond the computer and into a vast range of everyday routines and activity" (2014). More specifically, the app shared new agency with the user. Music fans stumbled over new sounds in their daily lives. They chose the songs they liked. The app let users identify songs without risking derision from a know-it-all friend or snobby record store clerk. It enhanced the serendipity of a musical encounter.

Since 2002 Shazam's algorithm, database, and interface have all evolved. Shazam updates the Apple version of its app monthly—about average for an iOS top-seller (Kumura 2014). Shazam's song library grows daily, aiming to catch up with emergent music scenes. The company makes deals with other music services, its employees seek out new songs, and independent musicians submit their own (Thompson 2014). The app's software now includes music recommendations and tie-ins with consumer brands as well as push notifications that congratulate users on their last "shazam" ("Nice find!"). Label executives use Shazam's data to make production and marketing decisions. While these features make Shazam more commercial and profitable, they also constrain user agency and work against the accidental aspects of music discovery that users like Mary H. found so compelling.

Shazam's chronology echoes a key dilemma in music history: new styles that emerge informally in local music scenes often end up co-opted by the commercial pop mainstream (Straw 2004). This also plays out the ethical divide in the software industry between corporations that use proprietary algorithms and those open source developers who freely distribute their code (Kelty 2008), particularly as the recognition algorithm has transitioned from a relatively open music identification hack to a proprietary user surveillance infrastructure. Over the course of its history, Shazam's database gradually excised any reference to music outside corporate cloud services. The app's interface, which once delegated agency to listeners, now almost completely curates user experience. The Shazam experience that has evolved is designed less for listeners to discover new songs and more for music producers and brands to discover new markets.

The Algorithm

Shazam was a breakthrough algorithm for its time. When the app launched in 2002, the ability to identify music via audio listening was still mostly theoretical. The most widely used "music recognition" solutions analyzed and tagged audio files on personal computers as metadata or imported CDs into iTunes (Vara 2005). Conversely, Shazam recorded ambient noise on cell phones designed to filter out everything but the loudest vocal frequencies. When the founders tried to listen to these audio recordings, they could not recognize even their favorite songs (Wang 2006).

To overcome this limitation, software engineer and Shazam's cofounder Avery Wang created an algorithm that extracted just a few highest and lowest points from a given recording to create and store a "fingerprint" for every track. Every time any user tried to "tag" a song, the Shazam server received a fifteen-second recording, extracted its fingerprint, and matched it to a relevant record in the database (Wang 2003). Wang's algorithm accounted for the noise, the low recording quality, and the ever growing database. Contrary to the "black box" practice for commercial algorithms (Gillespie 2013), Wang detailed how Shazam's music recognition worked in a 2003 conference proceedings article.

Shazam's popularity spurred algorithmic creativity. In the six months after the iOS store launch in 2008, the app's reach extended from 20 million users to 35 million in sixty countries (Shazam 2009). Several software engineers drew on Wang's article to post online do-it-yourself manuals on music recognition (laplacian 2009; Jacobs 2010; Surdu 2011). Developers ported the algorithm to other programming languages and operating systems. Dan Ellis, professor of electrical engineering at Columbia University, collaborated with the Echo Nest, the creators of an open source music recommendation algorithm, and MusicBrainz, a community-maintained music metadata database, to launch Echoprint, an open alternative to Shazam, in 2011 (Ellis et al. 2011; Echo Nest 2011).

Open source programmers aimed to give musicians and fans control over the range of music available for recognition. "The catalog of current commercial music resolvers is unnecessarily limited," Echoprint's creators pointed out. "You cannot submit your band's music to Shazam." An open algorithm would free developers to create apps that allow users to "'convert' a local catalog of music into any streaming or music service" (Echo Nest 2011). Independent musicians worldwide would submit their tracks to a public "fingerprint" repository, linked to a database that crowdsourced song metadata. New apps would use the algorithm to link fans to a vaster and more diverse sonic universe.

Almost immediately, these DIY projects came under fire from patent owners. By 2009, Shazam's founders, Wang included, left the company, to be directed by investors who sold away the app's core patents. Dutch programmer Roy van Rijn, creator of an open Java version of Shazam, was the first to receive cease-and-desist patent infringement letters. Protests erupted on Reddit, Slashdot, and other hacker sites (van Rijn 2010). "Music fingerprinting," Echoprint cocreator Brian Whitman expressed from the protesters' point of view, "should be a service that every developer can rely on" (Andrews 2011). But in 2014 Spotify purchased Echo Nest and took Echoprint's database offline, making its algorithm unusable. Aggressive patent enforcement halted work on open alternatives to Shazam.

Instead, Shazam's proprietary algorithm evolved to give record companies and commercial brands new ways to probe users. Within months of the iOS store launch, Shazam delivered its first weekly popularity charts based on user data to record labels (Manjoo 2009). Today Shazam's dynamic "heat maps" trace the popularity of a new single over time, which producers use to detect emerging hits and sign artists (Thompson 2014). The app also ventured beyond music. At the time Echoprint went public, Shazam piloted tie-ins with TV commercials: a recognized tune would pop up on phones with a link to a vendor website (Arthur 2015). Now users can identify images with their phone camera wherever they see a Shazam icon and QR code and get augmented reality effects and special offers in return.

This feature creep transformed Shazam into an effective surveillance and marketing engine. For business executives, this is progress, but from the perspective of the open source community, it represents a loss. An open algorithm became proprietary. Users and musicians have little control over the diversity of sounds available for recognition, and the app's "bloatware" foists content and products on users. "Serendipity" is no longer Shazam's most appropriate keyword.

The Database

Shazam required a music fingerprint library when large-scale digital music collections did not yet exist. The four founders first tested the app with their favorite songs (Wang 2006). Then they scaled up with a barter deal. Ex-CEO Andrew Fisher described how Shazam employees, day and night, digitized vinyl records from a major music retailer, Entertainment UK (EUK). In return, Wang got to experiment with 1.6 million fingerprints (Albanesius 2012). Together with small music assets Shazam had acquired, the database had 1.8 million songs in 2003 (Wang 2003; 2006). But ripping vinyl and CDs even for personal use became legal in Britain only three years later (BBC 2006). When Shazam workers copied the entire EUK distribution catalog of records issued by major and indie labels, they may well have been breaking the law.

Shazam's initial database thus occupied the borderlands between the legal and illegal music spheres. In the first decade of the twenty-first century, many digital recordings circulated via alternative infrastructures: MP3 blogs of digitized antique vinyl records, online avant-garde audio archives like UbuWeb, and other venues outside of official distribution channels. We might call these recordings "gray music"—analogous to the "gray literature" of technical manuals or white papers, for example—produced and circulated outside official publishing channels (Gitelman 2014, 115–16). Informally distributed gray music, like Shazam's digitized vinyl, moved in and out of the "pirate" status, depending on the vagaries of perpetually revised intellectual property law.

Shazam still works to encompass some marginal music scenes. Its algorithm has improved in recognizing electronic and South Asian music. In 2013 it purchased Beatport's catalog of underground club music, to tag "dubstep, tech house, trance, and other specialist subgenres" (Fischer 2012; Sherburne 2013). The same year, the company partnered with Indian streaming service Saavn to identify "genres like Bhangra, Devotional, Ghazals, Carnatic and Indipop" (SN 2013). Self-producing musicians can now submit their own tracks through digital distribution companies, including CD Baby, DistroKid, and TuneCore.

At a closer look, however, Shazam seeks out not music cultures but paying customers. While a wealthy global diaspora may have led Shazam to incorporate South Asian music genres, most African music scenes remain outside of its purview. In 2014 smartphones took up only 18 percent of the African cellular market overall (Raile 2014), but 40 percent of the South African market (Macauley 2015). Mindful of this digital divide, Shazam partnered with the South African advertising company AdVine but not with streaming services in other African countries, such as Kenyan Mdundo, Nigerian iRocking, or Senegalese MusikiBi (Shazam 2016).

Shazam shuns uncommodifiable scenes, because it has become a key player in the corporate music sphere. It has partnered with major online stores and streaming services—iTunes, Google Play, Spotify, Deezer, and others—that operate according to strict licensing restrictions. These transnational cloud-based services supplant the gray music infrastructure, decimated by takedown software such as Google's Content ID or by FBI raids like the shutdown of Megaupload's cloud service that stored MP3 bloggers' libraries (Bennett 2012). Now Shazam identifies only licensed content. Eighty percent of its income comes from a cut from songs sold and streamed after they have been tagged in the app (Thompson 2014).

Shazam's willful blindness becomes especially misleading when record labels rely on the app to track emerging hits. As Lisa Gitelman points out about digital collections, "the most pervasive" omission is "the elimination of clues that point toward missing material" (2010, 32). With almost the entire African continent left out, Shazam's so-called world popularity charts are global in name only. Even in North America a significant share of independent musicians post tracks on SoundCloud or UbuWeb without signing up with digital distributors—that music does not make its way to Shazam.

The app's selective library, combined with user data, redraws the visible music landscape. Derek Thompson coined this process the "Shazam effect" in an influential 2014 *Atlantic* article: song-tagging data shows mass preference for a few top artists. Label executives welcome a view that shows only the songs they can turn into profit. Musicians, however, believe that reliance on data will make pop music more homogeneous. "If there is the opportunity to throw out the grey with the black, this is often done," sound artist Vicki Bennett argues, "narrowing down results to only mainstream or sponsored content" (2012). Instead of discovering gray music, Shazam makes it disappear.

The Interface

Shazam's interface grew simpler as mobile hardware improved. Its initial dial-up and text-message mode gave way to a Java applet in 2004. Users pressed dial keys on their cell phones to navigate a small screen (Wang 2006). An iconic blue pulsating "Listening" button arrived with the touch screen in the 2008 iPhone app. It gave users the pleasure of identifying a song with a single gesture. As the licensed music sphere expanded, Shazam's interface grew more complex. More buttons cropped up offering ways to stream or buy the identified tune. The Java app offered ringtones, while the first iPhone app offered iTunes downloads and YouTube videos. Now Shazam links to streaming services and displays music recommendations based on tagged songs. These links let fans explore music further but only in a commercial online environment.

The latest interface options push the commercial environment on music fans. Users can long-press the "Shazam" button to turn on automatic recognition of all songs in the vicinity. Push notifications announce new album releases based on user history. These options take away accidental discovery. TV advertising tie-ins, image recognition, and augmented reality effects seek to structure users' physical environment, expecting them to follow instructions on a billboard or in a TV announcement. The new and improved Shazam provides "richer, deeper brand experiences" (Macauley 2015), at least to its business and corporate partners.

But lay reviewers of the app see this transformation in another light. Of all comments currently available in the app store, none praise the app for its marketing features. One music fan used Siri as an interface to recognize songs without opening Shazam and its recommendations. "I don't see the point," he wrote in a typical comment (@disqus_I9SPCKW30J 2016). For another user, the feature-laden app was too slow: by the time it opened, any song he tried to tag was already over (blunden 2016). In 2016 Shazam released a Lite version of the app. Designed for areas with spotty internet access, it does nothing but identify songs and save tunes for later recognition. Within four days of release, a hacker bypassed the region limit of the Lite app and got thanks from music fans who downloaded his version (blunden 2016). These active users preferred the original pleasure of identifying ambient music with just one touch or, with Siri, one phrase.

Avant-garde musicians and their hard-core fans have measured Shazam against informal interfaces for discovering sounds in everyday life. Vicki Bennett remembers the clunky but seductive routes to gray music's "gift economy" before the commercial services arrived:

> A web search for an obscure artist heard on the radio will take you to a blog telling you about them, sharing out-of-print material, with tags linking to related areas. An adjacent column will have links to 25 other websites and radio stations with similar interests. There then follows a wonderful odyssey into hidden and often forgotten sonic worlds. (2012)

Many of these online music scenes have disappeared as college radio stations closed and MP3 blogs went offline, together with "blog rolls" of recommended links.

But other alternative sonic worlds and interfaces have become prominent. Urban Dictionary listed the term "unshazamable" (spelling varies) as early as 2009. It connoted rare and coveted material, as in "Wow! That song is unshazamable . . . can I copy that cd?" (B lav 2009) With only one "Shazammable show" on its grid, cult freeform station WFMU airs music ranging from early twentieth-century phonograph cylinders to local demo tapes (Greg from Bloomfield 2014). "Un-Shazam-able" DJ Jonathan Toubin draws thousands to his club dance parties, where he spins forgotten 45s from the 1950s and 1960s (Fischer 2012). Entire genres and styles of music can be unshazamable, as user @Wyntonav mocks on Twitter: "When you Shazam African music you get 'we didn't catch that" (2016). In this light, historical, grassroots, live, and non-Western gray music genres gain in cultural value, because Shazam fails to recognize and data-mine them, but they also become harder to discover and diffuse to new listeners as a result of their exclusion from the database.

Conclusion

By corporate standards, Shazam has become an app juggernaut. It has over a billion downloads, it has a number of partnerships with major online services, and it integrates with mobile operating systems. Its user data plays a key role in music production decisions. Yet several important groups have interpreted the app's story as decline: open source software developers; avant-garde musicians and DJs; active app users; as well as hardcore fans of experimental, out-of-print, and non-Western music. They discern a decay of the app's original purpose: loss of serendipity, user agency, and experimentation. For these observers, Shazam's chronology represents the ways corporate digital economies have encroached upon the spheres where open source software and gray music circulate.

Taken together, these appraisals suggest new directions for critical histories of "mundane software." They zero in on the exploratory stages of app development. Shazam may have been typical in requiring some version of music "piracy" at an early stage: Spotify incorporated MP3 files from the Pirate Bay website to grow its initial database and test its recommendation algorithms (Andy 2017).

Open music fingerprint experiments represent political ideas that are still resonant today (Kelty 2008) arguments for making all cultural works available for
discovery. An algorithm, or an app, can have different political and aesthetic
valences depending on people and institutions taking it up.

In calling for a vaster sonic universe, Shazam's critics shift the focus from
individual sonic works to the diverse spheres and scenes where they circulate (Gaonkar and Povinelli 2003). Scholars have analyzed digital music as a
"fluid" and "conflicted" commodity whose status, value, and meaning change
with formats, circulation, and use (Morris 2015; Sterne 2012). From the perspective of "unshazamable" scenes, "discoverability" means more today than a
work's commodity status or digital format (McKelvey 2016). A rare 45 spun at
a club has more in common with an MP3 posted for free on a blog than with
a vinyl record that is also available on iTunes. The MP3 on that blog also has
more in common with an MP3 traded for cash over Bluetooth in Mali (Shimkovitz 2012) than with a track streamed for free on Spotify.

Shazam's critics imagine a digital economy that would allow all musicians
to make a living wage, against maximum profits for the few promised by Shazam. This may seem a nostalgic, utopian idea. The cultural caché of "unsha-
zammable" genres echoes punk and indie rock musicians' disdain for selling
out and the memory of gift economies of the early online music era. In the
meantime, Shazam continues to hone its app to take advantage of marketing
partnerships and user data, focusing on top-selling acts at the expense of independent and nontraditional artists. The future of music apps looks pretty grim
from the point of view of the corporate cloud, with its visions of a "deeper
brand experience" (Macauley 2015).

But scholars of mundane software need not follow the corporate gaze.
Several appified services bring independent music closer to a sustainable digital economy. "Unshazamable" WFMU and UbuWeb offer an app. Bandcamp, a pay-what-you-can music store with its own app, lowers the barrier
to entry into the commercial music sphere for self-published artists. On
Bandcamp, MP3 blog *Awesome Tapes from Africa* releases music from the region, paying 50 percent of profits to African artists. "The circle of influence
and inspiration has been critical to music's evolution since humans began
singing songs," blog creator Brian Shimkovitz explains. "Musicians growing
up in Africa are aware of what is happening and deserve to be a part of it"
(2012). These "appified" scenes and services offer a sense of history and a
sense of place. And while all songs (best-selling or not) offer a sense of place

and history, it is services like Shazam that circulate those sounds in ways that either foster those connections for listeners or just play them out in the hope of gathering clicks and data.

Note

1. Apple bought Shazam in December 2017. As of March 2018, as *Appified* went to print, the app still worked as described in this article. The European Commission was reviewing the deal for possible ways the sale could "adversely affect competitive practices" (Gibbs 2018).

iMaschine 2
Music-Making Apps and Interface Aesthetics
VICTORIA SIMON

App: iMaschine 2
Developer: Native Instruments
Release Date: November 2015 (Current Version: 2.2.0)
Category: Music
Price: $9.99
Platform: iOS
Tags: productivity, update culture, music-making, work flow
Tagline: "Make Music Anywhere"
Related Apps: Traktor DJ, Figure

"Did you press a button and something awesome happen? That was a productive minute."
—Andrew Higgins, former head of Native Instruments Mobile
App Division

iMaschine 2 is a composition app developed by the Berlin-based music-technology company Native Instruments that enables users to quickly sketch out musical ideas or create full-length tracks. On iMaschine 2's website, potential customers are presented with a promotional video that depicts a typical day in the life of a Los Angeles–based music producer, "TK." The video follows TK as he creates an entire music track from start to finish. In the morning, he constructs a drum rhythm while leisurely sipping his cappuccino at a busy coffee shop. After this, he walks down the street absorbed in the musical arrangement process the app offers, during which he creates a melody. Toward the end of the day, TK eagerly returns home to export his project into Native Instruments' larger flagship music production suite, Maschine. This marks the final step in the work flow of the track's completion, where TK can now fine-tune the song he casually produced throughout the day on the app.

The video of TK conveys the message that iMaschine 2 is an efficient conduit for music-making that seamlessly permeates his mundane daily routines. The app provides a way to channel and capture TK's musical creativity regardless of where he happens to be at any given moment and even in the absence of his full studio. The features of the app's user interface (UI) highlighted in the video are those that expedite the process of creative musical practice and showcase the app's value in assisting the user toward reaching the ultimate goal: the production of professional-sounding electronic music that fits within the musical genre expectations of the app. Seen through the marketing video, iMaschine 2 provides users with the freedom to effortlessly produce music outside the confines of a recording studio and in the absence of any music gear. The app liberates the user from the tedium of musical production. It streamlines the musical process so that the blunders and mistakes a user might otherwise make in their desktop music software setup will not disrupt the natural flow and rhythm of their day. iMaschine 2 indeed provides a valuable service to many users. Yet the benefits of productivity, time management, and flexibility of mobile musical labor embodied in apps such as iMaschine 2, I argue, come at the cost of some significant processes of musical expressivity: creative agency, discovery through failure, and habituation.

Once the mainstay of professional recording studios, music software— digital tools designed for music production with a computer—first entered the homes of amateur and professional musicians, producers, and DJs with personal computers in the 1990s. As a result, Paul Théberge argues, music software forged a distinct relationship between the user, music technology, and the process of musical composition more broadly; making music, then, became increasingly intertwined with the practice of consumption (1997). Following Théberge, the question now is, If music technologies alter the structure of musical practice, place music-making into a new relationship with consumer

society, and prompt new ideas about what music is and can be, what if anything changes with the use of music-making apps?

Based on personal interviews conducted with the app developer and members of the app development team at Native Instruments, combined with an
analysis of the UI of the app itself, I show how the appification of music
production not only shifts the tools with which music is made but also promotes different values involved with music-making. The UI features that the
app's interface brings to the foreground produce norms surrounding musical
production and establish which types of musical functions and practices are
important. These features uphold the values of time management and achievement. In the app, creativity is defined by production, efficiency, and outcome
rather than by process. These values infiltrate the musical composition process
and aesthetic outcome as well as the marketing discourse that encircles the
app. I argue that while the app format streamlines the musical process, the
logic of productivity embodied in the app's UI trades creative agency for user
efficiency and makes musical expression subject to the culture of updates and
iterative design that pervades app ecosystems. Like many other music apps
on the market, the app is designed to foster the feeling of accomplishment
without the blunders and "mistakes" that would otherwise attend music production.

Instruments of Productivity

While iMaschine 2 can be found in the app store under the category of "Music" and the subcategory of "Producers & DJs," it is in fact a productivity tool.
Productivity is not only a category in the app store; it is also an ideology that
infiltrates many categories, including apps for creative expression. As Melissa
Gregg explains:

> Productivity apps facilitate the pleasure of time management, which is ultimately
> the pleasure of control. Productivity techniques deliver an enhanced relationship
> to time by focusing only on what is important, maximizing opportunities for
> optimal work "flow." In technology design, the ultimate user experience hinges on
> securing this state of uninhibited flow as quickly as possible. (2015c)

Gregg proceeds to define productivity apps as those that maximize the user's
output while minimizing the user's labor. They do so by directing the user's

attention to the most essential tasks and create the conditions for optimal efficiency. Similarly, in Matthew Fuller and Andrew Goffey's *Evil Media*, the authors argue that the techniques involved in app culture are those that center on "extracting the most value from minimal finger swiping movements" (2012, 55). These scholars interact with a wider Latourian conception of science and technology studies, where humans and technologies are part of hybrid sociotechnical assemblages, with humans delegating or translating activities to technology that were once the province of human labor and action (Akrich and Latour, 1992).

In iMaschine 2's interface, the ideology of productivity that Gregg describes and the techniques of efficiency that Fuller and Goffey articulate suffuse the app's functions, UI architecture, and time-saving operations. This process exemplifies what Lev Manovich calls the "deep remixability" of software, where previously distinct forms of media exchange properties and modes of expression—sharing working methods and fundamental techniques (2013, 46). Music apps such as iMaschine 2 are a hybrid between productivity tools and music software. As a result, they embody a particular form of musical practice unique to the app format that serves to broaden the app's user base by translating the techniques that are specific to musical software and hardware into the working methods, gestures, and UI conventions familiar to all users of the iOS platform.

Like many productivity apps, iMaschine 2's website highlights the app's efficient work flow. For example, in describing the app's "Arranger" feature, the website calls attention to the various ways users can build a track with just "the swipe of a finger" and the ease with which the user can create sound patterns "on the fly." These are the features that will ostensibly optimize time management and hasten the speed of musical production. In iMaschine2's marketing materials, musical practice is subsumed into the mundane rhetoric of time management and work flow. The app's website states, for instance, that the app "introduces color coding for fast visual organization of your projects" (iMaschine 2). The promotional materials negotiate the meaning of music making by painting the activity as any other form of work. It depicts the app as a technological assistant in the service of getting the task done (Gregg 2015c). The task is to efficiently corral the user toward the ultimate end goal: a complete musical track.

In the service of time management—what Gregg describes as preserving the "state of flow" (2015)—the app (or the developers behind the app) makes significant aesthetic decisions on behalf of users to eliminate mistakes. The app's work flow—that is, the formal sequence of activity it undertakes to produce a musical track—simplifies the process of making music and delegates some of the more tedious parts of the process to computers in an effort to reduce errors and increase productivity. However, while this could be seen as a welcome utility to overcome a creative impasse, or to avoid the time-consuming process otherwise involved in producing a full musical track, this technique of the interface also forgoes the creative benefits of musical errors. In this case, "error" is defined as touching a musical note outside of a specific key signature, or playing a rhythmic pattern that does not fit squarely within the quantized confines of a standard metric scheme. "Error" is going outside of the strictly defined musical conventions embedded within the app. The music is quarantined into a set of musical assumptions, constrained within a discrete, quantifiable rhythm and traditional key. For instance, the "Smart Play Keyboard," a feature of iMaschine 2, reduces the chance that the user will make a mistake by playing a note outside of a given key signature.

The emphasis on efficiency and reduction of errors fulfills many of the expectations furnished by the iOS Human Interface Guidelines online document that Apple provides for developers. Specifically, iOS places significant emphasis on balancing the user's need to feel in control of decision making in the app; at the same time, it prevents them from the possibility of making choices that will result in outcomes the user may not want. Apple's guidelines for developers clearly state, "The best apps find the correct balance between enabling users and avoiding unwanted outcomes" (Apple 2017). In explaining the philosophy behind the design of iMaschine 2, the developer Scott Hobbs says, "We want to let producers capture what they have in their mind as quickly as possible" (author interview, April 10, 2016). This statement speaks to the app's role in enabling users and augmenting their creative agency. By contrast, the app's role in automating creative decisions on behalf of efficiency is perhaps best encapsulated in how the website promotes the Smart Play Keyboard. It advertises "When you use Smart Play, the keyboard in iMaschine 2 maps to the notes of a scale, so every note you play sounds right—because it is" (iMaschine 2).

Beyond eliminating errors, iMaschine 2's interface further attempts to avoid "unwanted outcomes" by encoding musical qualities that the designer assumes the user will ultimately desire, thereby saving the user the time and labor of having to make and implement too many decisions. For example, the UI encodes musical assumptions through the use of default settings. As José van Dijck argues, "Defaults are not just technical but also ideological maneuverings; if changing a default takes effort, users are more likely to conform to the site's decision architecture" (2013, 32). This observation is especially applicable to software such as apps that are accompanied by the cultural expectation of an effortless and efficient UI, a broader trend in digital media that trades fidelity of sound and creative expression for convenience and efficiency (see, for example, the long history of audio technologies described in Jonathan Sterne's *Mp3: The Meaning of a Format* (2012).

In the case of iMaschine 2, when the user opens the app, there are preloaded samples for each instrument that are typical of sounds one would find in the genre of house music. This combined with the default tempo of 125 BPM (beats per minute) ensures that without much effort the user can produce a polished and complete track that conforms to the standard conventions of the genre. The use of preloaded samples and genre-specific default settings is not unique to iMaschine 2 but can be found in many music production apps. While desktop commercial music software also has default settings, apps for music production reduce the complexity of the interface, assume that users want to avoid errors in musical practice, and automate significant creative decisions in the name of efficiency.

iMaschine 2's focus on productivity, defined by the creation of a full-length musical track in a short amount of time, is a feature that stitches together the niche professional user and the larger novice demographic in order to expand the overall user base and increase downloads. For professional users, productivity is appealing because it allows them to maximize their musical output and expedite their work flow. For novice users, the app allows them to produce music without having to invest a significant amount of time learning the technical details that would otherwise be required to produce a piece of music in their preferred genre. Mobility is another highlighted feature of the app that is intended to maximize the user base. Its appeal is that it lets the user record musical ideas when inspiration strikes. Above all, the app accomplishes this by reducing the complexity of functions typically found in desktop music software, transforming them into efficiently managed steps that the user is directed to take, and increasing the speed with which users can create music that typifies the musical genre of the app.

Andrew Higgins, the former head of Native Instruments' Mobile App Division, states that the norm for user habits with apps is to spend a short amount of time engaged with the interface, with the expectation that the user will accomplish something through the app with little to no effort. Apps, according to Higgins, should be "immediate and fun" (author interview, March

5, 2016) and designed for users with short attention spans. The goal of apps, as he sees it, is to optimize the speed it takes for the user to produce a full-length professional-sounding musical track in a way that also evokes the feeling that the finished product is the result of the user's creative input. Productivity and the UI techniques of efficiency appeal to both user bases through the feeling of accomplishment.

Updates and the Patina of Efficiency

Productivity is also the justification for the culture of updates and iterative design that surrounds the app format. According to Apple's developer guidelines, the apps that they feature in the App Store are either new apps or those with significant updates. Since its initial release in November 2015, iMaschine 2 has been updated seven times, with various tweaks to its functionality and features. Although sold as a separate product, the app itself is one large update of the original iMaschine app, which can still be found in the App Store. While more empirical research is needed on how consistent updates shape user perceptions of an app, the culture of updates infiltrates the overall representation of the creative process in iMaschine 2, ostensibly empowering users while making their creativity within the app dependent on and subject to updates that are beyond their control.

In *Updating to Remain the Same*, Wendy Chun asserts, "Users have become creatures of the update." She states that "change for the sake of change," even if it is not a change for the better, characterizes the current cultural moment, wherein users are addicted to updates in part because they signal that their software is consistently being cared for (2016, 2). In the interviews with the developers at Native Instruments, there was an overall sentiment that users have an affective attachment to updates due to the expectations Apple sets for the user and the broader economy of the iOS platform through the normalizing of consistent updates.

Hobbs explains how this process becomes normalized as an expectation for app developers to "continuously deliver." He says, "As soon as you stop updating your app on a regular schedule, users sense this and they start to feel they, and the product, have been abandoned." He theorizes that the sense of abandonment users feel when they don't receive updates is the result of the set of expectations that Apple has built around the app market—namely, that "software is cheap and plentiful and it's constantly on the cutting edge." Furthermore, within the iOS platform, an app's visibility is predicated on a user rating system. Hobbs is accountable for keeping the app above four stars at all times, lest the app disappears from visibility in the app market. Within this system, Hobbs discloses that "users literally hold stars for ransom" if they do not receive a desired update to the features in the app's UI. Therefore, the update culture that is perpetuated through the algorithms of the iOS app store rating system is internalized as a design strategy that has implications for both the labor of the designers and for the users.

The normalization of updates in turn infiltrates the design, work flow, and architecture of the app, and makes musical practice subject to updates and new versions. Like other apps in the app store, the developers of iMaschine 2 have designed its interface so that it can be updated and changed on a regular basis while maintaining consistent overall work flow. As the envisioned user expects their app to constantly be on the cutting edge and efficient, developers use the delivery mechanism of progressive disclosure, delivering innovations bit by bit. This user experience ostensibly makes people feel that they are efficient and that the app is responsive to their needs.

Progressive disclosure is a design strategy used by many app developers stemming from the belief that the UI of apps should not overwhelm the user with lots of features at once; rather, they should have the most common features up front and disclose less central features and functions progressively in a sequence. This ensures novice users will not be overwhelmed by multiple functions, while at the same time satisfying more advanced users who expect more complex features, serving both Apple and Native Instruments' economic goals of appealing to both niche and broad audiences. In an interview with Peter Siciliano, vice president of software development at Native Instruments, explained that because updates and minor interface changes are the norm in the app world—also known as "iterative design"—people expect a progressive unveiling of features (author interview, March 30, 2016). With apps, the user expectations and design philosophy merge in the very process of versioning. In the process of developing apps, the desired result is never a finished product.

In order to fulfill the expectation for updates, developers make changes to apps periodically, each justified as making the app more "efficient." More often than not, the updates arrive in a banal form, from support for different ways to browse for sounds, to support for Bluetooth headphones, or they can be more substantial, such as adding a new audio effect to the functionality of the app. The updates provide the discursive patina of efficiency and productivity. This has repercussions for the music production process and for the subjectivity of those who use music production apps.

Chun argues that habit is linked to creativity. When we are habituated into an activity through repetition, "it [also] gives us the time and space needed to be truly creative, for without habit there could be no thinking, no creativity, and no freedom" (2016, 6). As updates change and disrupt the UI of music apps, they have the potential to prevent the user from habituation into the interface. While some users perhaps crave updates as a general process, this does not mean they are habituated into the interface features that get updated. The purportedly sought-after updates that the user desires in order to feel that the app is responsive may in some ways undermine habits for making music.

For example, while a musician might learn and become proficient with a traditional musical instrument such as a piano through years of repetition, were the piano to consistently undergo tweaks and changes to its functionality, it would be significantly more difficult for playing it to become second nature. Writing about the traditional process of learning to play an instrument, Richard Sennet (2008) notes that through repetition, these "hard won movements" become ever more deeply ingrained in the body and form muscle memory. Apps such as iMaschine 2 are not intended for this form of musical mastery through repetition of bodily techniques, since they are designed for users to immediately produce music on the go. At the same time, periodic updates to the interface pose a tension with this form of musical entrainment and have the potential to erode the habit necessary for the familiarity and fluidity with a musical instrument.

In apps such as iMaschine 2, updates to the interface are advertised as a precondition for creativity rather than a disruption to the habituation required for mastering a musical instrument. According to iMaschine 2's website, "All new features give you a fresh burst of inspiration." Updates, therefore, have the potential to become a source of dependency for creativity. The marketing of updates conveys the message that they enhance the user's flow of musical expression and production, freeing the user from a creative roadblock, and facilitating the efficient accomplishment of the desired outcome: a full-length musical track. With the justification of higher efficiency through "fresh bursts of inspiration," in the form of additional creative features or as minor tweaks to the interface and technical improvements, users are meant to feel that the developers are responding to their needs and not abandoning them, turning them into more powerful and efficient musicians who are up to date on the technological innovations of the music technology market. However, constant updates also have the potential to prevent users from generating the familiarity and rhythm necessary for achieving the ease of musical creativity and output the app promises.

Conclusion

In the case of iMaschine 2, the appification of musical practice leaves open the question of what is lost and gained as musical practice is infused with the norms of UI design for apps. As I have shown, the logic of productivity and efficiency, combined with the normalization of updates, has influenced what developers can do with their products and subsequently mediate the user's creative process. Native Instruments markets iMaschine 2 as a tool of empowerment, freeing the professional user from the confines of the isolated recording studio, enabling the freedom and flexibility of mobile musical labor, and enhancing the user's overall musical productivity. Indeed, music apps such as iMaschine 2 provide the user with many benefits. The app's focus on efficiency through the prevention of mistakes and errors, as well as its default settings, opens up the potential for cultural participation to a wide range of people, those who own and know how to operate an iPhone but may not have musical experience. It does so by making it easy for users to produce a complete track that conforms to the standard conventions of a narrow range of genres.

At the same time, the conventions of UI create a norm of musical practice that values accomplishment and productivity over musical values such as expressivity, experimentation, or improvisation. The constraints of the interface disavow the forms of mistakes and errors that frequently lead to musical innovation, while consistent updates potentially prevent the user from habituation. In the process, the app's UI, as well as the way it is marketed, promotes the idea that musical creativity should be measured by production, efficiency, and outcome, a notion that is consistent with the ideology of app culture but one that remains at odds with notions of creative agency, discovery through failure, habituation and other practices that are valuable to making music.

Here: Active Listening System
Sound Technologies and the
Personalization of Listening

MACK HAGOOD

App: Here: Active Listening System
Developer: Doppler Labs Inc.
Release Date: January 2016 (Current Version: 1.5.36)
Category: Music
Price: Free (with purchase of $199 or $299 earbuds)
Platforms: iOS, Android
Tags: hearables, noise cancellation, augmented reality,
   orphic media
Tagline: "Transform the way you hear the world."
Related Apps: RJDJ; Inception–The App, Muzik Connect,
   White Noise

One morning in 2016, I drove through my university's campus and everything sounded wrong. As I pulled into a parking spot in front of the building where I work, the landscape was its familiar self but the soundscape was not. When I killed the car engine, the jingle of my keys repeated itself. When I got out of the car, slammed the door shut, and turned toward the building, I heard the *thunk* of the car door once, twice, three times in rapid succession. Weirdly, these sonic afterimages of "car door" remained in front of me even though my back was now facing the car. I then heard my own footsteps echo beneath me as I entered the building and walked toward my office. My entire world sounded like it had been remixed by dub reggae producer Lee "Scratch" Perry, drenched in a combination of reverb and delay. The mismatch between my visual and aural surroundings made me feel slightly ill.

But I wasn't ill. I was using a new app-enabled technology called Here: Active Listening, the product of a successful Kickstarter campaign. According to the website of its creator, Doppler Labs Inc., Here consists of "two wireless [ear]buds and a smartphone app [to] adjust the volume, EQ, and effects of the world around you" (2015). In other words, as I opened the door to my office to begin my day's work, Here was enabling me to apply the techniques and technologies of a recording studio engineer to the sonic space I entered. In an aural application of two buzzed-about trends in digital technology—"augmented reality" and "wearables"—Here relocates echo, reverb, distortion, and filtration from sites of musical production to the very locus of audition, screwed securely into the user's own ear holes.

"Augmented reality" (AR) is a type of virtual reality (VR) technology that integrates the virtual into the user's phenomenological experience of the "real world" rather than enveloping the user in a virtual environment; "wearables," or wearable technology, refers to digitally enabled personal accessories and clothing. The former offers the possibility of better integrating information and computation into everyday life—for example, by visually overlaying relevant data on the user's surroundings; the latter promises to integrate the user's everyday activities, whereabouts, and bodily function into informatic flows for new kinds of self-awareness and control. The two come together in technologies such as Google Glass, the wearable computer and optical head-mounted display that allows users to see overlaid search results and to record their surroundings, among other things. As Google Glass's rather notorious "glasses" form reinforces, AR is usually understood by the public as visual technology. However, Glass's primary input method is actually voice command and, in fact, the *sonic* applications of AR have long been considered and prototyped by designers.

With the arrival of the Here app and earbuds, augmented auditory reality is truly coming into fruition, prompting us to consider the changing relations between sound, the digital, and everyday life. As discussed in the introduction to this book, "software forges modalities of experience—sensoriums through which the world is made and known" (Fuller 2003, 63). Listening through dub-like effects radically reshaped my sonic world for a day, creating a disorienting schism between my sense of hearing and my other senses—one that I was in no hurry to experience again. However, it is actually Here's subtler and more mundane forms of sonic augmentation—like targeted noise-cancellation—that have the greatest potential to reorient our relationships to others and the world.

By 2018, Doppler Labs was out of business, the victim of strategic errors and poor battery life. Yet, as a *Wired* reporter put it, "Doppler's core idea—that in-ear computers are the next frontier—has permeated the industry." By inserting an "appified" logic of choice between our ears and the sounds around us, products like Here will potentially change not only what we hear but also how we listen. Taking the critical perspective encouraged in this collection, this study of Here: Active Listening examines the difference between experiencing the spaces we move through daily and interfacing with them through a mobile app.

Hearables and the Silicon Sonic Turn

As a sonic "wearable," Here is an excellent example of a new product category known by a terrible portmanteau: *hearables*. Coined in 2014 by engineer, entrepreneur, and self-proclaimed "Wireless Evangelist" Nick Hunn, the term "hearables" is meant to signify the earbud's evolution—thanks to advancements in processor miniaturization, battery efficiency, and wireless protocols—into a "smart," wearable, human-computer interface (HCI). Though popular, fitness wristbands provide limited interactivity between apps and users, while tiny-screened smart watches and their apps have so far seemed to garner a collective "meh" from all but the hard-core geek, fitness enthusiast, or Apple fashionista. Two-way earbuds and headphones, on the other hand, are ubiquitous among smartphone users, while innovations in machine learning, speech recognition, and voice synthesis are opening new possibilities in verbal human-machine communication. Voice-based HCI also offers the added advantages of hands-free, no-screen, mobile computing, allowing users to keep their eyes on their surroundings (though their ears and attention might be elsewhere). According to Hunn, "We'll be spending over $5 billion on Hearables" by 2018. "The ear is the new wrist," he enthuses (2014).

Hearables are not a stand-alone phenomenon but part of a larger silicon sonic turn that seems to be taking place in the cultural imagination and everyday use of computers. Consider the proliferation of devices, services, and popular representations such as the introduction of Siri in late 2011; the starring role of the talking OS Samantha in *Her* in 2013; the rise of Beats headphones as a status brand; Amazon's introduction of its "smart" listening/speaking Echo speaker in late 2015; and the proliferation of virtual assistants like Microsoft's Cortana, Samsung's Bixby, and Google Assistant. In a sense, information technology seems to be returning to its telephonic origins at AT&T, but thanks to better natural language processing and text-to-speech algorithms, the voice on the other end of the line is finally the talking computer of early sci-fi dreams.

The silicon sonic turn has implications not only for HCI but also for the very notion of apps themselves: if the essence of appification is the creation of small applications with simple, discrete, mundane functions (Morris and Elkins 2015), these applications may be increasingly subsumed as aspects of, for example, Siri's "personality" and capabilities. In other words, if many of our interactions with mobile and cloud-based computers turn from screens to conversations, many "apps" may simply become invisible "skills" that we add to Siri's toolkit. The fantasy of always-on, available-everywhere, friction-free, mobile computing truly comes into fruition when the phone and its software are hidden in our bag or pocket and the computer simply becomes a voice in our head or home.

An Interface with the Audible Environment

I purchased Here not because I was interested in the sonic future of HCI but because I wanted to hear the future of computing as an interface with the audible world. Here is both an expansion and a refinement of a new media practice already embedded in the ostensibly "dumb" technology of noise-canceling headphones. Previously I have argued that noise-canceling headphones fit Lev Manovich's definition of "new media" as *devices that use an interface to access a database of content* (2001); the power button on these headphones functions as an on-off interface with the aural world, turning it into a database of content that can be selectively accessed (Hagood 2012). Active noise-canceling headphones go beyond the sound-masking capabilities of the Walkman and other similar technologies in that they mediatize the sonic world in order to cancel it out: using tiny microphones and signal processing, they produce an out-of-phase copy of the audible environment in an attempt to negate its phenomenological existence.

With Here, this interface moves well beyond the on-off binary to offer more sophisticated customization. Does a concert need a little bass boost? Can do. Want to eliminate the sound of your roommate talking on the phone so you can study? No problem. You can even turn up the volume on a conversation in a noisy restaurant. The noise-canceling Here earbuds completely fill the ear canal so that (theoretically) sound comes to the user electronically instead of through the air, allowing for control via a Bluetooth-connected smartphone. The mobile app's clean, white screens and simple sliders seem to suggest that controlling the sounds in the air around us can—and should—be as easy as setting your home air temperature with a Nest wireless thermostat.

What is the difference between experiencing the spaces we inhabit and interfacing with them in this way? To begin answering this question, we first have to define "interface." An interface is a technology that allows us to access and manipulate information: a keyboard, touchscreen, graphical user interface, microphone, speaker, or software application (Manovich 2001). Good interfaces are "transparent" and "user-friendly," while bad interfaces are "opaque" and "non-user-friendly" (Norman 2004). Many have pointed out the problems in this commonsense definition, however. The more "transparent" an interface seems, the more layers of code and complexity its programmers have successfully hidden; the more "user-friendly" an app is, the more choices and complexity its programmers have removed from the user (Turkle 1995). The more an interface seems to be a simple window on the world, the more opaque it really is, allowing for only certain kinds of information and certain kinds of user experiences.

In other words, an interface is not so much a thing as it is a process in which some aspect of the world becomes sensible and controllable through standardized representations and operations. An interface seems transparent or natural when, as Lisa Gitelman writes, people "forget that they are heeding norms and standards at all" (2008, 7). But when we do pay attention to these standards, keeping in mind that programmers are fallible individuals with specific backgrounds and biases, we can see how the norms of the interface are ideological as well as informational (Nakamura 2009).

In order to critically examine how Here works as an ideological interface with the sonic world, we might think in terms of a double mediation. First, our sensory experiences are already mediated—both by our sense organs and by our sense-making, which is profoundly shaped by our cultural conditioning and habits. Which sounds we find pleasant, beautiful, annoying, ugly, or fearsome is influenced, in part, by ideology. There is a second mediation at work when we put in the Here earbuds and use the app, because an interface always presents a simplified version of reality, a reduction of the real, constructed to better facilitate computer-based control. "Ideology gets *modeled* in software" precisely in this reduction of the vast, messy, unrepresentable reality of things to a simplified, controllable interface (Galloway 2012, 52). Therefore, a mobile sound app such as Here is an interface in which the ideological listening of users is amplified, filtered, muted, or mixed with the ideological listening of app developers. The interface and its user unconsciously (and consciously as well, through design) privilege some kinds of sounds, music, and even voices over others, turning up the volume on some while "fixing," masking, or canceling out others—reducing the messy diversity of the aural world.

Perhaps the most fundamental ideology embedded in Here is that our listening should always be "in control." The Here app consists of three main screens, all organized around the control of one's sonic surroundings: "Volume" (noise cancellation and amplification), "Live EQ" (graphic equalization and effects), and "Filters." The "Filters" screen offers preset combinations of Here's volume, EQ, and effects capabilities, divided into two main sections: "Tune In" and "Tune Out."[1] Offering presets such as "60s Sizzle" and "Dirty South," "Tune In" allows concert and club goers to activate combinations of aesthetic choices common to particular genres of popular music—in effect second-guessing and overriding the choices of the musicians, DJs, and sound engineers in live settings. "Tune Out," on the other hand, is organized around the control of acoustic spaces the user might not want to hear at all: "Airplane," "Train," "Large office," "City," and the like. These filters use varying combinations of noise cancellation and masking ("white noise") to facilitate "tuning out" one's surroundings.

The clean design and clear parameters of the app seem to suggest that whether we seek pleasure, calm, or concentration, the simple solution lies in manipulating the sound around us. The app renders the audible world as a database of sounds, sorted into two categories: worth hearing and not worth hearing. Sounds worth hearing can always be improved for individual taste via the "Tune In" filters; sounds not worth hearing are not merely ignored but also canceled out or masked with white noise. As an interface, Here prioritizes individual choice above hearing the same thing as others or hearing the unexpected by chance; rather than changing our reactions to sounds, we are encouraged to change the sounds themselves.

As for the earbuds, the first-generation, Kickstarter-funded model I used was really a $199 pair of digitally enhanced earplugs, offering no recorded music playback or telephony. When I used Here in my office, it struck me as little more than a proof-of-concept device—a gimmicky luxury item for "club kids," live music fans, and "early adopter" technophiles. In contrast, the $299 second-generation product, called Here One, sounds like the hearable future. As it allows for music streaming, podcast listening, phone calls, and the use of Siri and Google Now, Here One promises to fulfill the hearable promise of true augmented reality, merging the "virtual" and "real" in one app. Here One is an interface for managing mediated and ambient sound as one continuous, controllable flow of information, whether it emanates from a distant server or the air around you.

Orphic Mediation

While it may be tempting to characterize Here's digital virtualization of listening as unprecedented, as hinted at above, sound has long played underappreciated roles in the history of virtual and augmented realities. In a 1997 survey of augmented reality technologies and applications, engineer and AR pioneer Ronald T. Azuma defined AR as consisting of three characteristics:

(1) Combines real and virtual
(2) Interactive in real time
(3) Registered in 3-D (Azuma 1997, 356)

While noting the primacy typically granted to vision in AR, Azuma also suggested:

> AR could be extended to include sound. The user would wear headphones equipped with microphones on the outside. The headphones would add synthetic, directional 3-D sound, while the external microphones would detect incoming sounds from the environment. This would give the system a chance to mask or cover up selected real sounds from the environment by generating a masking signal that exactly canceled the incoming real sound. While this would not be easy to do, it might be possible. (361)

In actuality, by 1997 active noise cancellation was far more than a possibility, having long been in use in aviation headsets—and Bose QuietComfort noise-canceling headphones were soon to make their debut on the ears of business travelers. Azuma's assertion is nevertheless helpful, as it clarifies how the now mundane and ubiquitous noise-canceling headphone *is* a weak form of augmented reality: it combines the real (ambient sound) and the virtual (its electronic reproduction) in real time to change the user's perception of the three-dimensional soundscape.

Both QuietComfort and Here alike *fabricate a desired sense of space through sound* in what I refer to as "orphic mediation." Like Orpheus, who defeated the mind-capturing sound of the Sirens with the musical sound of his lyre, orphic media users fight sound with sound to control the space they inhabit, rendering it suitable to their aesthetic, emotional, and cognitive demands (Hagood, 2019). Orphic media need not be digital—an old vinyl LP of nature sounds can transform a bedroom for purposes of sleep or concentration, for example. In fact, it could be argued that the acoustic construction of virtual realities long predates digital VR—think of the audiophile of the 1950s or '60s, entering a virtual concert hall in his living room via his "hi-fi" system (Keightley 1996; Doyle 2005). It has also been argued that the telephone constructs "a third, virtual place" between interlocutors and "probably represents the first virtual world in the contemporary sense of the term" (Boellstorff 2008, 36). A final example of predigital "analog virtuality" through sound technology might be the "imagined community" of listeners that one becomes a part of through listening to AM or FM radio programs. We can't perceive this community directly, yet its imagined presence gives radio listening a feeling of "liveness" not found in listening to our own collection of recorded music (Douglas 2004). In fact, the cultural logics of a mobile sound app like Here can be traced back not only to noise-canceling headphones and the Walkman but also to the comfy bubble of sonic choice that an automobile and its car radio provide (Bijsterveld et al. 2013).

Digital Music Production: A Harbinger of Sonic AR

Nevertheless, the advent of digital technology has revolutionized the production and transformation of sounds and sonic spaces, particularly in the world of music. Many of the studio engineer's tools—filtration, reverb, echo, computer synthesis, and other technologies—moved from the studio to the stage, embedded in musicians' digital keyboards, effects pedals, and DJ mixers themselves. By the 1990s, scholars of music and culture were grappling with the fact that music-making was now "driven not so much by the vibrations of membranes, chords, hard surfaces, or molecules of air but rather primarily by the manipulation of electrical impulses"; these scholars realized that the logics and practices embedded in these digital musical technologies required cultural, political, and economic analysis (Greene 2005, 1). One major consequence of the digital was to inject a kind of consumption practice into the heart of music production—for example, a keyboardist might record a melody as a digital sequence and then spend the next half hour scrolling through digital menus, listening to her notes played back in the guise of various prerecorded or programmed virtual instruments and effects. What had changed was more than the musician's instrument—it was the very practices and meanings of being a musician (Theberge 1997).

If music technologies have actually been ahead of the virtual curve, their social history may hint at the power and problems that augmented reality writ large may bring. On one hand, musicians gained unprecedented abilities to trigger virtual sounds and acoustic spaces, from realistic emulations of acoustic instruments and concert halls to previously unimaginable digital permutations; on the other hand, the use of these virtual instruments and effects exploded the necessary relationship between a specific performer, performance, instrument, acoustic space, and the resultant sound. This proliferation of virtual choice and control set the stage for a social crisis of musical authenticity, as seen in debates over technologies such as Auto-Tune and machine music: What is a musician (Auslander 2005)? What is an authentic performance (T. Taylor 2014)?

Doppler Labs' website promised (to my ears, overpromised) "real-world sound control," the power to "selectively choose what you want to hear and remove what you don't." If some other developer delivers on this promise of the perfect orphic interface, it will complete the ongoing deconstruction of the necessary relationship between specific listeners, acoustic spaces, social interactions, and resultant perceptions of sound. If a hearable revolution comes, the power and dilemma of the digital musician will belong to every listener with earbuds. For better or worse, every sound will become a choice.

Conclusion

As indicated by its name, Here: Active Listening is meant to inject a measure of proactive choice into a sensory modality long seen as "passive" and in need of technological improvement (Schmidt 2000; Sterne 2003). However, in this chapter I have argued that the sonic interface is not simply an augmentation of reality but also a reduction of the sonic world—one that processes the sound around us according to the utilitarian ideology embedded in its code. Instead of being specific to the space and moment of its origin and the embodied presence of its perceiver, sound becomes media content that we model, manipulate, and transform to better suit us. Rather than experiencing and engaging with sound as it happens to emerge, Here users interface with it as information, consuming sound via a standardized menu of amplification, cancellation, and enhancement. Like any technological interface with our senses, Here amplifies our powers of perception and attention in some ways while reducing them in others (Ihde 1979, 21). If interfaces encourage us to hear only what we want, our tolerance and ability for "open listening" may be diminished as a result.

The social impact of the sonic interface will depend upon whether hear-ables live up to the hype and achieve the technological sophistication and widespread use their proponents expect. And they might: due to the success of Bose, Beats, and others, earphone R&D spending is way up, portending greater powers of sonic customization in the future. Miniaturization may al-low for continuous use of tiny, almost invisible devices in the ears—achieving the dream of "earlids" at last. As one tech journalist exclaimed on a podcast discussion of Here, "I want to live in my own filter bubble!" He gleefully an-ticipated never hearing a crying baby again (Elgan 2016).

Perhaps my own experiments with Here hint at the nature of a hearable future. It was strange how quickly I adapted to the echo and reverb—while they made conversation next to impossible, they actually made working at my computer easier. My office is usually a sonically distracting space, but hallway voices were so distant now that they didn't capture my attention. The hard part came when I stopped using Here after half a day. The world suddenly sounded—and felt—too immediate, like it was pressing in on me. Had the earbud batteries not died, I might have scrolled through the Here app to find just the right setting to help me transition back to my normal hearing, which for a half hour or so sounded like a bad preset that needed some tweaking.

Note

1. While I used Here, there was a third promotional filter category, "Coachella," which referred to the music festival where Doppler Labs focused much of its marketing energy.

 Casual/Games

RuPaul's Drag Race Keyboard
Affect and Resistance through Visual Communication

KATE MILTNER

App: RuPaul's Drag Race Keyboard
Developer: Snaps Media Inc.; © 2015 Viacom International
Release Date: April 2015 (Current and Final Version: 1.5)
Category: Entertainment
Price: Free
Platforms: iOS/Android
Tags: RuPaul's Drag Race, GIFs, emoji, gender
Tagline: "Now Sissy that Talk!"
Related Apps: Giphy Keys, Jiffmoji, Kimoji, VH1
     Entertainment

"Now sissy that talk! Using Drag Race emojis (*sic*) is easier than lip-synching for your life. Just cut and paste, gurl! Are you a size queen? Drip your convo in eleganza with these oversized stickers. Oh yes, she betta DO!" So began the installation instructions on the RuPaul's Drag Race Keyboard (RPDRK), a mobile application that allowed the user to pepper their text conversations with a series of customized emoji, stickers, and animated GIFs.

RuPaul's Drag Race is a competition-based reality TV show on Viacom's Logo network, the self-labeled "gay epicenter of culture" ("About", 2015). Combining elements of *America's Next Top Model* and *Project Runway*, *RuPaul's Drag Race* pits a cast of drag queens against one another in the quest to win the title of "America's Next Drag Superstar" along with a cash prize of one hundred thousand dollars, a vacation, and a large supply of cosmetics. It is one of the most profitable and highly rated titles for the network (Baron 2015), so when it came time to promote the seventh season, Logo partnered with branded keyboard company Snaps Media to create a custom keyboard app for the show.1 The free RPDRK app contained a series of custom *emoji*, stickers, and animated GIFs that featured the show's most beloved cast members and memorable moments. Although the app was discontinued in 2017, its initial

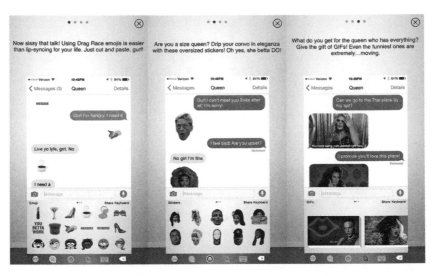

Installation instructions for the RuPaul's Drag Race Keyboard

success after its 2015 launch was still notable: the keyboard was downloaded over one million times in its first twenty-four hours, and it became a trending search in Apple's App Store ("Solutions," 2015). As of May 2015, almost seven hundred thousand emoji, stickers, and GIFs had been sent through the app (ibid.). Furthermore, the RPDRK connected with users outside of the show's audience and community: the app's user base is approximately 50 percent women, far exceeding Snaps's demographic expectations given the dominantly male viewership of the show (Austin Bone interview, May 1, 2015). Part of the explanation for the keyboard's success has to do with socio-technical practices. Austin Bone, the executive in charge of the keyboard's development, explained that Snaps Media was simply capitalizing on preexisting behavior. The use of emoji and GIFs in text messaging has become a widespread cultural phenomenon. According to Bone, the keyboard was a way for Logo to "empower" their fans with custom emoji and facilitate the behavior of those who were already sending each other funny GIFs and images of the show's contestants via text message (Austin Bone interview, May 1, 2015).

However, most branded keyboards don't take off in the way that the RPDRK did: the metrics for Snaps's other crown jewel—a branded keyboard for the Comedy Central hit *Broad City*—paled in comparison to the RPDRK. Furthermore, the app resonated so much with its users that they began using a hashtag, #rupaulsdragracekeyboard, to document their use of the app in situ on Instagram and Twitter. While the app might have originated as a marketing ploy, the RPDRK clearly struck a cultural chord, particularly since at least part of its user base was comprised of women who may not even have been fans of the show itself. This chapter argues that the app's cross-audience appeal can be partially explained by the fact that it was rife with transgressive potential that was uncommon for messaging apps at the time. An examination of the app using visual analysis and the walkthrough method (Light, Burgess, and Duguay 2016) revealed that in its specific curation and display of *Drag Race* content, the RPDRK both enabled and encouraged the deployment of drag femininity in everyday mobile-based conversation. While the mainstreaming of drag culture occurred long before the app's creation, the ways in which the app was used to subvert hegemonic norms by cisgender heterosexual women in particular offers some insight into why the performances of "reality queens" (Gamson 2013) are so appealing in our contemporary moment, whether on TV or as affective punctuation and communicative proxies in text messaging apps. As a (short-lived) keyboard app, the RPDRK exemplified the idea of "mundane software" (Morris and Elkins, 2015; Morris and Murray, this volume); however, the ways in which the app trafficked in resistant modes of gendered self-representation and communication were anything but mundane.

Gender Trouble? The Complicated Subversiveness of Drag

Drag and feminism have not always been comfortable bedfellows. Judith Butler noted that a feminist critique of drag centers on the perception that drag is fundamentally misogynist because it is based upon the degradation and ridicule of women (1993, 126–27). The assertion that drag culture (and gay culture more generally) has "a misogyny problem" has experienced a recent resurgence in popular discourse (see Donovan 2017). The crux of the drag-as-misogyny argument is that male-to-female drag is a form of punching down, so to speak; rather than ridiculing the straight men who have harassed them, drag queens send up women instead (Murphy 2014). Similarly, the use of terms like "fishy"—which refers to the purported odor of women's genitalia—to describe queens who are "suspiciously convincing" is offered as evidence that gay men view women as objects of disdain (Tabberer 2017). While it may be true that "a lot of queer men have work to do" in this arena (Lang 2017, para. 5), arguing that drag *as a practice* is misogynist because some gay men or drag artists are misogynist ignores some of the more complicated, subversive, and potentially liberating forces in play.

In "The Aesthetic of Drag," Daniel Harris addresses the perception of misogyny in drag performance. Drag, he explains, "involves a ritual descent into the morass of American vulgarity" (1995, 71). Drag dissects and reflects, funhouse mirror–style, the "egregious tastes of the homosexual's bigoted opponents," from Moral Majoritarians to mall rats (71). Rather than mistakenly interpreting drag as misogynistic, Harris argues that it is really "complacent heterosexuals in general" who are being taunted (71). The representation of femininity in drag performance does not necessarily reflect a love of womanhood, but a desire to deconstruct the binary categories of gender (Barrett 1995). The assertion that drag is misogynist is partially an assertion that what is being represented on stage as "woman" is, in fact, a valid representation of women. As such, the critique of drag as misogynistic somewhat misses the point; the intent of the drag queen is not to pass as a woman, but to exaggerate the qualities that are associated with women and complicate gender representations (Mann 2011, 795). Drag queens draw on stereotypes of what it means to be feminine, and in doing so, they "highlight, provide commentary on, and often challenge prevailing ideologies" (794).

In fact, it is the appropriations of the drag queen that are the source of her political power; drag is "an appropriation that seeks to make over the terms of domination" (Butler 1993, 137). Drag queens reverse the power dynamic between the marginalized and the mainstream by becoming the source of the "gaze" and subjecting the bigoted heterosexual to the scrutiny that she experienced as a "deviant" (Harris 1995, 72). However, the most radical politics of the drag queen are not necessarily in her outward lampoonery, but her acceptance of herself in the face of a society that despises and rejects her. Aesthetic terms such as "fabulous," whose repeated usage may seem like effeminate verbal punctuation to outsiders, is "an entirely ideological expression" that signifies full self-acceptance in the face of social bigotry (69). Arguably, it is this politics of resilience that is an entry point for heterosexual cis women when it comes to embracing and performing drag femininity. Perpetual self-negation is at the heart of hegemonic femininity: women are expected to always put others first, diet until they are shadows, sacrifice their careers for the "good" of their families, internalize any "unpleasant" affect (such as anger) so as not to upset others, and so on. To reject these forms of erasure and instead engage in full self-acceptance in the face of unrelenting, unattainable expectations is a deeply political move.

This is not the only way that heterosexual women can deploy drag politics, however. Women, too, are subject to a heterosexual, patriarchal gaze; women, too, are severely constrained by gender binaries. Furthermore, when women (whether straight, gay, or trans) blatantly or aggressively push back against hegemonic expectations, they are often met with violence. However, when cisgender heterosexual women deploy forms or representations of drag femininity, it is usually not seen as subversion or resistance against oppressive gender norms. Thanks to reality TV, the once avant-garde and often risky performances of queens (Newton 1972) have become standard (if not entirely apolitical) fare, and the resulting proliferation of "reality queens" in female-targeted media means that performances of "queeny" femininity by cis straight women can be written off as expressions of fandom or cultural knowledge. While this makes drag a safer route to resistance for women, unfortunately this is not the case for LGBTQ+ individuals for whom any display of non-normative femininity (drag or otherwise) comes with considerable risks outside of LGBTQ+ spaces and places.

The Rise of "Reality Queens"

Joshua Gamson (2013) notes that despite the violence and ridicule that effeminate boys and men face in the off-television world, they have become fixtures in reality television. He argues that the dominance of queens who "tell you what not to wear and how to decorate your house, judge your taste or exhibit their own, [and] decide if you're the winner" is thanks to market forces and "the exigencies of television," rather than broader sociocultural shifts (Gamson 2013, 52–53). The explosion of reality TV in the late 1990s and early 2000s brought new economic exigencies and thus new formats of shows, such as style makeover and fashion-based competition shows, that were welcoming to gays and lesbians (Dovey 1998). The queeny-ness of the reality TV landscape helped lay the groundwork for the introduction of drag into that particular milieu, as did the new era of mass celebrity that accompanied the explosion of *Keeping Up with the Kardashians*, *Real Housewives*, and the *Bachelor/ettes* onto the scene. As Harris explains, "Modern drag is rooted in the culture of mass celebrity. . . . It is an eccentric by-product of our increasingly intense involvement with popular entertainment" (1995, 66). What better way to offer a critique of "reality" celebrity than to create a competition show that both literally and figuratively shows the backstage of reality TV? Launched in 2009,

RuPaul's Drag Race is part of the broader reality TV trend of "dualcasting" programming to heterosexual women and gay men (Sender 2007) and was central to Logo's strategy of "gaystreaming" LGBTQ content to straight women as a method for garnering higher ratings (Ng 2013). By taking drag out of the gay bar and putting it on national television, *RuPaul's Drag Race* was accused of producing "a more normalizing view of drag performance and competition" (Edgar 2011, 136). However, while the drag offered by *RuPaul's Drag Race* may not be genderfuck or other more radical forms of the genre, it is hard to argue that the show fails to operate in incendiary ways; after all, the judgment criteria for the queens is their c.u.n.t: charisma, uniqueness, nerve, and talent. As Gamson argues, *RuPaul's Drag Race* "is no assimilationist fare" (2013, 54), and it this undercurrent of mutiny and subversion that was actively deployable—and deployed—in the RuPaul's Drag Race Keyboard.

"F*ck All You Bitches": Transgressive Affect in the RuPaul's Drag Race Keyboard

The emojis, stickers, and animated GIFs in the RPDRK were, in a word, hilarious. They were also over-the-top, bawdy, grotesque, and snarky. In short, they were everything hegemonic femininity is not, and this was arguably the core of their appeal for the women who used the app. They provided color, both literal and figurative, to a technical environment where emojis, the dominant affective "digital companions" (Stark and Crawford 2014), had only recently added nonwhite skin tones in response to public outcry. Furthermore, the RPDRK offered a broader spectrum of performative options. As Luke Stark and Kate Crawford have argued, traditional emoji are inherently conservative and offer "new opportunities for digital expression, but only if you're speaking their language" (2014, 1). In this language, even the euphemisms are sanitized: in 2015 Instagram banned the eggplant emoji from appearing in search results because of its phallic associations (Bonnington 2015), and the peach emoji was redesigned in 2016 so that it looked less like a rear end (Dickson 2016). Not so with the RuPaul's Drag Race Keyboard emoji: despite the fact that there are only seventeen of them, one is, literally, a pink furry box (subtlety be damned). While it may be true that the Unicode emoji have "nothing much to say about our political impasses" (Stark and Crawford 2014, 1), the same cannot be said about those in the RPDRK.

Emoji and other sticker applications on commonly used apps such as Face-

Stickers from the RuPaul's Drag Race Keyboard

book provide a range of "negative" emotions, including sadness, annoyance, fatigue, and frustration, but they do so in a safe and sanitized way. The stickers and GIFs provided by the RPDRK, on the other hand, are downright carnivalesque (Bakhtin 1984), subverting dominant culture through humor and chaos and invoking the rage of the subordinated. In his explanation of the shift in drag culture in the aftermath of the Stonewall riots, Harris describes how the default facial expression of the drag queen became the "ferocious baring of the fangs" characteristic of the "drag screech" (1995, 69). The drag screech is on full display in the RPDRK, along with a host of other outlandish facial expressions and exaggerated missteps. These blatant failures of femininity are not held up as moralistic warning signs, but instead as the gleeful embrace and embodiment of, as the queens would say, the hot mess. The femininity on display here is the opposite of the "cool girl," all good times and sangfroid; instead, it shows what femininity could be if (and arguably is when) it fully lets go: hungry, desirous, playful, funny, angry, bitchy, messy, fabulous. The difference between standard emoji and the options available on the RPDRK is something that users have commented on; as one Instagram user gushed, "#rupaulsdragracekeyboard has made my life so much easier . . . your queens deliver your mood better than any emoji." In another post, the same woman simply posted a large photo of one of the drag screech stickers with no comment, much to the appreciation of her Instagram followers.

"I Can Speak Drag!" Subtle Subversion on #rupaulsdragracekeyboard

The hashtag #rupaulsdragracekeyboard is used on Instagram and Twitter, where users—most of whom identify as female—document the conversations they have on the keyboard and the different ways they can, in the words of one user, "speak drag." Perhaps unsurprisingly, these conversations are not exactly prim and proper. One example includes a conversation between two girls discussing a carrot in an explicitly sexual way. "Lmao don't stick it in," says the first; "I won't haha I'm eating it," says the other, only to be met with a GIF of a drag queen throwing some considerable shade. Another young woman shared how she used the GIFs of Bianca Del Rio (the season 6 winner) and Michelle Visage (one of the judges) to reject an unworthy suitor. "Used the new @rupaulsdragrace keyboard today to turn down an annoying boy," she commented. "@michellevisage and @thebiancadelrio thanks for the help lol." Her friend responded "OMG THIS IS PERFECT," and the young woman concurred: "It's literally the best way to reject a guy without coming out and saying 'ew no.' LOVE IT! You should download it and bask in it's [*sic*] wonderfulness!"

While not all of the posts on #rupaulsdragracekeyboard blatantly referenced masturbation with vegetables or the gleeful rejection of men, most of them demonstrated that the specific content of the app offered its users the opportunity to express themselves in a unique and engaging manner. One woman posted several times on the hashtag; each post was a picture of her and her friend exchanging over-the-top *Drag Race* GIFs with no text. Each of these posts had the comment "Thank god for the #rupauldragracekeyboard" followed by the hashtag "#expressyourself." Another woman posted a picture of a friend responding to her new hair color with the comment "Oewh fancy!!" followed by a GIF of a drag queen tossing her hair in a glam, defiant way. The caption read, "when you get new hurr and your best friend gets a new iPhone." The playfulness of these and other posts on the hashtag reflect the gratification of engaging on one's own terms—a pleasure that is frequently denied to even the most powerful women. This is, perhaps, what the RPDRK afforded on a micro level for its cis, straight, female users: the ability to express one's fabulousness regardless of size, shape, or color and a means of celebrating affect that society deems impolite, unladylike, and inappropriate. The posts on the hashtag are only a few of the hundreds of thousands (if not millions) of interactions that took place on the app over its lifespan, but the fact that they were publicly documented indicate that they were meaningful to their users on some level. These posts also underline that the affective affordances of the RPDRK clearly served an unfulfilled need in the technological landscape where it thrived.

"Making it work against society": Concluding Thoughts

The fact that the RPDRK was removed from the iTunes store and Google Play at some point in 2016 or 2017 probably says more about the economic priorities of Viacom's marketing department than it does about the app itself; in all likelihood, Viacom didn't want to pay to keep the app compatible with perpetually updated operating systems. Or perhaps they realized that they didn't *need* to pay: since the last version of the app was integrated with massive GIF search engine Giphy, the app's previously exclusive content was now easily searchable and available on Giphy's own mobile app. As of September 2017, a search of "RuPaul's Drag Race" on Giphy returned almost twenty-three thousand GIFs; it seems that the show's fans are fully "empowered" without a proprietary keyboard, and the queens continue to live on in conversations across the globe.

However, the initial success of the keyboard and the use of the queens as affective proxies by cis straight women also raises some slightly different political questions than those of the resistance discussed in this chapter. In a 2017 article for *Teen Vogue*, English literature scholar Lauren Michele Jackson argued that when black people become the go-to choice for nonblack users to engage in affective hyperbole in online environments, it operates as a form of "digital blackface." She explains:

> These GIFs often enact fantasies of black women as "sassy" and extravagant, allowing nonblack users to harness and inhabit these images as an extension of themselves. . . . We are your sass, your nonchalance, your fury, your delight, your annoyance, your happy dance, your diva, your shade, your "yaas" moments. (para. 12)

Jackson's argument raises several questions about the politics of *Drag Race* GIF usage. To start, the relationship between gay culture, drag culture, and black culture is complicated. While some have argued that the presence of phrases like "shade," "truth tea," and "yaaaas" is a co-option of black womanhood (see Mannie 2014), others argue that the presence of black vernacular in contemporary gay and drag cultures connects back to the drag balls of the 1980s, and that the use of these phrases reflects the "mimicking of gay men from another era, who found a way to forge an identity amid difficult circumstances" (D'Addario 2014, para. 4). This debate notwithstanding, Jackson's argument raises another question as to whether or not the cisgender heterosexual women using *Drag Race* GIFs are engaging in a type of "digital dragface," so to speak. What does it mean that individuals from a comparatively privileged identity category are using representations of a marginalized group to perform specific kinds of affect and emotion that they might not otherwise feel comfortable expressing? Furthermore, many of the most popular *Drag Race* GIFs have queens of color in them; of course, part of this is because the show's cast is quite diverse and RuPaul herself is black. Nevertheless, the racial identity of the queens certainly complicates the issue—there very well may be multiple types of transgression going on here, and not all of them in service of subverting the patriarchy. There aren't any simple answers to these questions; as always, representations are fraught with intersectional considerations and tensions. Nevertheless, Jackson's argument certainly highlights how purportedly mundane media such as GIFs and GIF keyboards are in fact deeply political and operate at the crossroads of identity and power.

In his behind-the-scenes look at the reunion for *RuPaul's Drag Race* season 2, Rich Juzwiak commented that "making it work against society, drab backgrounds and oneself . . . is what being a drag queen is all about. These are the flowers fighting through the asphalt of our culture" (2010, para. 52). It is likely this spirit that is so appealing and inspiring to the women who choose to represent themselves and their emotions with images and GIFs of the *Drag Race* queens. Complicated and fraught though it may be, the RPDRK (and the GIFs that sustain its legacy) allows users across a spectrum of genders and sexualities to extend a metaphorical middle finger to the patriarchy and, as if instructed by RuPaul, sashay away.

Notes

1. A keyboard app allows mobile phone users to insert a variety of image formats into text message conversations. It installs itself in the messaging software of a mobile device operating system and is accessible as a keyboard within the text messaging software. Popular keyboard apps include Bitmoji and Tenor.

DraftKings
Daily Fantasy Sports Leagues, Legality, and Shifting Mobile Spaces
JASON KIDO LOPEZ

App: DraftKings
Developer: DraftKings LLC
Release Date: December 2013 (Current Version: 3.12.0)
Category: Sports
Price: Free
Platforms: iOS, Android
Tags: fantasy sports, gambling, games
Tagline: "The Game Inside the Game"
Related Apps: FanDuel

The daily fantasy sports company DraftKings seemed to be everywhere in 2015. The business offers a website and an app that allows users to participate in one form of the popular trend of fantasy sports play, the "daily fantasy game" or "daily fantasy play"—an accelerated gaming mode where participants construct their own teams of real athletes and compete with one another by accruing points based on their team's actual sporting performances over different increments of time during a season (e.g., day, weekend, evening). After spending millions of dollars on advertising and garnering near-ubiquitous visibility on television, DraftKings also made news headlines in October 2015 when the company was sued for negligence, fraud, and false advertising. Rather than simply a popular mode of entertainment, some legislators conceived of daily fantasy sports as a form of problematic gambling that involved financial risk and potential financial reward. Subsequently, the DraftKings app was banned in some states—by 2016 there were ten states that expressly prohibited daily fantasy sports and thirty-eight considering a ban (van Natta 2016).

DraftKings offers free accounts, through which users can deposit money and use that money to join different competitions for financial prizes. Although one can access DraftKings' daily fantasy game through a computer or a mobile application, its appification increases its mobility. The app transforms any physical location—a park, a coffee shop, a classroom, etc.—into a space where one can risk money on the outcomes of sporting events. DraftKings transforms space both through the nature of the game and through the restrictions on where one can play it. For example, because of the prohibition in a handful of states, the app could restrict fans from the opportunity to engage more deeply with the sport they love. But because it is permitted in other states, the app also allows people to gamble money they may or may not have on a sports competition.

Tracing the history of the game of fantasy sports into its current space-changing, appified form helps us to explain both possibilities. Many aspects of the daily fantasy game are similar to earlier forms of fantasy sports that allowed small groups of friends and acquaintances to compete with each other over sports knowledge. DraftKings allows its players to engage in a similar competition, as well as risk money quickly and regularly from any location. Yet anti–daily fantasy regulations focus only on the latter, with the entire game being viewed as an opportunity for gambling that poses a social and moral threat. DraftKings, therefore, offers an opportunity to see how an app's space-altering affordances allow not only for a wide range of engagement from its users, but also a more narrow and one-sided view of that range from legislators. When considering legislation of a mobile app, it is vital to consider the app's history, its affordances, and how rules can be enforced on and through the software. This has implications not just for daily fantasy and gambling apps but also for any app that allows for quick, easy play and in-app purchases.

The Game of Fantasy Sports

Although DraftKings offers engagements with fantasy football, soccer, golf, e-sports, and other games, the history of fantasy sports is, to a degree, tied to baseball. As early as 1980, a group of baseball fans were participating in a fantasy league called "The Rotisserie League." This league became influential through a publication called *Rotisserie League Baseball: The Greatest Game for Baseball Fans since Baseball* (Waggoner 1984). The publication laid out the original rules of the game, which consisted of ten to twelve participants picking a unique team of twenty-three actual Major League Baseball (MLB) players, including fielding positions, starting pitchers, relief pitchers, and bench players. Participants were given an imaginary budget of $260 to spend on players, who were auctioned off during a mock draft. The competition ran the length of an MLB season, and the goal was to score points by having one's team score well in statistics such as home runs, runs batted in, strikeouts, and saves. Throughout the season, participants managed their team by selecting their starting lineups of the players they drafted, picking up undrafted players, making trades with other participants, and accommodating their injured players. The gameplay was designed to offer participants an opportunity to role-play a general manager of an MLB baseball team (Waggoner 1984, 4, Walker 2006, 331). Although there are currently many different permutations of fantasy baseball, this particular version largely served as their foundation and inspiration.

Baseball fans were drawn to fantasy sports because it allowed them to spend more time with the sport they loved. This devotion was important, since early forms of fantasy baseball were extremely time-consuming. Scoring had to be done by hand through totaling statistics of players' performances from a daily box score. Given that there were almost no fantasy resources available, only the most dedicated fans took the necessary time to research. Since fantasy allowed for competitive fandom (Halverson and Halverson 2008), it was not enough to know and enjoy baseball; fantasy players wanted to demonstrate that their knowledge was deeper than that of their league mates, and that took commitment.

Financial incentives, although present, did not play a primary role in early fantasy sports. For example, participants in the Rotisserie League played for money, but more attention was given to the fact that their league's winner got doused in Yoo-Hoo by his or her league mates (Waggoner 1984). Furthermore, approximately half of season-long fantasy players played for no money, as the "exchange of money from losers to winners . . . does not appear to be a primary aim" (Billings and Ruihley 2013a, 3). Money may have played some role in encouraging participants to play season-long fantasy (Mahan et al. 2012), and in some ways it could be compared to gambling (Bernhard and Eade 2005), but the most often discussed motivations to play generally focused on friendly competition, the love of the game, or other nonmonetary reasons.

Fantasy sports rose in popularity in the nineties as sports media companies began to develop fantasy resources and more opportunities to play. Companies like ESPN realized that fantasy sports increase sports fan activity and could be profitable for them (e.g., Dwyer and Kim 2011; Randle and Nyland 2008; Billings and Ruihley 2013b). This resulted in increased coverage of fantasy news, publications of game-play tips and strategies, and advertisement of opportunities to play a variety of versions of fantasy online. The advent of online fantasy leagues through a large sports media company was particularly important, because it made playing fantasy easier (for example, there was no need to calculate scores by hand) and gave sports media companies yet another service to offer sports fans.

What was once a niche game that had to be played offline with pen and paper became a fully digital and widely accessible pastime: "Like the Internet itself (interestingly, the tool that has enabled and simplified fantasy baseball's allure), fantasy baseball no longer belongs solely to an enthusiastic but socially disconnected cult (Bernhard and Eade 2005, 32). Thus, online fantasy became a popular and broadly available activity for sports fans, many of whom transitioned online as fantasy became popular and easy in digital form.

The Rise of Daily Fantasy Sports

As online versions of fantasy sports increased, so too did other forms of internet-based gaming. Online poker became increasingly popular from 2003 to 2006 (Silver 2011), and online sports betting was also on the rise (Lang 2016, 209). Online versions of poker and sports betting allowed players to skirt local geographically based laws against gambling. For instance, it might be illegal to gamble within the state of Utah, but one could gamble on a website run by a company located in a different state or country. The US government responded by tightening regulations through the Unlawful Internet Gambling Enforcement Act (UIGEA) in 2006. The UIGEA restricts banks from handling money involved in "betting or wagering" on the internet, which had an adverse effect on most opportunities to gamble online. The act also took special care to note that fantasy, under certain conditions, didn't count as betting or wagering (section 5362). There was fairly good reason to be unconcerned about fantasy, as the contests lasted months and it would be difficult, outside of high-stakes leagues, to lose too much money.

The popularity of fantasy, the ability to play online easily, and the explicit governmental permission to play it online for money set the grounds for the arrival in the late oughts for a new version of gaming: daily fantasy sports. DraftKings LLC launched in 2012 alongside other already existing fantasy sports companies (including, most importantly, FanDuel), and in 2014 Draft-Kings was the first to release a daily fantasy app (Rotogrinders 2016). Rather than running an entire year of a sports league, DraftKings offered a game that is much quicker; competitions can run over the length of a weekend, day, or even afternoon or night.

These games are similar to earlier forms of fantasy with a few adjustments to make the speed of the game work logistically. In baseball, for example, participants still pick a team of athletes, points are awarded for a similar set of statistical occurrences (strikeouts, home runs, etc.), and the player who draft-ed the team with the most points wins. However, in daily fantasy, the draft is done without knowing the lineup of the opposing fantasy teams, and the same MLB player can be drafted to multiple teams. Despite these differences, the daily game bears a strong resemblance to both the original niche and the popular online season-long games. These games enable similar sorts of engage-ments with fandom and sports knowledge as the daily version.

Between 2012 and 2015, daily fantasy became wildly popular, and Draft-Kings and FanDuel became locked in a race to corner the market (van Natta 2016). Major advertising campaigns and deals with sports leagues and media companies fueled the spread of daily fantasy sports and its mainstream visibil-ity. But as the game's popularity increased, concerns about its alignment with gambling practices intensified as well. To some state legislators, daily fantasy seemed to create a popular and worrisome way to gamble legally.

Daily Fantasy as Gambling

The version of fantasy gaming that the DraftKings app enabled resembled gambling in many ways, particularly in its interface and business model. For example, the app's interface design and features seem primarily to emphasize money and winnings rather than, say, sports fandom. Upon opening the app, users arrive on a home screen that prominently displays the amount of money that has been deposited through a credit card or PayPal account. When the user tries to join a contest, the cost of entry and amount of prize money are first highlighted. The list of which contests are currently "live" or still in play foregrounds the money the user is currently on track to win (if any). Finally, when the user wins money, a notification alert shows up. If enabled, a notification will pop up even if the app is closed.

Not only is there an emphasis on winning money in the design of Draft-Kings, but the economics supporting the app are also very similar in structure to poker. In the game of poker, participants are charged a "rake" just for the opportunity to play the game (Hayano 1982). In DraftKings there is a similar cost to play. For instance, you can play a game where three players all pay a $1.00 entry fee and the winner gets $2.70. The remaining $0.30 is the rake, which goes to the company. There are many different types of games on Draft-Kings that vary by number of players, entry fee, and payout, but the rake is applied in non-free games in order for the company to profit.

Furthermore, there are many similarities between the play of DraftKings and contemporary slot machines. Slot machines have moved away from using physical money and allow for a high play rate (Schüll 2005, 67). DraftKings similarly encourages quick and easy play, with games starting nearly every hour that can be joined for an electronic entry fee. Other features that slot machines use to promote gambling can also be found on the app, such as multiple plays and payout adjustment to player preference (Schüll 2005, 69–70), since it is possible to submit the same lineup for multiple competitions and the payouts of those competitions range from twenty-five cents to millions of dollars.

These connections to gambling have led some states to debate the legality of daily fantasy, and even to start prohibiting it. They reason that if it is like gambling, then it should be regulated like gambling. Yet not all forms of gambling warrant state intervention, given that most states allow some form of slot machine gambling (Schüll 2012, 5) as well as horse racing and lotteries (Lang 2016). If this is the case, the DraftKings app must offer a problematic form of gambling. Some of the arguments against daily fantasy specifically are that it is easily available to vulnerable populations and that the game is unfair.

Concerns about vulnerable populations are connected to the qualities of daily fantasy already mentioned; it is a game that is quick to play and easily accessible. If children or those struggling with addiction get hold of a game like this, its speed and ease could potentially contribute to problematic play. Furthermore, the game also offers unchecked play. Unlike most early forms of fantasy, users don't need to know anyone to join a competition. Users of DraftKings compete with complete strangers over a number of hours. Like sports betting and playing poker online, it is possible to play in an anonymous, isolated fashion. Those who are concerned about DraftKings see it as an easy, instantaneous, and unsupervised way to lose money, and they think that vulnerable populations need to be protected from this kind of play. Indeed, many new media technologies are already framed within similar discourses of addiction, so it makes sense that DraftKings' play and incorporation of money only fuel more concern.

The arguments that the game is unfair rest on accusations that DraftKings doesn't give most of its players an honest chance to win. Not only was it discovered that employees of DraftKings and FanDuel were using proprietary information to win money on each others' sites, but more experienced players with advanced technology were found to have a large advantage over average players (van Natta 2016). In sum, the DraftKings app seems to allow unsupervised, vulnerable populations the opportunity to lose money quickly and easily in a likely unfair game.

Regulating and Policing Daily Fantasy

The view that DraftKings enables problematic play is reflected in the app itself. For example, the bottom of the app's home screen contains a link titled "Responsible Gaming," which connects users to information about the National Center for Responsible Gaming, as well as a set of options for setting spending limits, taking a time out, self-excluding, or establishing parental controls. Yet these features are not enough for some states, and legislation against the game has moved forward.

One difficulty of legislating against DraftKings has been that it is a mobile, mediated game rather than a game that happens in a static physical space. Gambling regulations have traditionally been enacted upon geographically specific locations, with states restricting where and how one can gamble (Lang 2016). The enforcement of these gambling regulations then requires policing of the physical spaces where gambling is or is not allowed. Given that this policing becomes impossible for a mobile game that is continuously playable while moving through physical space, the responsibility of enforcing anti–daily fantasy regulations falls to the app itself.

After users download and open DraftKings, a pop-up notification appears and asks permission to access the user's location. In smaller type, the notification states, "Due to varying geographical restrictions on contest participation and cash prizes, location services must be ALLOWED." Until location services are allowed—which enables DraftKings to use Wi-Fi, GPS, or GSM signals to locate the user (DraftKings 2016)—it is impossible to do anything else on the app. Once enabled, users are free to peruse the app, select a contest to play in, and draft a team. However, if the user is found to be in a state that has made it illegal to play DraftKings (or in some cases if legislation is pending), the user is prohibited from submitting a lineup and fully joining the competition, and a pop-up notification explains that there are "prize restrictions in the state."

The legislation against DraftKings, then, not only applies to the app but also is instantiated through the app. Such implementation of policy through technology is not new. Jeremy W. Morris, for example, looks at how cultural and technical protocols like patents take on regulatory-like roles. Technical protocols are "what capabilities devices do and do not have" which in turn shape "the features and attributes of media technologies" (Morris 2013, 216). Similar types of protocols are in place for copyright law. Tarleton L. Gillespie also notes a shift away from seeking legal enforcement, with designers instead proactively constructing technologies in ways that prevent misuse. If such efforts are successful, they "would regulate every single user automatically and without the bureaucracy of enforcement and adjudication; it would anticipate and preempt infringement" (Gillespie 2006, 652). We can see such preemptive measures built into DraftKings, since it forces users to submit to the surveillance of their geographic location or not use the app at all. DraftKings takes up the role of surveillance and enforcement on the state's behalf.

Furthermore, to protect minors from playing the game, the app again regulates itself. The company's website notes that "in order to register an Account you are required to provide your full name, address, date of birth, email address and telephone number," and some individuals may need to provide additional documentation such as uploading a picture of a valid identification (DraftKings 2017). Play on the DraftKings app, therefore, demands surveillance of geographic location as well as age.

The Legality of Altering Spaces

Although technologies may be commonly designed to prohibit unwanted behaviors, DraftKings offers an important case for asking how legal regulations are now shaping the design of technologies and affecting the freedom of its users. With regard to gambling regulations, we can see a fundamental shift from regulating physical spaces toward regulating users and technologies. The deployment of these specific laws is premised on the understanding that app usage can change any space into a gaming/gambling space, which has traditionally required regulation and supervision by the state.

A space that is supervised in this way by technological protocols ends up not only limiting use of a particular technology, but also "frustrat[ing] the agency of the users" (Gillespie 2006, 653). Although DraftKings furnishes users with a wide variety of reasons to play daily fantasy, it also can limit their agency regardless of those reasons. All players are surveilled, and if they are found to be located in a state that prohibits play, all play is restricted. Using DraftKings for monetary gain, fun, and competition all get treated the same. Just as the app enables all of these experiences, in some cases it also restricts them.

In the view of some state legislatures, DraftKings is both the cause of a problem and its solution. But to view the app as nothing but an unfair and dangerous game ignores the other experiences that are possible. When one looks at how the game of daily fantasy is constructed by DraftKings, it is possible to see its roots both in the origins of the fandom and enthusiasm of fantasy sports and in the common ways that gambling is monetized. Thus, DraftKings' transformation of space is a double-edged sword: betting can occur anywhere and so can expressions of fandom. Unfortunately, much of the self-governance of the app due to state regulation privileges the former at the expense of the latter.

This investigation of how apps are legislated reveals the importance of considering the full spectrum of users and uses—including an app's technological and cultural affordances, as well as the implications of restrictions for both app design and users. Studying apps like DraftKings is about more than just understanding the shifting parameters of gaming practices as they are digitized; it also requires addressing a new and murky area of technologically supported power that governing bodies wield to shape and enforce policy based on troubling assumptions about users.

Probing this set of issues is important not just for fantasy sports and sports in general (for example, sports betting and horse racing apps), but for a wide variety of gaming apps—especially those that offer in-app purchases. Many games that offer in-app purchases, such as *World Series of Poker*, *Candy Crush*, and *Pokémon Go*, might offer addictive play that encourages quick, easy, and unsupervised spending. This isn't to say that these games ought to receive the same attention from legislators, but their ability to alter space into a gaming space and to encourage spending can be compared to DraftKings. As with DraftKings, any legislation and policing of gamers must be sensitive to the history of the games, how they are played, and by whom.

Kendall & Kylie
Girl Affects, Celebrity, and Digital Gaming in Millennial Culture

JESSALYNN KELLER AND ALISON HARVEY

App: Kendall & Kylie
Developer: Glu Mobile
Release Date: February 2016 (Current Version: 2.6.0)
Category: Game
Price: Free
Platforms: iOS, Android
Tags: casual gaming, girl culture
Tagline: "BE YOURSELF, EXPRESS YOURSELF!"
Related apps: Kim Kardashian: Hollywood; Stardom:
    Hollywood, Clash of Clans; Hotel Dash; Britney Spears:
    American Dream; Nicki Minaj: The Empire

brb . . . Kendall and Kylie Jenner are texting.

If you've downloaded and played Kendall & Kylie, the Jenner sisters' recently released app, you've likely also received a message in the middle of dinner from them, reminding you that a gift box awaits or that one of the celebrity siblings likes your in-game social media post. These interactions with the younger half-sisters of Kourtney, Kim, and Khloe Kardashian, or rather their avatars, are the foundation of this mobile game. Much like the rise to fame of the Kardashian family following the success of their reality television series *Keeping Up with the Kardashians*, the app's main quest is the creation of a highly customizable star on the rise who gains celebrity status through the cultivation of social media "followers," both non-player characters and others playing the game. Since its release in February 2016, the free app has been installed by over six thousand new users a day (Think Gaming 2016) and at the time of this writing has been reviewed on the iTunes app store nearly fifty thousand times, with an average ranking of 4.5 stars out of 5.

Kendall & Kylie follows other highly successful apps released by the Kardashian-Jenner clan, most notably, Kim Kardashian: Hollywood (KK:H)

(Glu Mobile, 2016), which earned Kardashian $74 million in 2014 (Reisinger 2016). Also designed by Glu Mobile, Kendall & Kylie reimagines many of the same themes and game mechanics found in KK:H for millennial fans, including the pursuit of fame via publicity work, the styling of digital bodies, and constant social networking with friends, followers, professional contacts, and love interests. What is different from the game based on older sister Kim is that in Kendall & Kylie fame is acquired exclusively through your character's own savvy use of social media. As a player, you are asked to produce beauty vlogs (video blogs) with Kylie, post selfies with Kendall, and answer texts from a range of racially diverse suitors. Rather than collecting fans through third-party celebrity gossip and news tweets as in Kim Kardashian: Hollywood, you constantly engage with your in-game mobile phone as you pursue ever increasing follower goals through posts about activities you undertake in the game. In this sense, Kendall & Kylie gamifies the building of a digital reputation that is found in apps like Klout, making playful a privileged practice of selfhood in neoliberal capitalism (Hearn 2010).

Through the successful completion of tasks, such as producing an online makeup tutorial, you will earn energy (needed to complete further tasks), cash, and glittering pink "k-gems," the premium currency of the game that allows you to buy exclusive designer clothes, funky accessories, and trendy manicures. All of these things, of course, are necessary accoutrements for navigating the app's simulation of the fame game, which includes a global range of photo shoots, appearances, and other gigs in Los Angeles, New York, Montreal, and Paris. As with KK:H, Kendall & Kylie is categorized on Google Play as part of the "Adventure" genre,[1] and is centered on digital work in various immaterial forms, including promotional, entrepreneurial, glamour, and affective labor.

It may seem easy to dismiss Kendall & Kylie as another frivolous Kardashian-Jenner branded product that brings together the most derided elements of celebrity culture and casual gaming and builds on their established empire of media and consumer products. Indeed, until very recently, within digital game studies casual gaming has been largely overlooked or denigrated, a tendency that researchers argue serves to marginalize feminine participation in play (Vanderhoef 2013), leave unexamined the cultural values embedded in these games (Anable 2013), and obscure the reality that casual games are played by both male and female audiences (Eklund 2015). Despite challenges to assumptions made about the supposedly minimal temporary, financial, and attentional investment of casual game players (Juul 2010), there remains a stereotype that casual game play is cheap, low status, and culturally insignificant, representing the feminization of games culture to the detriment of its opposite: hardcore play (Taylor 2012). Yet dismissals of casual (often mobile) games negates both the economic and cultural impact of titles such as Kendall & Kylie and their celebrity girl creators/influencers, as well as an important history of girls' culture that, we argue, informs the app.

In this chapter, we contribute to the growing body of work exploring the design and expressive meaning of casual mobile games. As part of the growing corpus of celebrity game apps, Kendall & Kylie exemplifies a contemporary expression of girl culture via mobile media and digital games. In addition to challenging the tendency to dismiss casual games as insignificant beyond their ability to provide a temporary distraction, we offer a consideration of a game clearly designed with a girl audience in mind that does not simply relegate them to the passé category of "girl game."2 We make the argument that this game app is significant because of how its design indicates key shifts in the relationship between girl culture, contemporary celebrity, digital play, and mobile media, providing players with the opportunity to engage with a fun, fantasy-based contemporary girlhood that is nonetheless realized through highly feminized labor. While we argue that Kendall & Kylie functions as part of a lengthy history of girl culture via its mobilization of a girl affect (Swindle 2011), we maintain that the app is distinctively part of millennial digital culture, reconfiguring the immaterial labor that arguably defines digital culture as a necessary part of millennial girlhood. In this sense, the affects attached to girlhood, most notably fun, is married to the immaterial and temporally bound labor of publicity, glamour, and fame—a relationship that is "appified" in Kendall & Kylie. Thus, while the app may momentarily disrupt the masculinization of tech cultures through its embrace of teenage girldom, it also problematically prescribes feminized work in the "digital reputation economy" (Hearn 2010) as *the* way to achieve successful feminine selfhood—leaving little room for the articulation of feminine subjectivities that do not achieve "likes," "followers," and, ultimately, fame.

Mermaid Hair and BFFs: Girldom and Girl Affects in Digital L.A.

Kendall & Kylie is . . . *fun*. And while the summary of the game play sounds remarkably banal and akin to everyday practices of social media performance on other apps, Kendall & Kylie goes beyond click-based work, seamlessly linking the investment of your time with the ability to express yourself through avatar customization. And indeed, this is likely a large part of its widespread appeal. Online reviews and commentary describe the app as "weirdly addictive" (Szubiak 2016), "delightful" (Rees 2016), and "super-fun and surprisingly creative" (Fanning 2016). In *Tech Times* Lauren Keating (2016) contends that "dressing your avatar is so much more fun [than in KK:H], because players can try out more interesting hair color choices, like the mermaid blue trend, along with more signature looks the girls are known for rocking, including their latest collection from [California clothing retailer] PacSun." And *New York Magazine* writer Kevin Fanning (2016) assesses the app as "a tongue-in-cheek exploration of the way social media opens new opportunities for us, and a creative toolbox that actually, against all odds, makes social media fun again." This is an interesting contrast to the cultural reception of its spiritual antecedent Kim Kardashian: Hollywood, which has been called "evil and vile" (Curry 2014). The distinction appears to be linked to the parodic pleasure to be taken in the social-media-based game mechanics in Kendall & Kylie—it is not a game that takes itself too seriously.

Indeed, the game's reviews hint at the affective tenor of the app; it's fun because of its exaggerated, yet self-aware, construction of the Jenners's (implausible to most) world. You can tap on the screen to mimic the infamous Kardashian "duckface" in a concert selfie with Kendall or pay 499 k-gems to purchase a (gorgeous!) branded Nicole Miller flowy maxi-dress to show off at a party at WeHo's (West Hollywood) The Beverly. The incorporation of such cultural practices and material objects—the duckface, the playful abbreviation, and the designer dress, for example—into the app can be read as a knowing wink to players who recognize such practices and objects as both performative fantasy and representative of the excess associated with much-derided feminized celebrity culture (see Holmes and Negra 2011). However, the app invites us to momentarily suspend our critiques and *enjoy* documenting our ever changing hair color with stickers proclaiming "Perf" on our newsfeed or choosing a new coat for our weekend ski trip to Jackson Hole.

While there has been some insightful writing on how the Kendall & Kylie app functions as a "surprisingly creative playground for experimenting with narratives online" and the relationship between social media and our performative selves (Fanning 2016), little commentary has discussed how the app is distinctively *gendered* and *aged*. In other words, Kendall & Kylie trades in objects of *teenage girldom*: low-pressure casual dating, instant ramen and popcorn, texts, selfies, crop tops, and multicolored hair. To Monica Swindle "girldom" refers to the objects that "circulate girl throughout the culture by girls and others as well as by language and cultural representations of girlhood created by marketers, the media, feminism, and other groups who control the means to creating discourse" (2011, para. 29). Girldom, in this sense, is far-reaching; objects may include commercial products such as sparkly nail polish, as well as aesthetics, relationships, movements, rhetorical modes and affects that circulate the notion of "girl" (para 31).

Here, Swindle (2011) theorizes girl as an affect in late capitalist Western culture that circulates between and "sticks" (Ahmed 2004) to particular bodies and objects. She writes, "Girls themselves are not necessarily the objects of the affect girl, though they often generate and circulate this feeling, and many women, when they feel girl are . . . recalled back into the experience of girlhood" (Swindle, para. 16). Using this framework, Swindle complicates girl as a biologically based category anchored in a narrative of progress (becoming woman) and encourages us to consider how girl moves and generates feelings, such as fun, happiness, and power. However, Swindle acknowledges how this "feeling of girl" remains a privilege that some bodies (nonwhite, disabled, non-Western, fat, transgender) cannot access easily, an important reminder that points to the ways girl affects are situated within larger structures of power and inequality.

Yet Kendall & Kylie is fun, in part, because the app grants access to feeling girl, even for those, like the chapter's authors, that are well beyond their teenage years. And while it is of course possible to critique the game for drawing on stereotypical ideas of a privileged, feminized girlhood modeled after the Jenner sisters own images (white, slim, able-bodied, cisgender), we're taking a different perspective here. By theorizing Kendall & Kylie's use of girldom and girl affects as fun, we're following Mary Celeste Kearney's (2015) challenge to reconsider how girl affects such as sparkle can be deployed in the traditions of queer camp culture, challenging the patriarchal dismissal of femininity while reframing them as performative and playful. We may even consider these girl affects as momentarily disrupting the masculinity of tech culture, opening up potential space for not just feminized games but for girl and women app developers and digital media producers as well.[3]

From Paper Dolls to Digital 'Dos: Reconfiguring the App as Part of Girl Culture

Although it's tempting to suggest that Kendall & Kylie is doing something wholly new and innovative based on the unique hardware and software affordance of mobile casual games and the contemporary style of the game's interface, we want to instead contextualize our discussion of girldom and girl affects in relation to a history of girl culture in Western societies. Catherine Driscoll (2002) distinguishes "girl culture" from "girl market" suggesting that the former refers to the circulation of the things that girls can do, be, have, and make, while the latter denotes the advertising, sales, or commercial discourse guiding who buys what popular cultural products. Indeed, while Kendall & Kylie's aesthetics (cute and casual interfaces), language (heavy on the "lol speak"), and early twenties characters all suggest youth, there are neither statistics about who is playing the game nor data on actual players' experiences of the game. Without these empirical insights, however, it is still possible to examine how the game constructs a fantastical affective experience of girlhood, anchored in the historical practices of girl culture yet tied to the expectations of immaterial labor that are so prevalent in contemporary digital culture.

Indeed, Kendall & Kylie incorporates different historical practices of girl culture into a digital space. Doll play, including paper dolls and Barbie, is often about experimentation with fashion and style, facilitating practices ranging from mix-and-match outfitting to impromptu home haircuts that have left beloved toys with lopsided bobs and the occasional Mohawk. These practices have been a part of girls' digital culture for several years, with online paper dolls and fashion-themed world-building games popular among girls (see Willett 2007, 2008; Stein 2009). Kendall & Kylie draws on these longstanding practices, repurposing them to a mobile landscape that assumes a near-constant access to one's phone. Thus, there is a distinctive temporal dimension to the app that distinguishes it from other girl culture play.

In this sense it is worth considering how apps such as Kendall & Kylie extend girls' culture into increasingly convergent mobile media and digital gaming practices and how the norms of this culture shape girl culture in turn. The feminized, low cultural status accorded to both girl culture and casual games allows for an easy marriage. However, we are more interested in the ways both girl culture and casual games operate through distinct ideological ideas about feminine labor. As Aubrey Anable (2013, 8) argues in her analysis of *Diner Dash*, casual games are inextricably bound up in our work cultures, as they are designed around the organization, management, and rhythm of time that we devote to our labor. These games hide the work of affect from the player, representing "fantasy workspaces" where women are successful entrepreneurs engaging in doing what they love, with well-defined, always-rewarded tasks in safe work environments. Indeed, Anable's argument can be contextualized within contemporary convergence culture more broadly, which extends women's affective labor into women's media use and reception practices (Ouellette and Wilson 2011).

Likewise, Kendall & Kylie is an "invest/express" game that primarily involves the coordination of time, people, products, and resources—all of which are in short supply (Chess 2016). The player may express herself through hairstyles, clothing, home decorating, and dating partners, but she must first and foremost discipline her time. Because tasks must be completed in a set period, the player requires strict time management outside of the game to address the constraints of the in-game clock, in a virtual context where time is literally money. All tasks in this game are time-delimited and require set amounts of energy to complete. However, you will exhaust your energy before you can complete the task, and must wait for the energy to replenish. You will therefore have to leave the game but also remember to return before the task ends. In our own play, we have set alarms to remind us to return to the game, to ensure time does not run out, to leave us with a poorly performed task, and to suffer the harsh judgment from in-game avatars that comes with it. In this sense, the app demands not only time management but also multitasking, a practice that is both feminized in the tradition of domestic labor and aligns with the flexible, "always on" neoliberal worker. Interestingly, the app is designed to be played in portrait view, lending itself to easy multitasking play in a way that KK:H's landscape view makes difficult, appealing to always-online users who want to check their other app notifications during play and keep an eye on their energy replenishment's progress.

This tight control over time in play is how free-to-play or "freemium" game apps earn the kind of profits we see within the celebrity genre, with players paying a premium to bypass time-intensive tasks to progress, be it to the million-follower mark in Kendall & Kylie or the "A-List" in Kim Kardashian: Hollywood. As Anable notes of the way these games are played, "The maniacal and rapid tapping and clicking of the player to complete a timed task is a highly visible form of work on a smooth machine that is designed to conceal our labor and to conceal the digital processes that structure our lives" (2013). Quite simply then, these games naturalize work as play (Chess 2017) in a manner that contradicts notions that play is the opposite of work. The targeting of this genre toward the girl market denotes an opportunity for mobile game developers to provide younger audiences engagement with ideological notions of feminized work and labor in contemporary life, while simultaneously recognizing that feminized labor (everything from babysitting to various fan practices) has always been a part of girl culture (Swindle 2011). In Kendall & Kylie the neoliberal and postfeminist discourse of individualist meritocracy to achieve recognition meets a computational structure *making fun from labor*, creating a perfect simulation of contemporary digital work for girls.

Conclusion: The Neoliberal Constraints of Celebrity Apps

The development of other celebrity-inspired apps, such as the forthcoming Glu Mobile–designed Taylor Swift app, suggests a longevity of this genre that previous games designed for girls (such as Ubisoft's Imagine collection and the Barbie series) could not sustain. As research into mobile games indicates (Ramirez 2015), the appification of games allows for new affective appeals to players, through their marketing, design, and payment structures. Constant updates, seasonal events, and push notifications create a constant feedback loop with players, pulling them back to play in ways that a console-based game could not. Indeed, this chapter offers only an introductory and partial analysis of this genre yet points to how the logic of these apps stems from their location within a wider neoliberal media landscape where the labor of developing—and then selling—a self is promoted to millennial girls as a primary route to success in the digital economy (see Banet-Weiser 2012). Yet the implications of this require unpacking. First, Kendall & Kylie socializes girls into the logic of a gendered neoliberal capitalism, offering no alternatives to the persistent grind of feminized work on the self. While the fun girl affects of the game momentarily temper or mask the routinized labor that undergirds Kendall & Kylie, this normalizing of neoliberal capitalism constrains girls— not only temporally and physically to their phones (at least if you want to be a "winner" in the game) but also in terms of the feminine subjectivities deemed desirable both today and into the future.

Second, then, we must consider which feminine girlhood subjectivities are valued within media like Kendall & Kylie and how digital "likes" might reproduce hegemonic values that privilege normative bodies, feminized careers, and individualized selves. Indeed, the app's yoking of girlhood to celebrity suggests that doing millennial girlhood correctly means finding success as an individual star within the narrow confines of fashion, beauty, media, and fitness industries. Once again, labor becomes part of one's feminine identity, yet only particular kinds of work are idealized, what Anita Harris (2004, 23) calls a "glamour worker subjectivity." In this sense, girls are also constrained in how they can imagine their future, imaginings that may shape real-life decisions about education, careers, and relationships. And while it is of course impossible to draw definitive conclusions about the implications of these ideologies without having studied girl users of the Kendall & Kylie app, the concerns we highlight here point to the ongoing need to interrogate how digital engagement operates alongside ideologies that are distinctively gendered, raced, classed, and aged.

Kendall & Kylie represents a cultural text at the intersection of feminized celebrity, casual gaming, and girl culture, all of which tend to be accorded low cultural status and subject to higher scrutiny related to legitimacy, authenticity, and value. However, as the mechanics of the game indicate, nuanced attention is required to perceive how the heteronormative beauty norms of the game and the stereotypical vision of Western girlhood are not the only critique worth making about this game. While these arguments are valid, we chose not to engage with them substantially here. Instead, we suggest the need to move beyond these common critiques of girl's cultural texts, celebrity culture, and casual mobile games in order to better understand how new models of digital engagement operate as ideological systems, particularly in relation to immaterial labor. In the case of Kendall & Kylie, this would include a deeply affective simulation of feminized work and how success functions in the social media era.

Notes

1. Classifications based on genre tend to be fraught with disagreement in game studies, and Google Play does not clarify the attributes of each category of game it deploys. While the placement of Kendall & Kylie under "Adventure" rather than "Casual" may seem curious, this is likely simply a marketing strategy, as it is developers/publishers who suggest the category for their game. Given the sheer volume of "Casual" game apps, this might be a tactic for Glu Mobile to make their games more visible in the Google Play store.

2. This category refers to a collection of games developed in the 1990s designed to cultivate audiences of young female players in a male-dominated domain. Such games have been roundly dismissed as a means of essentializing the relationship between digital play and masculinity and through stereotypical game design features acting as "'special' feminized playthings that escort girls to their proper place in the gender order" (de Castell & Bryson 1998, 232).

3. In July 2016 American business magazine *Forbes* featured Kim Kardashian on its cover alongside the headline "The New Mobile Moguls." The issue suggests a potentially interesting shift away from the association of masculinity with digital production, though

of course ongoing and indeed very recent controversies related to gender-based discrimination in high-tech workforces such as Google indicate the need to be cautious about these promises and attuned to systemic marginalization of women in this domain.

Neko Atsume
Affective Play and Mobile Casual Gaming
SHIRA CHESS

App: Neko Atsume: Kitty Collector
Developer: Hit-Point Co. Ltd.
Release Date: October 2014 (Current Version: 1.10.5)
Category: Games
Price: free-to-play, offers in-app purchases
Platform: Android, iOS
Tags: cats, cuteness, virtual pets
Tagline: "Leave goodies and snacks out in your backyard and wait for the cats of the neighborhood to crowd."
Related Apps: KleptoCats, Tamagotchi Classic, Pokémon Go

The Internet Is for Cats

According to the vast "knowledge base" of the internet, Sigmund Freud once said, "Time spent with cats is never wasted." While there is no proof that Freud actually said this (and there are historical documents insinuating that in reality Freud did not even *like* cats),[1] it seems unsurprising that this questionable sentiment of Freud's ambiguous love of cats would proliferate online. Because regardless of the pithy lyrics from off-Broadway musicals, the internet is for cats. Between a deep obsession with "Grumpy Cat," the propagation of memes from the website icanhaz.cheezburger.com, and a YouTube full of cats doing generally kittenly things, it seems that cats are an inescapable force of digital culture, Freud be damned.

Given that, Neko Atsume: Kitty Collector and its ensuing popularity as an app should not have caught anyone by surprise. Neko Atsume refers to itself as a "cat collector," which is quite literal as far as app descriptions are concerned. The game-play is simple. The player begins with a small amount of in-game currency (gold and silver sardines). Sardines can be used to purchase food and toys for virtual cats. Players can place several different kinds of toys within a small space and put out varying qualities of cat food. The player is told to leave the game for a while, and when they return, smiling cats almost magically appear. Most of the cats that show up are fairly ordinary: cats with names like Snowball, Fred, Pickles, and Socks curl up and snuggle on your smartphone screen, pleased with the offerings. Sometimes they play in cute animations, and sometimes they just take little naps.

If you place specific special items in the game, you might get visits from "rare" cats. For example, if you put out a cowboy hat, you might get a visit from Billy the Kitten, a baseball brings Joe DiMeowgio, a cardboard train cutout brings Conductor Whiskers, and a cat habitat shaped like a restaurant lures in Sassy Fran. The game revels in this strange combination of reality and fantasy. As each cat visits, the player can take photos, which get recorded in a "Catbook" (the collecting part). After the food runs out, the cats wander off until you refresh the food, although they leave gold and silver sardines, which the player can use to replenish food and buy more toys and habitats. The player can expand the yard (although only to a limited extent). When the cats visit enough times, they bring the player "Mementos"—generally meaningless items, but the kind of gifts that cats might bring a person they liked, such as acorns, feathers, coins, and bottle caps. The game is deeply submerged in the aesthetic of cute.

However, as simple and obvious a game as this might be in a lot of respects, it did take game developers, internet communities, and global culture by surprise. CNN referred to it as "addicting" (Luu et al., 2015). *The Telegraph* referred to it as a "classier angry birds" (Horton 2016), despite the fact that the two games couldn't have any less in common. Mikki Halpin, an editor from the feminist newsletter *Lenny* asserts, "I don't meditate or anything, but I think that Neko Atsume is really fucking Zen" (Halpin 2016). Even the game's developer, Yutaka Takazaki, has admitted uncertainty as to how the game got as popular as it has. In one interview he states, "Honestly speaking, I don't understand why it became so popular" (Horton 2016). Yet it has not only been popular—it has been award-winning. Among other awards, in 2016 it was awarded as one of the Top 5 Mobile Games of 2015 by GameStop. The game has inspired a trove of cute Neko Atsume cat-themed products too numerous to catalog and inspired a live event at a cat café in Japan.

But the question remains: what underlies this game's seemingly viral popularity, and what can it tell us more broadly about how apps reframe our relationships with software, feline-focused or otherwise. The game's success is strongly linked to its ability to deploy a kind of affect on players—it relies entirely on our emotional needs and emotional entanglements, our desire to feel things, and our desire to take care of things, even when those "things" exist only as mere pixels. While we might find it unsurprising that affect plays such an important everyday role in our (very) emotional worlds, it would seem to be far more jarring to recognize that we might desire an emotional relationship with a smartphone app. Yet, unlike other games that play with digital affect, Neko Atsume is pressure free—we can call on it for emotional catharsis but can abandon it easily and without repercussion. In what follows, I identify this strange mode of digital affect as "conceptual affect" linking its cute aesthetic to the idea that the affect is never quite real. It is because of the game's conceptual affect and strange tension between emotional desire and distance that makes Neko so successful. Similarly, through Neko Atsume we can better understand the emotional resonances that occur in seemingly small apps.

Video Games and the Labor of Affect

Neko Atsume, at its core, is ruled by affect, or our emotional relationship with the game and its critters. Affect is always about personal emotional entanglements and our very human desires to be touched emotionally by things. Many scholars and game designers have considered the various ways that video games can make us feel things. Katherine Isbister (2016) asserts that because of a combination of game flow and player agency, video games have the ability to have more emotional impact than any other medium. Elsewhere she claims that this emotional impact is primarily performed through the characters—the relationship between the characters in the games and how they are able to relate to the player in intimate ways (Isbister 2006). Game designer David Freeman refers to in-game emotional engineering as "emotioneering," explaining, "People go to films, watch TV shows, and listen to music that moves them. Emotion will be one of the keys to the mass market in games as well. Thus, from the point of view of economics, emotioneering in games is good business" (Freeman 2004, 11). Similar to Isbister, Eugénie Shinkle (2008) argues that video games are more equipped to create emotional responses than other forms of media because of combinations of sensory feedback with location and movement, necessitated by gaming interfaces. All of this is to say that games make us feel things; they create spaces where affect is a potential byproduct and motivator toward play.

Affect, however, is messy. Often, affective and emotional forms of labor end up being forms of work traditionally assigned to women in non-virtual spaces such as jobs like nursing, flight attendants, and waitressing. Arlie Russell Hochschild coined the term "emotional labor" to refer to labor that "requires one to induce or suppress feeling in order to sustain the outward countenance that produces the proper state of mind in others" (Hochschild 1983, 7). This labor is institutionalized in a way that makes the purveyor of that labor essentially powerless. Similarly, as Kylie Jarrett (2013) notes, "affective labor" tends to refer to similar forms of labor activities in the domestic sphere, that are often unaccounted for as a kind of immaterial labor. Jarrett argues that this kind of affect can slip into digital spaces, ultimately fueling the digital economy. Our emotions are central to the capitalist machine. Within all of this, one can agree that (1) video games have a particular power to resonate with the emotional lives of players, and (2) women, in particular, might have a complicated, labor-intensive relationship with that degree of emotionality.

However, while affect and emotional labor are certainly core to Neko Atsume, the game lacks any inherent gendering, in terms of its design. Elsewhere I have argued that we can understand gender and game design on a continuum supported by several factors, including aesthetics, game-play attributes, and general issues of accessibility (Chess 2017). Using this scale, the game seems to be relatively gender neutral in its style and execution. For example, the aforementioned news articles on the game were written by both men and women, and the game's popularity seems to reinforce its rather broad appeal regardless of gender, age, and ethnicity. The game seems to offer a kind of affective experience for almost anyone.[2] While the neutrality of the game is notable unto itself—we often hear more about video games that are deliberately designed for masculine audiences—Neko Atsume manages to be fairly agnostic in its appeal.

The affect in Neko Atsume, in many ways, can be likened to other virtual pet games, such as *Nintendogs*, *FooPets*, *GoPets*, or *Kinectimals*. Even classic semi-physical toys such as Tamagotchi and Furby are predecessors of the Neko Atsume experience. In these games, the player is given virtual pets that they must feed, play with, and groom. However, with most virtual pet games there are repercussions for a lapse in player engagement. If you ignore a Tamagotchi, it will eventually get sick and die. If you ignore a Nintendog, it will get surly and act out in less than desirable ways. The games push the player toward a kind of constant engagement and touching of screens. The emotional and affective labor is built into these games in such a way that the player knows that if they do not fulfill their promises to their virtual pet, their affect will be lost and they will ultimately become an abusive owner. Because we cannot possibly play with a Tamagotchi forever, it must eventually die—and likely at our own hand. Most virtual pets, then, are digital embodiments of guilt and predictors of neglect. The virtual pet is, necessarily, a pet that will eventually be forgotten in almost every case.

However, the cats of Neko Atsume are entirely different beasts. They are digital strays that wander into our virtual backyard, and we feed them until we get bored. Not feeding them does not mean the cats will die. Not providing them with warm heating pads doesn't mean they will freeze. When we don't feed them for a while the cats disappear, but when we put out new food they always come back, just as happy as they were before. We are left to assume that these strays are doing just fine, and they aren't our responsibility—although they certainly appreciate us when we try. It is a caregiving game, but one without strings or commitment. Neko Atsume offers the pleasures of affect, but removes the overlay of Tamagotchi guilt. No matter what the player does, the cats will be fine.

The Aesthetics of "Cute"

At their core, the cats of Neko Atsume are subsumed in a tidal wave of cute. The draw of the game, the pull that keeps us coming back and visiting our pixeled, faux-furred friends, is a promise of cuteness. And, for sure, the game delivers. In a blog post from 2009, Ian Bogost distinguishes between cute aesthetics that are characterized by inability (or partial inability) fostered by a kind of parental protectiveness versus Japanese depictions of cuteness (which he argues are of the Sanrio, "Hello Kitty" variety) that focus less on inability and more on appearance (Bogost 2009).

There is a childishness inherent in the cute aesthetics and design of Neko Atsume. The simplicity of dropping food into a bowl (which never needs to be cleaned) implies a kind of "sweet" labor that emulates reality without messiness. The drag-and-drop of the spatial aesthetic lacks any real labor or even mindful game-play. The game exchanges affect for cuteness, with its smiling, easy-to-please cats reveling in these aesthetics, simplifying all cat behaviors down to eating, sleeping, and playing. In reality, we don't even ever get to see the cats eat—food is consumed when we aren't looking.[3] The cats on the screen, no matter what they are doing, are necessarily smiling with big, penciled curves. Their cuteness is to alert us that we have done something right, although the game design dictates that no wrong can really be done. The worst we can do to the Neko Atsume cats is to not show up, or (perhaps) to uninstall the game. But because someone else on some other screen will always feed the cats, because they are not our responsibility, their cuteness is central to our desire to nurture them as well as a form of positive feedback that we are doing things right.

However, part of the weirdness of the game lies within that aesthetic of cute. Those of us who are cat lovers (and I will admit to being one myself) rarely are drawn to these pets for their "smiling" disposition. A cat may be contented, for sure, but their real-world behavioral conceits tend to be more prone toward aloofness and passive aggressive snuggling.[4] Cats are certainly capable of being cute, but that cuteness is rarely related to a smiling disposition. In this way, the smiles of Neko Atsume seem to make them distinct from actual cats. After all, even the ever optimistic title character from Hello Kitty denies us actual smiles. The in-game cuteness of Neko Atsume cats seems to be an essence, but one that is devoid of the emotional roller coaster inherent in actual cat ownership. Like little Cheshires, these cats *always* smile. And this lack of reality helps us keep the affect from ever feeling too real or too much a part of our actual lives.

Specific aspects of the game seem to drive this aesthetic even further, rewarding the player's time with improbable cuteness. It is impossible to resist a smiling round-faced cat in the first place, but one wearing a conductor's uniform (Conductor Whiskers) or a little waitressing outfit (Sassy Fran) seems almost too much to bear. The animations, the movements of the cats are small and mostly inconsequential. At many points when the player opens the screen (after having put down the proper amount of food), a cat might be tossing a ball back and forth with itself or playing with a kick toy. The latter involves the cat on its back, displaying its tiny anus for the player to revel in. And, somehow, even the Neko Atsume cat anus manages to be cute. If Mikki Halprin is correct and the game really is "Zen," then part of that zen-ness lies within the reward system. There is no failure, only cute little rewards. The cuteness makes all of the affect positive. It is in the "cute" nature of Neko that we can see childish and simple affect that is devoid of real responsibilities, yet also guilt.

The Problem with Cats You Can't Pet

The cuteness of Neko Atsume is what brings players in. It is our payment in full for our affective labor. It is also what keeps us from feeling that it bears too much emotional weight on our actual lives. Because, of course, in real life the pleasure of pet ownership is beyond feeding them or just leaving out toys for them—a pet owner typically wants the more visceral aspects of petting, prodding, and playing. While having a fictional cat smile contentedly is nice, stroking a cat's soft fur while it purrs is nicer. As with all virtual pets, herein lies the problem: we can't touch our Neko Atsume cats. We are shown their adorable smiles—so we know that they are pleased. We get gifts from them—so we know they probably like us. But beyond that our affect is entirely one-sided. We give but never receive affection. The Neko Atsume cats construct a kind of desire that they can never quite fulfill.

This returns us to the idea of "conceptual affect." After all, Neko Atsume makes one wonder if the affect a player feels for a digital cat that is incapable of mirroring those emotions is real affect. Real affect implies a subject of emotional largesse, and it is difficult to conceive of that in a digital pet. Because we are performing labor for virtual animals that are never fully capable of feeling (or responding properly to our feelings) the affect is ultimately hollow. It has all of the structural conceits of affective labor sans actual emotional content.

The pleasures offered by the Neko Atsume app are so slight, so delicate, that if you blink you might miss them. It forces the player to seek the very smallest validations and minuscule forms of attention and subsequently exploding meaning into them. *Tabitha gave me random seeds, so she must love me a lot and appreciate all that I've done for her. Pickles gave me five gold sardines, so he knows that the Cat Condo was meant for him.* In Neko Atsume all affection is conceptual, and we internalize the small gestures of our digital strays as love. But that affection also elicits more desire—the things the cats offer is never quite enough, which is why we keep playing. In this way, we can think of conceptual affect as a necessary part of our "app-iness." Conceptual affect is a way to manage our emotions on a small scale, reaping only positive emotional experiences.

And this, of course, is the problem with cats you can't pet. When cats are coded down to pure behaviors—and only positive behaviors, at that—we lose some of the essence of their catness. At the same time, we are trading on our affect for fleeting moments and the persuasive power of cuteness. We get exactly the amount of affirmation that we infer into the app experience—it supplies short-form validation with no strings attached.

Conclusion

One day, soon after I had begun playing Neko Atsume, I realized that one of my real cats was underfoot. She clearly wanted to be fed. I gave her the side-eye. "I understand your needs, but I have pretend cats to feed, too. Can you give me five minutes? It's very important that I finally get to see Joe Di-Meowgio in action."[5] This is not, of course, to infer that we are ignoring the real for the virtual—I did ultimately feed my cat. However, this moment does remark upon the neatness of the affective experience offered by Neko Atsume. The cats in the game offer only sanguine responses, no matter how often they are fed. They require no cleanup after their meals, and despite their positively adorable anuses, we do not have to clean a virtual litter box. They do not get sick and cannot die. The experience is pure pleasure, but it is also impossibly sterile.

It is within this sterility that we can understand the cultural obsession with the game and perhaps even its importance in our app-filled lives. The app promises a release system for the affect we feel every day, all around us. We are overwhelmed with emotions, stimulated by feels in real life and the internet. We cannot always disclose those emotions nor have an easy outlet for their release. The conditional nature of conceptual affect allows us to sneak emotion into our everyday, quietly and without fanfare. Certainly we can't touch our Neko Atsume cats, but we always know that if we do a certain set of things, we will make them happy. We may not always be able to express our feelings and devotion to our loved ones, our friends, our colleagues, or even our actual cats. Neko Atsume provides a space where we can steal away for moments, get our affect on, and know that nothing we do can possibly be wrong.

Notes

1. According to the Freud Museum of London's website, this quote is fairly dubious, as they cited a letter to his friend Arnold Zweig in which he asserts, "I, as is well known, do not like cats" (Freud Museum London).

2. Except, perhaps, for Freud.

3. The single exception to this is the overweight cat, Tubbs, who occasionally comes in and steals all of the food in a bowl in one shot.

4. This is by no means meant to be dismissive about the lovely qualities of cats. Please don't tell my cats that I wrote this.

5. My cat is not now, and has never been, impressed with Joe DiMeowgio no matter how many times I place my iPhone in front of her.

References

"About Logo TV". 2015. *Logo TV*. http://LogoTV.com/about.

Acker, Amelia, and Brian Beaton. 2016. "Software Update Unrest: The Recent Happenings around Tinder and Tesla." In *Proceedings of the 49th Hawaii International Conference on System Sciences (HICSS)* January 5–8. Koloa, Hawaii. doi: 10.1109/HICSS.2016.240.

ACLU. 2015. "ACLU Releases Tool to Hold Law Enforcement Accountable." ACLU, November 13. www.acludc.org/en/node/43399.

ACLU. 2016. "ACLU Apps to Record Police Conduct." www.aclu.org/feature/aclu-apps-record-police-conduct.*Adam Walsh Child Protection and Safety Act of 2006 published by 109th Congress Public Law 248*. www.gpo.gov/fdsys/pkg/PLAW-109publ248/html/PLAW-109publ248.htm.

Adams, Douglas. 1980. *The Restaurant at the End of the Universe*. New York: Pan Books.

Adams, Vincanne, Michelle Murphy, and Adele E. Clarke. 2009. "Anticipation: Technoscience, Life, Affect, Temporality." *Subjectivity* 28, no. 1: 246–65.

Addison, Joseph. 1811 [1711]. Entry No. 25, Thursday March 29. In *The Spectator, Eight volumes*. London: Whittingham and Rowland. 29–30.

Adelman, Lori. 2011. "Can Technology Help Prevent Sexual Assault and Dating Violence on College Campuses?" *Feministing.com*. July 18. http://feministing.com/2011/07/18/can-technology-help-prevent-sexual-assault-and-dating-violence-on-college-campuses/.

Ahmed, Sara. 2004. *The Cultural Politics of Emotion*. Edinburgh: Edinburgh University Press.

Akrich, Madeleine, and Bruno Latour. 1992. "A Summary of a Convenient Vocabulary for the Semiotics of Human and Nonhuman Assemblies." In *Shaping Technology/Building Society: Studies in Sociotechnical Change*, edited by Wiebe E. Bijker and John Law, 259–64. Cambridge: MIT Press.

Albanesius, Chloe. 2012. "Shazam Shifts from Music to TV, Ad Discovery." *PCMag UK*. June 5. http://uk.pcmag.com/apps/63233/news/shazam-shifts-from-music-to-tv-ad-discovery.

Al Jazeera America. 2013. "Is 'Stop and Frisk' Street Harassment?" *The Stream*. http://america.aljazeera.com/watch/shows/the-stream/the-stream-officialblog/2013/8/20/is-stop-and-friskstreetharassment.html.

Anable, Aubrey. 2013. "Casual Games, Time Management, and the Work of Affect." *Ada* 2. http://adanewmedia.org/blog/2013/06/01/issue2-anable.

Anand, N., and Richard A. Peterson. 2000. "When Market Information Constitutes Fields: Sensemaking of Markets in the Commercial Music Industry." *Organization Science* 11, no. 3: 270–84.

Ananny, Mike. 2015. "From Noxious to Public? Tracing Ethical Dynamics of Social Media Platform Conversions." *Social Media + Society* 1, no. 1. doi: 2056305115578140.

Anderson, Chris. 2010. "The Web Is Dead? A Debate." *Wired.* August 17. www.wired. com/2010/08/ff_webrip_debate.

Anderson, Monica. 2015. "Key Takeaways on Mobile Apps and Privacy." *PEW Internet Research.* www.pewresearch.org/fact-tank/2015/11/10/key-takeaways-mobile-apps.

Andrejevic, Mark. 2005. "The Work of Watching One Another: Lateral Surveillance, Risk, and Governance." *Surveillance & Society* 2, no. 4: 479–97.

Andrejevic, Mark. 2006. "The Discipline of Watching: Detection, Risk, and Lateral Surveillance." *Critical Studies in Media Communication* 23, no. 5: 391–407.

Andrejevic, Mark, and Mark Burdon. 2015. "Defining the Sensor Society." *Television & New Media* 16, no. 1: 19–36. doi: 10.1177/1527476414541552.

Andrejevic, Mark. 2017. "To Preempt a Thief." *International Journal of Communication* 11: 879–96.

Andrews, Michelle, Jody Goehring, Sam Hui, Joseph Pancras, and Lance Thomswood. 2016. "Mobile Promotions: A Framework and Research Priorities." *Journal of Interactive Marketing* 34: 15–24.

Andrews, Robert. 2011. "Listen Up, Shazam; Hundreds of Rivals Are About to Bloom." *Gigaom.* June 23. https://gigaom.com/2011/06/23/419-look-out-shazam-hundreds-of-rivals-are-about-to-bloom.

Andy. 2017. "Spotify's Beta Used 'Pirate' MP3 Files, Some from Pirate Bay." *TorrentFreak.* May 9. https://torrentfreak.com/spotifys-beta-used-pirate-mp3-files-some-from-pirate-bay-170509.

Ankerson, Megan Sapnar. 2015. "Social Media and the 'Read-Only' Web: Reconfiguring Social Logics and Historical Boundaries." *Social Media+ Society* 1, no. 2. doi: 2056305115621935.

App Annie. 2010. "Introducing App Annie—Now in Open Beta!" App Annie Company Blog, appannie.com. March 2 (retrieved via WayBack Machine).

App Annie. 2016. "App Annie Mobile App Forecast: The Path to $100 Billion." www.appannie.com/insights/market-data/app-annie-releases-inaugural-mobile-app-forecast.

App Annie. 2018. "App Annie 2017 Retrospective." www.appannie.com/en/insights/market-data/app-annie-2017-retrospective/#download

App Annie TV. 2016. "App Annie Case Studies: Etermax." www.youtube.com/watch?v=5vmusDAgI_0.

Apple. 2008. "iPhone App Store Downloads Top 10 Million in First Weekend." Apple Newsroom. July 14. www.apple.com/pr/library/2008/07/14iPhone-App-Store-Downloads-Top-10-Million-in-First-Weekend.html.

Apple. "Design Principles—Overview—IOS Human Interface Guidelines." https://developer.apple.com/ios/human-interface-guidelines/overview/design-principles.

Apple. 2018. "See Send." App Store. https://itunes.apple.com/ca/app/see-send/id556069712?mt=8.

Apple. 2016. "Tinder." iTunes Store. Retrieved June 28, 2016, from https://itunes.apple.com/au/app/tinder/id547702041?mt=8.

Arnold, Matthew. 2011. "Devices & Diagnostics: The App Avant-Garde." *Medical Marketing and Media.* www.mmm-online.com.

Arthur, Charles. 2015. "Shazam—From Text-Based Service to $1bn App in 15 Years." *The*

Guardian. January 26. www.theguardian.com/technology/2015/jan/26/shazam-from-text-based-service-to-one-billion-dollar-app-15-years.

Aspers, Patrik, and Nigel Dodd, eds. 2015. *Re-Imagining Economic Sociology*. Oxford: Oxford University Press.

Athique, Adrian. 2016. *Transnational Audiences: Media Reception on a Global Scale*. Cambridge, UK: Polity Press.

Auletta, Ken. 2009. "Searching for Trouble: Why Google Is on Its Guard." *New Yorker*, October 12. 46–56.

Auslander, Philip. 2005. "At the Listening Post, or, Do Machines Perform?" *International Journal of Performance Arts and Digital Media* 1, no. 1:5–10.

Azuma, Ronald T. 1997. "A Survey of Augmented Reality." *Presence: Teleoperators and Virtual Environments* 6, no. 4: 355–85.

Bakhtin, Mikhail. 1984. *Rabelais and His World*. Vol. 341. Bloomington: Indiana University Press.

Baldwin-Philippi, Jessica. 2015. *Using Technology, Building Democracy: Digital Campaigning and the Construction of Citizenship*. Oxford: Oxford University Press.

Banaji, Shakuntala (2015) "Behind the High-Tech Fetish: Children, Work and Media Use across Classes in India." *International Communication Gazette* 77, no. 6: 519–32.

Banet-Weiser, Sarah. 2011. "Branding the Post-Feminist Self: Girls' Video Production and YouTube." In *Mediated Girlhoods*, edited by Mary Celeste Kearney, 277–94. New York: Peter Lang.

Banks, John. 2012. "The iPhone as Innovation Platform: Reimagining the Videogames Developer." In *Studying Mobile Media: Cultural Technologies, Mobile Communication, and the iPhone*, edited by Larissa Hjorth, Jean Burgess, and Ingrid Richardson, 155–72. New York: Routledge.

Baron, Steve. 2015. "'RuPaul's Drag Race' Renewed for Season 8 by Logo." *TV By The Numbers*. March 20. http://tvbythenumbers.zap2it.com/2015/03/20/rupauls-drag-race-renewed-for-season-8-by-logo/377626.

Barrett, Rusty. 1995. "Supermodels of the World, Unite! Political Economy and the Language of Performance among African-American Drag Queens." *The Language and Sexuality Reader*, edited by Deborah Cameron and Don Kulick, 151–63. New York: Routledge 2006.

Barrientos J. 2012. "Cell Phones Making Campus Emergency Systems Obsolete." *Bakersfield Californian*. March 25.

Barzilai-Nahon, Karine. 2008. "Toward a Theory of Network Gatekeeping: A Framework for Exploring Information Control." *Journal of the American Society for Information Science and Technology* 59, no. 9: 1493–1512.

Bayer, Joseph B., Nicole B. Ellison, Sarita Y. Schoenebeck, and Emily B. Falk. 2016. "Sharing the Small Moments: Ephemeral Social Interaction on Snapchat." *Information, Communication & Society* 19, no. 7: 956–77. doi: 10.1080/1369118X.2015.1084349.

BBC. 2006. "UK Music Fans Can Copy Own Tracks." *BBC News*. June 6. http://news.bbc.co.uk/2/hi/entertainment/5053658.stm.

"Become a Tasker." n.d. *TaskRabbit.com* https://www.taskrabbit.com/become-a-tasker.

Bennett, Vicki. 2012. "Collateral Damage." *Wire*. February. http://thewire.co.uk/articles/8439.

Bennett, W. Lance, and Jarol B. Manheim. 2006. "The One-Step Flow of Communication." *Annals of the American Academy of Political and Social Science* 608, no. 1: 213–32.

Bergstrom, Kelly. 2011. "'Don't feed the troll': Shutting Down Debate about Community Expectations on Reddit.com." *First Monday* 16, no. 8. http://firstmonday.org/ojs/index.php/fm/article/view/3498/3029.

Berlant, Lauren. 2006. "Cruel Optimism." *differences: A Journal of Feminist Cultural Studies* 17, no. 3: 20–36.

Bermejo, Fernando. 2009. "Audience Manufacture in Historical Perspective: From Broadcasting to Google." *New Media & Society* 11, nos. 1–2: 133–54.

Bernhard, Bo, and Vincent Eade. 2005. "Gambling in a Fantasy World: An Exploratory Study of Rotisserie Games." *UNLV Gaming Research & Review Journal* 9, no. 1: 29–42.

Bhaduri, Saugata. 2016. "Gaming." In *Digital Keywords: A Vocabulary of Information Society and Culture*, edited by Benjamin Peters, 140–48. Princeton Studies in Culture and Technology. Princeton: Princeton University Press.

Bijsterveld, Karin, Eefje Cleophas, Stefan Krebs, and Gijs Mom. 2013. *Sound and Safe: A History of Listening behind the Wheel*. Oxford: Oxford University Press.

Bilge, Sirma. 2013. "Intersectionality Undone: Saving Intersectionality from Feminist Intersectionality Studies." *Du Bois Review* 10, no. 2: 405–24.

Bilge, Sirma. 2014. "Whitening Intersectionality: Evanescence of Race in Intersectionality Scholarship." In *Racism and Sociology: Racism Analysis Yearbook 5*, edited by Wulf D. Hund and Alana Lentin,175–205. Berlin: Lit Verlag/Routledge.

Billings, Andrew, and Brody Ruihley. 2013a. *The Fantasy Sport Industry: Games within Games*. London: Routledge.

Billings, Andrew, and Brody Ruihley. 2013b. "Why We Watch, Why We Play: The Relationship between Fantasy Sport and Fanship Motivations." *Mass Communication and Society* 16, no. 1: 5–25.

Birkinbine, Benjamin, Rodrigo Gomez, and Janet Wasko, eds. 2016. *Global Media Giants*. London: Routledge.

Bivens, Rena, and Amy Adele Hasinoff. 2018. "Rape: Is There an App for That? An Empirical Analysis of the Features of Anti-Rape Apps." *Information, Communication & Society.* 21 no. 8: 1050–67. http://dx.doi.org/10.1080/1369118X.2017.1309444.

B lav. 2009. "Unshazamable." *Urban Dictionary*. www.urbandictionary.com/define.php?term=Unshazamable.

blunden. 2016. "Shazam Lite—No Region Restriction." *Android Development and Hacking*. October 10. http://forum.xda-developers.com/android/apps-games/app-shazam-lite-region-restriction-t3477858.

Boellstorff, Tom. 2008. *Coming of Age in Second Life: An Anthropologist Explores the Virtually Human*. Princeton, NJ: Princeton University Press.

Boghossian, Emily. 2016. "How I Made It: What Lands in the This.cm Newsletter, and Why." Airmedia.org. March 16. http://airmedia.org/how-i-made-it-what-lands-in-the-this-cm-newsletter-and-why.

Bogost, Ian. 2009. "A Theory of Cuteness." *Gamasutra*. www.gamasutra.com/blogs/IanBogost/20090812/85200/A_Theory_of_Cuteness.php.

Bogost, Ian. 2011a. "'Gamification Is Bullshit.'" *The Atlantic*. August 9. www.theatlantic.com/technology/archive/2011/08/gamification-is-bullshit/243338.

Bogost, Ian. 2011b. "What Is An App? A Shortened, Slang Application." Ian Bogost Blog. www.bogost.com/blog/what_is_an_app.shtml.

Bogost, Ian. 2012. *Alien Phenomenology, or, What It's Like to Be a Thing*. Posthumanities 20. Minneapolis: University of Minnesota Press.

Bonnington, Christina. 2015. "Instagram Won't Let You Search for THAT Emoji." *Refinery 29*. April 28. www.refinery29.com/2015/04/86409/instagram-has-already-banned-search-of-one-emoji.

Bosker, Bianca. 2013. "Siri Rising: The Inside Story Of Siri's Origins — And Why She Could Overshadow The iPhone." *Huffington Post*. January 22, 2013. www.huffington-post.com/2013/01/22/siri-do-engine-apple-iphone_n_2499165.html.

Bouchard, Anthony. 2016. "Link Your Spotify Account with Shazam to Enable Full Song Previews & More." *iDownload Blog*. February 8. @disqus_I9SPCKW30J. 2016. Comment to Bouchard 2016. www.idownloadblog.com/2016/02/08/how-to-link-shazam-with-spotify.

Boudreau, Kevin J. 2012. "Let a Thousand Flowers Bloom? An Early Look at Large Numbers of Software App Developers and Patterns of Innovation." *Organization Science* 23, no. 5:1409–27. doi: 10.1287/orsc.1110.0678.

Boudreau, Kevin, and Andrei Hagiu. 2009. "Platform Rules: Multi-Sided Platforms as Regulators." In *Platforms, Markets, and Innovation*, edited by Annabelle Gawer, 163–91. Cheltenham, UK: Edward Elgar Publishing.

Bourdieu, Pierre. 1990. *Photography: A Middle-Brow Art*. Stanford, CA: Stanford University Press.

Bourdieu, Pierre. 1991. *Language and Symbolic Power*. Translated by John B. Thompson. Cambridge, UK: Polity Press.

Bourne, Emily. 2001. "China Mobile Hopes to Emulate i-Mode Success with Monternet." *Total Telecom. April 25.*

Bowker, Geoffrey C., and Susan Leigh Star. 2000. "Invisible Mediators of Action: Classification and the Ubiquity of Standards." *Mind, Culture, and Activity* 7, nos. 1–2: 147–63. doi: 10.1080/10749039.2000.9677652.

boyd, danah, and Nicole B. Ellison. 2007. "Social Network Sites: Definition, History, and Scholarship." *Journal of Computer-Mediated Communication* 13, no. 1: 210–30.

Bratton, Benjamin H. 2016. *The Stack: On Software and Sovereignty*. Cambridge: MIT Press.

Brennan, Teresa. 2014. *The Transmission of Affect*. Ithaca, NY: Cornell University Press. Kindle Edition.

Brown, Sam. 2014. "Designing the New Foursquare." *Medium*. August 6. https://medium.com/@sambrown/designing-the-new-foursquare-8f8788d366f0#.t2xy5kyku.

Brownstone, Sydney. 2013. "An App to Help the Government Stop Street Harassment." FastCompany blog. August 20. www.fastcompany.com/3016176/an-app-to-help-the-government-stop-street-harrasment.

Brunton, Finn. 2013. *Spam: A Shadow History of the Internet*. Cambridge: MIT Press.

Bumiller, Kristin. 2008. *In an Abusive State: How Neoliberalism Appropriated the Feminist Movement against Sexual Violence*. Durham, NC: Duke University Press.

Burgess, Jean. 2012 "The iPhone Moment, the Apple Brand, and the Creative Consumer: From 'Hacka- bility and Usability' to Cultural Generativity." In *Studying Mobile Media:*

Cultural Technologies, Mobile Communication, and the iPhone, edited by Larissa Hjorth, Jean Burgess and Ingrid Richardson, 28–42. New York: Routledge.

Burgess, Jean, and Joshua Green. 2013. *YouTube: Online Video and Participatory Culture*. 2nd ed. Oxford: Wiley.

Burston, Jonathan. 2003. "War and Entertainment: New Research Priorities in an Age of Cyber-Patriotism." In *War and the Media: Reporting Conflict 24/7*, edited by Daya Kishan Thussu and Des Freedman, 163–75. London: Sage.

Bury, Rhiannon, and Li Johnson. 2015. "Is It Live or Is It Timeshifted, Streamed, or Downloaded? Watching Television in the Era of Multiple Screens." *New Media & Society* 17, no. 4: 592–610.

Butler, Andy. 2014. "Red Antler Tells Us More about the New Foursquare Identity." *Designboom | Architecture & Design Magazine*. August 8. www.designboom.com/design/red-antler-new-foursquare-identity-process-08-07-2014.

Butler, Judith. 1993. *Bodies That Matter: On the Limits of "Sex."* New York: Routledge.

Button, Deeanna M., Matthew DeMichele, and Brian K. Paynem. 2009. "Using Electronic Monitoring to Supervise Sex Offenders: Legislative Patterns and Implications for Community Corrections Officers." *Criminal Justice Policy Review* 20, no. 4) 414–36.

Byrne, Dara. 2013. "419 Digilantes and the Frontier of Radical Justice Online." *Radical History Review* 117: 70–82. doi 10.1215/01636545–2210464.

Cabral, Luis, and Gabriel Natividad. 2016. "Cross-Selling in the US Home Video Industry." *Rand Journal of Economics* 47, no. 1: 29–47.

Callon, Michel, ed. 1998. *The Laws of the Markets*. Oxford: Blackwell .

Campbell-Kelly, Martin. 2003. *From Airline Reservations to Sonic the Hedgehog: A History of the Software Industry*. History of Computing. Cambridge: MIT Press.

Capps K. 2014. People Just Won't Give Up on Awful Neighborhood Apps. *CityLab*. www.citylab.com/tech/2014/08/people-just-wont-give-up-on-awful-neighborhood-apps/375766.

Carlson, Matt. 2006. "Tapping into TiVo: Digital Video Producers and the Transition from Schedules to Surveillance in Television." *New Media & Society* 8, no. 1: 97–115.

Carpenter, Christopher, and Bree McEwan. 2016. "The Players of Micro-Dating: Individual and Gender Differences in Goal Orientations toward Micro-Dating Apps." *First Monday* 21, no. 5. http://firstmonday.org/ojs/index.php/fm/article/view/6187/5469.

Carson, Dan. 2016. "The Hillary Clinton App Is One of the Best Digital Organizing Tools Ever Made." *Medium*. October 13. https://medium.com/@okdan/the-hillary-clinton-app-is-one-of-the-best-digital-organizing-tools-ever-made-39628cd7ab91.

Centre for Equity Studies. *India Exclusion Report 2016*. New Delhi: Yoda Press. 2017.

Chen, Baoliang, 2015. "Meibu shouji 1 meiyuan mailu; Google Play ruhua qianjing buming" (The Unclear Picture of Google Play's Reentry into China). *163.com*. http://tech.163.com/15/0602/03/AR2SVN47000915BF.html.

Chess, Shira. 2016. "A Time for Play: Interstitial Time, Invest/Express Games, and Feminine Leisure Style." *New Media & Society*. doi: 10.1177/1461444816660729.

Chess, Shira. 2017. *Ready Player Two: Women Gamers and Designed Identity*. Minneapolis: University of Minnesota Press.

Chun, Wendy Hui Kyong. 2008a. "The Enduring Ephemeral, or the Future Is a Memory." *Critical Inquiry* 35, no. 1: 148–71. doi: 10.1086/595632.

Chun, Wendy Hui Kyong. 2008b. "On 'Sourcery' or Code as Fetish." *Configurations* 16, no. 3: 299–324.

Chun, Wendy Hui Kyong. 2011. *Programmed Visions.* Cambridge: MIT Press.

Chun, Wendy Hui Kyong. 2016. *Updating to Remain the Same: Habitual New Media.* Cambridge: MIT Press.

Chun, Wendy Hui Kyong, and Thomas Keenan. 2006. *New Media, Old Media: A History and Theory Reader.* New York: Routledge.

Citron, Danielle Keats. 2014. *Hate Crimes in Cyberspace.* Cambridge, MA: Harvard University Press.

Clark, Jessica, Nick Couldry, Abagail De Kosnik, Tarleton Gillespie, Henry Jenkins, Christopher Kelty, Zizi Papacharissi, Andrea Powell, and José van Dijck. 2014. "Participations: Dialogues on the Participatory Promise of Contemporary Culture and Politics—Part 5: PLATFORMS." *International Journal of Communication* 8: 1446–73.

Clark-Gordon, Cathlin V., Kimberly E. Workman, and Darren L. Linvill. 2017. "College Students and Yik Yak: An Exploratory Mixed-Methods Study." *Social Media + Society* 3, no. 2. https://doi.org/10.1177/2056305117715696.

Clayton, Nick. 2002. "SCOT: Does It Answer?" *Technology and Culture* 43, no. 2: 351–60. doi: 10.1353/tech.2002.0054.

Cockroft, Steph. 2016. "Pokemon Go is 'a Gateway for Paedophiles." *Daily Mail Online.* www.dailymail.co.uk/news/article-3688410/Pokemon-gateway-paedophiles-NSPCC-urges-makers-NOT-release-Britain-unless-guarantee-t-used-sexual-predators-lure-children.html.

Coleman, Gabriella. 2011. "Anonymous: From the Lulz to Collective Action." *The New Everyday: A Media Commons Project.* April 6. http://mediacommons.futureofthebook.org/tne/pieces/anonymous-lulz-collective-action.

Consalvo, M. 2002. "Selling the Internet to Women: The Early Years." In *Women and Everyday Uses of the Internet: Agency and Identity,* edited by Mia Consalvo and Susanna Paasonen, 111–38. New York: Peter Lang.

Consalvo, Mia. 2011. "Slingshot to Victory: Games, Play, and the iPhone." In *Moving Data: The iPhone and the Future of Media,* edited by Pelle Snickars and Patrick Vonderau, 184–95. New York: Columbia University Press.

Constine, Josh. 2015. "Yik Yak Systematically Downvotes Mentions of Competitors." *TechCrunch.* February 17. https://techcrunch.com/2015/02/17/yakgate.

Cooley, C. H. 1992/1902. Human Nature and the Social Order. Transaction Publishers.

Cox, Joseph. 2015. "Ashley Madison Hackers Speak Out: 'Nobody Was Watching.'" *Motherboard.* http://motherboard.vice.com/read/ashley-madison-hackers-speak-out-nobody-was-watching.

Cramer, Henriette, Mattias Rost, and Lars Erik Holmquist. 2011. "Performing a Check-in: Emerging Practices, Norms, and 'Conflicts' in Location-Sharing Using Foursquare." *Proceedings of the 13th International Conference on Human Computer Interaction with Mobile Devices and Services, HCI '11.* 57–66.

Craw, Victoria. 2014. "The Real Story behind Hugely Successful Dating App Tinder." March 17. www.news.com.au/finance/business/the-real-story-behind-hugely-successful-dating-app-tinder/story-fn5lic6c-1226856885645.

Crawford, Kate, Jessa Lingel, and Tero Karppi. 2015. "Our Metrics, Ourselves: A Hundred

Years of Self-Tracking from the Weight Scale to the Wrist Wearable Device." *European Journal of Cultural Studies* 18, nos. 4–5: 479–96.

Crook, Jordan. 2015. "Hate It or Love It, Tinder's Right Swipe Limit Is Working." *TechCrunch*. March 12. https://techcrunch.com/2015/03/12/hate-it-or-love-it-tinders-right-swipe-limit-is-working.

CrunchBase. 2016. "Yik Yak." www.crunchbase.com/organization/yik-yak#/entity.

Cumiskey, Kathleen, and Kendra Brewster. 2012. "Mobile Phones or Pepper Spray?" *Feminist Media Studies* 12, no. 4: 590–99.

Cunningham, Katelan. 2015. "14 Swiping Apps That Are Better Than Tinder." *Brit+Co*. www.brit.co/swiping-apps.

Curry, Ruth. 2014. "Towards a Unified Theory of Kim Kardashian: Hollywood." *Brooklyn Magazine*. September 10. www.bkmag.com/2014/09/10/toward-a-unified-theory-of-kim-kardashian-hollywood.

Curtin, Michael, Jennifer Holt, and Kevin Sanson, eds. 2014. *Distribution Revolution: Conversations about the Digital Future of Film and Television*. Oakland: University of California Press.

Dacome, Lucia. 2001. "Living with the Chair: Private Excreta, Collective Health and Medical Authority in the Eighteenth Century." *History of Science* 39, no. 4: 467–500.

Dacome, Lucia. 2012. "Balancing Acts: Picturing Perspiration in the Long Eighteenth Century." *Studies in History and Philosophy of Biological and Biomedical Sciences* 43, no. 2: 379–91.

D'Addario, Daniel. 2014. "Let's Talk about White Gays 'Stealing Black Female Culture.'" *The Cut*. July 15. www.thecut.com/2014/07/why-white-gays-steal-black-female-culture.html.

Dalla Costa, Mariarosa & Selma James. 1973. *The Power of Women and the Subversion of the Community*. Bristol, Eng: Falling Wall Press.

Daniels, Jessie. 2009. "Rethinking Cyberfeminism(S): Race, Gender and Embodiment." *WSQ: Women's Studies Quarterly* 37, nos. 1–2: 101–24.

Daston, Lorraine, ed. 2004. *Things That Talk: Object Lessons from Art and Science*. Cambridge: MIT Press.

Davenport, Thomas H., and John C. Beck. 2001. *The Attention Economy: Understanding the New Currency of Business*. Cambridge, MA: Harvard Business Press.

David, Matthieu. 2013. "Bertrand Schmitt, App Annie CEO Speaks about the Start of App Annie!" *Technode*. October 16. http://technode.com/2013/10/16/app-annie-ceo-betrand-schmitt-on-the-start-of-app-annie.

Dean, Jodi. 2005. "Communicative Capitalism: Circulation and the Foreclosure of Politics." *Cultural Politics* 1, no. 1: 51–74.

Dean, Jodi. 2010. "Affective Networks." *MediaTropes* 2, no. 2:, 19–44.

De Castell, Suzanne, and Mary Bryson. 1998. "Re-Tooling Play: Dystopia, Dysphoria, and Difference." In *From Barbie to Mortal Kombat: Girls and Computer Games*, edited by Justine Cassell and Henry Jenkins, 232–61. Cambridge: MIT Press.

de Certeau, Michel. 1988. *The Practice of Everyday Life*. Berkeley: University of California Press.

de Peuter, Greig, and Nick Dyer-Witherford. 2005. *Games of Empire: Global Capitalism and Video Games*. Minneapolis: University of Minnesota Press.

de Souza e Silva, Adriana, and Jordan Frith. 2012. *Mobile Interfaces in Public Spaces: Locational Privacy, Control, and Urban Sociability*. New York: Routledge.

Dewey, Caitlin. 2014. "The Many Problems with Sketchfactor, the New Crime Crowdsourcing App That Some Are Calling Racist." *Washington Post*. August 12. www.washingtonpost.com/news/the-intersect/wp/2014/08/12/the-many-problems-with-sketchfactor-the-new-crime-crowdsourcing-app-that-some-are-calling-racist.

D'heer, Evelien, and Cedric Courtois. 2016. "The Changing Dynamics of Television Consumption in the Multimedia Living Room." *Convergence* 22, no. 1: 3–17.

Dickson, E. J. 2016. "Apple's Peach Emoji No Longer Looks Like a Butt." *Glamour*. November 1. www.glamour.com/story/apple-peach-emoji-no-longer-looks-like-a-butt.

Didziokaite G., P. Saukko, and C. Greiffenhagen. 2017. "The Mundane Experience of Everyday Calorie Trackers: Beyond the Metaphor of Quantified Self." *New Media & Society*. https://doi.org/10.1177/1461444817698478.

Doctorow, Cory. 2012. "Lockdown: The Coming War on General-Purpose Computing." *Boing Boing*. http://boingboing.net/2012/01/10/lockdown.html.

Dodd, Nigel. 2014. *The Social Life of Money*. Princeton, NJ: Princeton University Press

Donovan, Charles. 2017. The Dangerous 'Gay Men Are Misogynists' Movement. *Huffington Post UK*. August 25. www.huffingtonpost.co.uk/charles-donovan/post_15730_b_17816344.html.

Doppler Labs Inc. 2015. "Here One Wireless Smart Earbuds--Get Your Here Buds." https://hereplus.me.

Dormehl, L. 2015. "How a Sarcastic AI Taskmaster Came to Rule the App Store." *Cultof-Mac.com*. www.cultofmac.com/318267/sarcastic-ai-carrot-apps-brian-mueller.

Douglas, Mary, and Aaron Wildavsky. 1983. *Risk and Culture: An Essay on the Selection of Technical and Environmental Dangers*. Los Angeles: University of California Press.

Douglas, Susan. 2004. *Listening In: Radio and the American Imagination*. Minneapolis: University of Minnesota Press.

Douglas, Susan J. 2010. *The Rise of Enlightened Sexism: How Pop Culture Took Us from Girl Power to Girls Gone Wild*. New York: Macmillan.

Dovey, Jon. 1998. *Freakshow: First Person Media and Factual Television*. Sterling, VA: Pluto Press.

Doyle, Peter. 2005. *Echo and Reverb: Fabricating Space in Popular Music Recording, 1900–1960*. Middletown, CT: Wesleyan University Press.

DraftKings. 2016. "Mobile Devices FAQ." *DraftKings.com*. www.draftkings.com/lp/geofaq.

DraftKings. 2017. "Frequently Asked Questions." *DraftKings.com*. www.draftkings.com/help/faq.

Driscoll, Catherine. 2002. *Girls: Feminine Adolescence in Popular Culture*. New York: Columbia University Press.

Droll, Tyler. 2015. "Making the World Feel Small Again." *Yik Yak* Blog, August 20. https://blog.yikyak.com/blog/making-the-world-feel-small-again.

Droll, Tyler. 2016a. "Introducing Profiles." *Yik Yak* Blog, July 27. http://blog.yikyak.com/blog/introducing-profiles.

Droll, Tyler. 2016b. "Introducing Handles." *Yik Yak* Blog, March 8. http://blog.yikyak.com/blog/introducing-handles.

Duggan, Maeve. 2013. "Photo and Video Sharing Grow Online." *Pew Research Center*.www.pewinternet.org/2013/10/28/photo-and-video-sharing-grow-online.

Duguay, Stefanie. 2016. "Lesbian, Gay, Bisexual, Trans, and Queer Visibility through Selfies: Comparing Platform Mediators across Ruby Rose's Instagram and Vine Presence." *Social Media + Society* 2, no. 2:1–12. doi: 10.1177/2056305116641975.

Duguay, Stefanie. 2017. "Dressing up Tinderella: Interrogating Authenticity Claims on the Mobile Dating App Tinder." *Information Communication & Society* 20, no. 3: 351–67. doi: 10.1080/1369118X.2016.1168471.

Dumit, Joseph. 2012. *Drugs for Life: How Pharmaceutical Companies Define Our Health.* Durham, NC: Duke University Press.

Dwyer, Brendan, and Yongjae Kim. 2011. "For Love or Money: Developing and Validating a Motivational Scale for Fantasy Football Participation." *Journal of Sport Management* 25: 70–83.

Echo Nest. 2011. "Announcing Echoprint." *Echo Nest Blog.* June 23. http://blog.echonest.com/post/6824753703/announcing-echoprint.

Edgar, Eir-Anne. 2011. "'Xtravaganza!': Drag Representation and Articulation in 'RuPaul's Drag Race.'" *Studies in Popular Culture* 34, no. 1: 133–46.

Edwards, Jim. 2014. "A 'Dark Pattern' in Flappy Bird Reveals How Apple's Mysterious App Store Ranking Algorithm Works." *Business Insider.* February 12. www.businessinsider.com/how-apple-app-store-ranking-algorithm-works-2014–2#ixzz3RErh65Py.

Edwards, Paul. 2003. "Infrastructure and Modernity: Force, Time, and Social Organization in the History of Sociotechnical Systems." In *Modernity and Technology,* edited by T. Misa, P. Bray, and A. Feenberg, 185–225. Cambridge: MIT Press.

Eklund, Lina. 2015. "Who Are the Casual Gamers? Gender Tropes and Tokenism in Game Culture." In *Social, Casual, and Mobile Games: The Changing Gaming Landscape,* edited by Tama Leaver and Michele Willson, 15–30. New York: Bloomsbury Academic.

Elaluf-Calderwood, Silvia, Ben Eaton, Carsten Sørensen, and Youngjin Yoo. 2011. "Control as a Strategy for the Development of Generativity in Business Models for Mobile Platforms." (conference paper) 15th International Conference on Intelligence in Next Generation Networks (ICIN 2011), October 4–7, Berlin, Germany.

Eler, Alicia, and Eve Peyser. 2016. "Tinderization of Feeling." *The New Inquiry.* http://thenewinquiry.com/essays/tinderization-of-feeling.

Elgan, Mike. 2016. Podcast no. 569. Transcript. July 6. *This Week in Tech.* https://twit.tv/posts/transcripts/week-tech-569.

Ellis, Dan, Brian Whitman, and Alastair Porter. 2011. "Echoprint: An Open Music Identification Service." In *Proceedings of the International Society for Music Information Retrieval Conference.*

Ellis, John. 2000. *Seeing Things: Television in the Age of Uncertainty.* London: IB Tauris.

Ellis, John. 2009. "Mundane Witness." *Media Witnessing: Testimony in the Age of Mass Communication,* edited by Paul Frosh and Amit Pinchevski, 73–88. Basingstoke: Palgrave Macmillan UK.

Elmer, Greg, and Andy Opel. *Preempting Dissent: The Politics of an Inevitable Future.* Winnipeg: Arbeiter Ring Press, 2008.

Epstein, Kayla. 2013. "New York Victim of Street Harassment? There's an App for That." *The Guardian.* August 20. www.theguardian.com/technology/2013/aug/20/hollaback-app-street-harassment-new-york.

Erevelles N., and A. Minear. 2016/2011. "'Unspeakable' Offenses: Disability Studies at the

Intersections of Multiple Differences." In *Disability and Difference in Global Contexts: Enabling a Transformative Body Politic*, by Nirmala Erevelles, 95–120. New York: Palgrave Macmillan.

Esteban-Bravo, Mercedes, Jose M. Vidal-Sanz, and Goekhan Yildirim. 2015. "Historical Impact of Technological Change on the US Mass Media Advertising Expenditure." *Technological Forecasting and Social Change* 100: 306–16.

Etherington, Darrell. 2013. "Snapchat Gets Its Own Timeline with Snapchat Stories, 24-Hour Photo & Video Tales." *TechCrunch*. October 3. http://techcrunch.com/2013/10/03/snapchat-gets-its-own-timeline-with-snapchat-stories-24-hour-photo-video-tales.

Etherington, Darrell. 2014. "Foursquare Can Now Satisfy Your Appetite at Home with Integrated Seamless and GrubHub Ordering." *TechCrunch*. January 30. http://techcrunch.com/2014/01/30/foursquare-can-now-satisfy-your-appetite-at-home-with-integrated-seamless-and-grubhub-ordering.

Fagan, Michael. E. 1976. "Design and Code Inspections to Reduce Errors in Program Development." *IBM Systems Journal* 15, no. 3:182–211. doi: 10.1147/sj.153.0182.

Fanning, Kevin. 2016. "Why Kendall and Kylie's New Game Is So Addictive." *NYMag.com*. March 11. http://nymag.com/thecut/2016/03/kendall-kylie-game-review.html.

Farman, Jason. 2012. *Mobile Interface Theory: Embodied Space and Locative Media*. New York: Routledge.

Federici, S. 2012. *Revolution at Point Zero: Housework, Reproduction, and Feminist Struggle*. Oakland, CA: PM Press.

Feldmar, Jamie. 2011. "Hollaback's Emily May Talks about Stopping Street Harassment." *The Gothamist*. June 30. http://gothamist.com/2011/06/30/hollabacks_emily_may_talks_about_st.php.

Fileborn, Bianca. 2014. "Online Activism and Street Harassment: Digital Justice or Shouting into the Future." *Griffith Journal of Law & Human Dignity* 2, no. 1: 32–50.

Fischer, Reed. 2012. "DJ Jonathan Toubin: I've Been Told I'm 'Un-Shazam-Able.'" *City Pages*. November 30. www.citypages.com/music/dj-jonathan-toubin-ive-been-told-im-un-shazam-able-6649572.

Fiske, John. 1998. "Surveilling the City: Whiteness, the Black Man, and Democratic Totalitarianism." *Theory, Culture & Society* 15, no. 2: 67–88.

Fitbit.com. n.d. Website. www.fitbit.com/home.

Florida Sexual Offenders and Predators. n.d. "Frequently Asked Questions." http://offender.fdle.state.fl.us/offender/FAQ.jsp#Question16.

Flückiger, Barbara. 2011. "The iPhone Apps: A Digital Culture of Interactivity." In *Moving Data: The iPhone and the Future of Media*, edited by Pelle Snickars and Patrick Vonderau, 171–83. New York: Columbia University Press.

Forbes. 2016. "The World's Billionaires, Brian Acton." June 7. www.forbes.com/profile/brian-acton.

Fortunati, Leopoldina, Mark Deuze, and Federico de Luca. 2014. "The New about News: How Print, Online, Free, and Mobile Coconstruct New Audiences in Italy, France, Spain, the UK, and Germany." *Journal of Computer-Mediated Communication* 19, no. 2: 121–40.

Foucault, Michel. 1975. *Discipline and Punish: The Birth of the Prison*. New York: Random House.

Fox, Bobby, John Senior, and Peter Thomas. 2010. "The Big Video Game: TV's Battle to Hold Online Viewers." *IEEE Communications Magazine* 48, no. 5: 52–56.

Fraser, Antonia. 1984/2011. *The Weaker Vessel: Woman's Lot in Seventeenth-Century England*. London: Hachette UK.

Freeman, David. 2004. "Creating Emotion in Games: The Craft and Art of Emotioneering." *ACM Computers in Entertainment* 2, no. 3: 1–11.

Freud Museum London. n.d. "Frequently Asked Questions." www.freud.org.uk/about/faq.

Frosh, Paul. 2006. "Telling Presences: Witnessing, Mass Media, and the Imagined Lives of Strangers." *Critical Studies in Media Communication* 23, no. 4: 265–84.

Fuller, Matthew. 2003. *Behind the Blip: Essays on the Culture of Software*. Brooklyn, NY: Autonomedia.

Fuller, Matthew. 2008. *Software Studies: A Lexicon*. Leonardo Books. Cambridge: MIT Press.

Fuller, Matthew, and Andrew Goffey. 2012. *Evil Media*. Cambridge: MIT Press.

Gajjala, Radhika, and Altman, Melissa. 2006. "Producing Cyberselves through Technospatial Praxis: Studying through Doing." In *Health Research in Cyberspace*, edited by Pranee Liamputtong, 67–84. New York: Nova Publishers.

Gajjala, Radhika. 2011. "Snapshots from Sari Trails: Cyborgs Old and New." *Social Identities* 17, no. 3: 393–408.

Gajjala, Radhika. 2016. "Circuits of Affect, Care, and Materiality." Grazier Lecture at University of Southern Florida. April 22.

Galloway, Alexander R. 2012. *The Interface Effect*. Cambridge: Polity Press.

Galloway, Alexander R. 2004. *Protocol: How Control Exists after Decentralization*. Cambridge: MIT Press.

Gamson, Joshua. 2013. "Reality Queens." *Contexts* 12, no. 2: 52–54.

Gannes, Liz. 2012. "Much-Cited Analytics Tool App Annie Raises $6M." *AllThingsD*. August 14. http://allthingsd.com/20120814/much-cited-analytics-tool-app-annie-raises-6m-series-b.

Gantz, Carroll. 2012. *The Vacuum Cleaner: A History*. Jefferson, NC: McFarland.

Gantz, Carroll. 2014. *Founders of American Industrial Design*. Jefferson, NC: McFarland.

Gaonkar, Dilip Parameshwar, and Elizabeth A. Povinelli. 2003. "Technologies of Public Forms: Circulation, Transfiguration, Recognition." *Public Culture* 15, no. 3: 385–97.

Gates, K. 2013. "The Cultural Labor of Surveillance: Video Forensics, Computational Objectivity, and the Production of Visual Evidence." *Social Semiotics* 23, no. 2: 242–60.

Gawer, Annabelle. 2014. "Bridging Differing Perspectives on Technological Platforms: Toward an Integrative Framework." *Research Policy* 43: p1239–49.

Gerlitz, Carolin, and Anne Helmond. 2013. "The Like Economy: Social Buttons and the Data-Intensive Web." *New Media & Society* 15, no. 8: 1348–65.

Gerlitz, Carolin, Fernando N. van der Vlist, Anne Helmond, and Esther Weltevrede. 2016. "App Support Ecologies. An Empirical Investigation of App-Platform Relations." *Infrastructures of Publics—Publics of Infrastructures*. The DFG Collaborative Research Centre Conference "Media of Cooperation." Artur-Woll-Haus, University of Siegen, Germany.

Gibbs, Samuel. 2018. "EU to Review Apple's Reported $400m Purchase of Music App Shazam." *Guardian*, February 7, 2018.

Gibson, R. K. 2015. "Party Change, Social Media, and the Rise of 'Citizen-Initiated' Campaigning." *Party Politics* 21, no. 2: 183–97. doi: 10.1177/1354068812472575.

Giddens, Anthony. 1991. *Modernity and Self-Identity: Self and Society in the Late Modern Age*. Stanford, CA: Stanford University Press, 1991.

Gillespie, Tarleton L. 2006. "Designed to 'Effectively Frustrate': Copyright, Technology and the Agency of Users." *New Media & Society* 8, no. 4: 651–69.

Gillespie, Tarleton L. 2010. "The Politics of 'Platforms.'" *New Media & Society* 12, no. 3: 347–64. doi: 10.1177/1461444809342738.

Gillespie, Tarleton L. 2013. "The Relevance of Algorithms." In *Media Technologies: Essays on Communication, Materiality, and Society*, edited by Tarleton L. Gillespie, Pablo Boczkowski, and Kirsten Foot, 167–94. Cambridge: MIT Press.

Gillespie, Tarleton L. 2015. "Platforms Intervene." *Social Media + Society* 1, no. 1. doi: 2056305115580479.

Gillespie, Tarleton L. 2017. "Governance of and by Platforms." In *SAGE Handbook of Social Media*, edited by Jean Burgess, Tom Poell, and Alice Marwick. London: Sage.

Gilmore J. N. 2015. "Everywear: The Quantified Self and Wearable Fitness Technologies." *New Media & Society* 18, no. 11: 2524–39. doi: 1461444815588768.

Gitelman, Lisa. 2010. "Welcome to the Bubble Chamber: Online in the Humanities Today." *Communication Review* 13, no. 1: 27–36.

Gitelman, Lisa. 2014. *Paper Knowledge: Toward a Media History of Documents*. Durham, NC: Duke University Press Books.

Glueck, Jeff. 2016. "Foursquare Ushers in a New Era." *Medium*. January 14. https://medium.com/foursquare-direct/foursquare-ushers-in-a-new-era-f52edb39af6#.supdtw9hn.

Goffman, Erving. 1959. *The Presentation of Self in Everyday Life*. Anchor Books edition. Garden City, NY: Doubleday.

Goggin, Gerard. 2011a. "Ubiquitous Apps: Politics of Openness in Global Mobile Cultures." *Digital Creativity* 22, no. 3: 148–59.

Goggin, Gerard. 2011b. "Going Mobile." In *Handbook of Media Audiences*, edited by Virginia Nightingale, 128–46. Cambridge: Blackwell.

Goggin, Gerard. 2012. "The Eccentric Career of Mobile Television." *International Journal of Digital Television* 3, no. 2: 119–40.

Goggin, Gerard. 2014. "Mobile Video: Spreading Stories with Mobile Media." In *Routledge Companion to Mobile Media*, edited by Gerard Goggin and Larissa Hjorth, 146–56. New York: Routledge.

Goggin, Gerard, and Mark McLelland, eds. 2017. *Routledge Companion to Global Internet Histories*. New York: Routledge.

Goggin, Gerard, and Rowan Wilken, eds. 2012. *Mobile Technology and Place*. New York: Routledge.

Goldhaber, Michael H. 1997. "The Attention Economy and the Net." *First Monday* 2, no. 4. http://firstmonday.org/ojs/index.php/fm/article/view/519/440.

Goldman, Alex. 2015. "The Writing on the Wall." Podcast audio. *ReplyAll*. January 14. https://gimletmedia.com/episode/9-yik-yak.

Goldsmith, Ben. 2014. "The Smartphone App Economy and App Ecosystems." In *The Routledge Companion to Mobile Media*, edited by Gerard Goggin and Larissa Hjort, 171–80. New York: Routledge.

Goldstein, Daniel M. 2005. "Flexible Justice: Neoliberal Violence and 'Self-Help' Security in Bolivia." *Critique of Anthropology* 25, no. 4: 389–411.

Golis, Andrew. 2014. "So What's Up with This?" *This. Blog*. August 17. https://blog.this.cm/so-whats-up-with-this-24ebb1f83dd9.

Golis, Andrew. 2015a. "Big News: This. Raises Money, Builds a Team and Loves It Some Links." *Tumblr*. August 11. http://golis.tumblr.com/post/126431776587/big-news-this-raises-money-builds-a-team-and.

Golis, Andrew. 2015b. "Introducing: This.—This. News." *This. Blog*. October 21. https://blog.this.cm/introducing-this-1633dfee851d#.kxo5q3src.

Golis, Andrew. 2015c. "This. Is a Social Magazine (and Also Download Our App)—This. News." *Medium*. May 20. https://blog.this.cm/this-is-a-social-magazine-and-also-download-our-app-8790a628d90#.koqachme1.

Goodman H. 1995. "Parents' Ivy Dreams Taking Root: The Start of the College Season Means a Tour of the Eastern Seaboard." *Philadelphia Inquirer*. August 8.

Google Play. 2018. "See Send." Play Store. https://play.google.com/store/apps/details?id=com.mymobilewitness.seesend&hl=en.

Graber, Diana. 2014. "Yik Yak App Makers Do the Right Thing." *Huffington Post*. March 26. www.huffingtonpost.com/diana-graber/yik-yak-app-makers-do-the_b_5029679.html.

Green, Joshua. 2012. "Corporations Want Obama's Winning Formula." *Bloomberg*. November 21. www.bloomberg.com/news/articles/2012-11-21/corporations-want-obamas-winning-formula#p2.

Greene, Paul D. 2005. "Introduction: Wired Sound and Sonic Cultures." In *Wired for Sound: Engineering and Technologies in Sonic Cultures*, edited by Paul D. Greene and Thomas Porcello, 1–22. Middletown, CT: Wesleyan University Press.

Greenfield, Adam. 2006. *Everyware: The Dawning Age of Ubiquitous Computing*. Berkeley, CA: New Riders.

Greg from Bloomfield. 2014. Comment to Ken Freedman and Andy Breckman. "Can You Beat Shazam?" *WFMU: Seven Second Delay with Ken and Andy*. May 28. www.wfmu.org/playlists/shows/55832.

Gregg, Melissa. 2015a. "Getting Things Done: Productivity, Self-Management, and the Order of Things." In *Networked Affect*, edited by Ken Hillis, Susanna Paasonen, and Michael Petit, 187–202. Cambridge: MIT Press.

Gregg, Melissa. 2015b. "Inside the Data Spectacle." *Television and New Media* 16, no. 1: 37–51.

Gregg, Melissa. 2015c. "The Productivity Obsession." *The Atlantic*. November 13. www.theatlantic.com/business/archive/2015/11/be-more-productive/415821.

Grewal, Dhruv, Yakov Bart, Martin Spann, and Peter Pal Zubcsek. 2016. "Mobile Advertising: A Framework and Research Agenda." *Journal of Interactive Marketing* 34: 3–14.

Grimes, Sara M. 2015. "Little Big Scene." *Cultural Studies* 29, no. 3: 379–400. doi: 10.1080/09502386.2014.937944.

Grimmelmann, J., and P. Ohm. 2010. "Dr. Generative or: How I Learned to Stop Worrying and Love the iPhone." *Maryland Law Review* 69: 910–53.

Grubin, Don. 1998. "Sex Offending against Children: Understanding the Risk." Police Research Series 99. Policing and Reducing Crime Unit. London. http://217.35.77.12/research/england/justice/fprs99.pdf.

Grusin, Richard. 2000. "Location, Location, Location: Desktop Real Estate and the Cultural Economy of the World Wide Web." *Convergence* 6, no. 1: 48–61.

Grusin, Richard. 2010. *Premediation: Affect and Mediality after 9/11*. Basingstoke, Eng.: Palgrave Macmillan.

Guerin, Jackie. 2016. "Companion: The Newest Safety Tool." *Odyssey Online*. April 11. www.theodysseyonline.com/companion-newest-safety-tool.

Haber, Chad. 2016. "Never Hillary Clinton on the App Store." *Apple App Store*. August 5. https://itunes.apple.com/us/app/never-hillary-clinton/id1131701133?mt=8.

Hackett, Robert. 2016. "Ashley Madison Hacking Victims Face a Big Decision." *Fortune*. April 20. http://fortune.com/2016/04/20/ashley-madison-data-breach-lawsuit-names.

Hagood, Mack. 2011. "Quiet Comfort: Noise, Otherness, and the Mobile Production of Personal Space." *American Quarterly* 63 (3): 573–89.

Hagood, Mack. 2019. *Hush: Media and Sonic Self-Control*. Durham: Duke University Press.

Halberstam, Judith. 2005. "Shame and White Gay Masculinity." *Social Text* 23.3–4 (84–85): 219–33.

Halegoua, Germaine R., Alex Leavitt, and Mary L. Gray. 2016. "Jumping for Fun? Negotiating Mobility and the Geopolitics of Foursquare." *Social Media + Society* 2, no. 3. doi: 10.1177/2056305116665859.

Halperin, David, and Valerie Traub, eds. 2009. *Gay Shame*. Chicago: University of Chicago Press.

Halpern, Daniel, Natalia Quintas-Froufe, and Francisco Fernandez-Medina. 2016. "Interactions between Television and Its Social Audience: Towards a Communication Conceptualization." *Profesional de la Informacion* 25, no. 3: 367–75.

Halpern, Orit. 2015. *Beautiful Data: A History of Vision and Reason since 1945*. Durham, NC: Duke University Press.

Halpin, Mikki. 2016. "Letter No. 30." *Lenny Email Newsletter*. April 19. http://mikkipedia.net.

Halverson, Erica, and Richard Halverson. 2008. "Fantasy Baseball: The Case for Competitive Fandom." *Games and Culture* 3, nos. 3–4: 286–308.

Hannah-Moffat, Kelly, Paula Maurutto, and Sarah Turnbull. 2009. "Negotiated Risk: Actuarial Illusions and Discretion in Probation." *Canadian Journal of Law & Society/La Revue Canadienne Droit et Société* 24, no. 3: 391–409.

Haraway, Donna. 2016. *Staying with the Trouble: Making Kin in the Chthulucene*. Durham, NC: Duke University Press.

Harper, Richard. 2012. *The Connected Home: The Future of Domestic Life*. Dordrecht, Germany: Springer.

Harris, Anita. 2004. *Future Girl: Young Women in the Twenty-first Century*. New York: Routledge.

Harris, Daniel. 1995. "The Aesthetic of Drag." *Salmagundi* 108: 62–74. http://salmagundimagazine.tumblr.com/post/37348041244/the-aesthetic-of-drag-an-essay-by-daniel.

Harshaw, Pendarvis. 2015. "Nextdoor, the Social Network for Neighbors, Is Becoming a Home for Racial Profiling." *Fusion*. March 3. http://fusion.net/story/106341/nextdoor-the-social-network-for-neighbors-is-becoming-a-home-for-racial-profiling.

Harvey, Sylvia, and Marko Ala-Fossi. 2016. "Eroding the Assets of Citizenship? From Broadcast to Broadband." *International Communication Gazette* 78, no. 4: 294–310.

Hasinoff, Amy Adele. 2015. *Sexting Panic: Rethinking Criminalization, Privacy, and Consent.* Urbana: University of Illinois Press.

Hayano, David. 1982. *Poker Faces: The Life and Work of Professional Card Players*. Berkeley: University of California Press.

Hayles, N. Katherine. 2005. *My Mother Was a Computer: Digital Subjects and Literary Texts*. Chicago: University of Chicago Press.

Hayles, N. Katherine. 2006. "Traumas of Code." *Critical Inquiry* 33 (Autumn): 136–57.

Hearn, Alison. 2010. "Structuring Feeling: Web 2.0, Online Ranking and Rating, and the Digital 'Reputation' Economy." *Ephemera: Theory & Politics in Organisation* 10, nos. 3–4: 421–38.

Heath Alex. 2015. "Use This App Whenever You're Walking Home Alone." *Tech Insider.* www.techinsider.io/companion-app-may-help-you-feel-safe-2015-8.

Heath Alex. 2016. "Foursquare's Location Data Is Way More Powerful Than People Realize." *Tech Insider*. January 1. www.techinsider.io/inside-foursquares-pilgrim-technology-2015-12.

He Jun. 2015. "Tengxun fouren yingyong bao hui 'dianda qike,' zhengxiang dazao zhongchuang pingtai" [Tencent's MyApp Denied Unfair Deals with App Developers, Competing to Build Mass Entrepreneurship Platforms]., *Securities Daily*. http://money.163.com/15/1026/06/B6R6UTB100253B0H.html.

Heldman, Caroline, and Lisa Wade. 2010. "Hook-Up Culture: Setting a New Research Agenda." *Sexuality Research and Social Policy* 7, no. 4: 323–33. doi: 10.1007/s13178-010-0024-z.

Helft, Miguel. 2010. "Charges Settled over Fake Reviews on iTunes." *New York Times*. August 26. www.nytimes.com/2010/08/27/technology/27ftc.html?_r=2&src=twt&twt=nytimestech.

Here One Wireless Smart Earbuds—Get Your Here Buds. 2015. https://hereplus.me.

Hestres, Luis. 2013. "App Neutrality: Apple's App Store and Freedom of Expression Online." *International Journal of Communication* 7: 1265–80.

Hillary for America. 2016. "FAQ: Hillary 2016 Mobile App." *Hillary for America*. July 26. www.hillaryclinton.com/page/mobile-faq.

Hillis, Ken. 2015. "The Avatar and Online Affect." In *Networked Affect*, edited by K. Hillis, S. Paasonen, and M Petit. Cambridge: MIT Press. Kindle Edition.

Hinchliffe, Emma. 2016. "Meet the Developer That's Making Apps for the Conservative Movement." *Mashable*. October 7. http://mashable.com/2016/10/07/ucampaign-conservative-apps.

Ho, Jocelyn. 2008. "Incest and Sex Offender Registration: Who Is Registration Helping and Who Is It Hurting?" *Cardozo Journal of Law and Gender* 14: 429–57.

Hochschild, ArlieRussell 1983. *The Managed Heart: Commercialization of Human Feeling*. New York: Penguin Books.

Hollaback! Halifax. n.d. "Our Values." https://halifax.ihollaback.org/about/our-values.

Hollaback! London. n.d. "About Us and FAQs." https://ldn.ihollaback.org/about.

Hollaback! London. n.d. "Share Your Story." https://ldn.ihollaback.org/share-your-story.

Holmes, Su, and Diane Negra, eds. 2011. *In the Limelight and Under the Microscope: Forms and Functions of Female Celebrity*. New York: Continuum.

Holt, Jennifer, and Sanson, Kevin, eds. 2014. *Connected Viewing: Selling, Streaming, and Sharing Media in the Digital Era*. New York: Routledge.

Horowitz, Emily. 2007. "Growing Media and Legal Attention to Sex Offenders: More Safety or More Injustice?" *Journal of the Institute of Justice & International Studies* 7: 143–58.

Horton, Helena. 2016. "The New, Classier Angry Birds? Cat Game Neko Atsume Takes Internet by Storm." *The Telegraph*. April 20. www.telegraph.co.uk/technology/2016/04/20/the-new-classier-angry-birds-cat-game-neko-atsume-takes-internet.

Hu, Elise. 2014. "Some Loyal Foursquare Users Are Checking Out after Swarm Spinoff." *NPR.Org*, July 30. www.npr.org/sections/alltechconsidered/2014/07/30/336531811/some-loyal-foursquare-users-are-checking-out-after-swarm-spinoff.

Humphreys, Lee. 2007. "Mobile Social Networks and Social Practice: A Case Study of Dodgeball." *Journal of Computer-Mediated Communication* 13, no. 1: 341–60. doi: 10.1111/j.1083–6101.2007.00399.

Hunn, Nick. 2014. "Hearables—The New Wearables." *Creative Connectivity*. April 3. www.nickhunn.com/hearables-the-new-wearables.

Huntemann, Nina B., and Matthew Thomas Payne, eds. 2009. *Joystick Soldiers: The Politics of Play in Military Video Games*. New York: Routledge.

Hutchby, Ian. 2001. *Conversation and Technology: From the Telephone to the Internet*. Cambridge, UK: Polity Press.

Hutchins, Brett. 2016. " 'The More Things Change, the More They Stay the Same': Path Dependency, Sports Content, and the Suppression of Innovation in Mobile Television." *Telematics and Informatics* 33, no. 2: 703–10.

Ihde, Don. 1979. *Technics and Praxis: A Philosophy of Technology*. Dordrecht, Germany: D. Riedel Publishing.

iHollaback. 2015. "Love and Revolution Ep. 6: 'Criminalization.'" Video. https://youtu.be/yUBGy2tdvfw.

iMaschine 2. n.d. *Native Instruments*. www.nativeinstruments.com/en/products/maschine/maschine-for-ios/imaschine-2.

Ingraham, Nick. 2013. "Apple Announces 1 Million Apps in the App Store, More Than 1 Billion Songs Played on iTunes Radio." *The Verge*. www.theverge.com/2013/10/22/4866302/apple-announces-1-million-apps-in-the-app-store.

Isbister, Katherine. 2006. *Better Game Characters by Design: A Psychological Approach*. Boston: CRC Press.

Isbister, Katherine. 2016. *How Games Move Us: Emotion by Design*. Cambridge: MIT Press.

IT home. 2015. "Tencent chuangye jidi you tian xin chengyuan; dazao fulianwang hangye zuida fuhuaqi" [Tencent's Incubation Base Welcomes New Members, Building the Biggest Internet Incubator]. *Techweb.com*. www.techweb.com.cn/news/2015–04–09/2141226.shtml.

Jackson, Lauren Michele. 2017. "We Need to Talk about Digital Blackface in Reaction GIFs." *Teen Vogue*. August 2. www.teenvogue.com/story/digital-blackface-reaction-gifs.

Jackson, Nicholas. 2011. "Exodus International's 'Gay Cure' App Failed to Turn Me Straight." *The Atlantic*. March 21. www.theatlantic.com/technology/archive/2011/03/exodus-internationals-gay-cure-app-failed-to-turn-me-straight/72783.

Jacobs, Bryan. 2010. "How Shazam Works to Identify (Nearly) Every Song You Throw at It." *Gizmodo*. September 24. http://gizmodo.com/5647458/how-shazam-works-to-identify-nearly-every-song-you-throw-at-it.

Jane, Emma A. 2016. *Misogyny Online: A Short (and Brutish) History.* Los Angeles: Sage.

Jane, Emma A. 2017. "Feminist Digilante Responses to a Slut-Shaming on Facebook." *Social Media + Society.* http://doi.org/10.1177/2056305117705996.

Jarrett, Kylie. 2008. "Interactivity Is Evil! A Critical Investigation of Web 2.0." *First Monday* 13, no. 3. http://firstmonday.org/ojs/index.php/fm/article/view/2140/1947.

Jarrett, Kylie. 2014. "The Relevance of 'Women's Work': Social Reproduction and Immaterial Labor in Digital Media." *Television & New Media* 15, no.1: 14–29. 10.1177/1527476413487607.

Jarrett, Kylie. 2015a. *Feminism, Labour and Digital Media: The Digital Housewife.* New York: Routledge.

Jarrett, Kylie. 2015b. "'Let's Express Our Friendship by Sending Each Other Funny Links Instead of Actually Talking': Gifts, Commodities, and Social Reproduction in Facebook." In *Networked Affect*, edited by K. Hillis, S. Paasonen, and M. Petit, 203–20. Cambridge: MIT Press. Kindle Edition.

Jay, Martin. 1988. "Scopic Regimes of Modernity." In *Vision and Visuality*, edited by Hal Foster, 29–50. Seattle: Bay Press.

Jenkins, Henry. 1992. *Textual Poachers: Television Fans & Participatory Culture*. Studies in Culture and Communication. New York: Routledge.

Jenkins, Henry. 2014. "Rethinking 'Rethinking Convergence/Culture.'" *Cultural Studies* 28, no. 2: 267–97.

Jewkes, Yvonne, and Wykes, Maggie. 2012. "Reconstructing the Sexual Abuse of Children: 'Cyber-Paeds,' Panic, and Power." *Sexualities* 15, no. 8: 935–52.

John, Nicholas A. 2013. "Sharing and Web 2.0: The Emergence of a Keyword." *New Media & Society* 15, no. 2: 167–82.

Johnson, Dennis W. 2016. *Democracy for Hire: A History of American Political Consulting.* New York: Oxford University Press.

Johnson, L. 2002. "Creating the Software Industry-Recollections of Software Company Founders of the 1960s." *IEEE Annals of the History of Computing* 24, no. 1: 14–42. doi: 10.1109/85.988576.

Johnson, Steven. 2010. "Rethinking a Gospel of the Web." *New York Times*. www.nytimes.com/2010/04/11/technology/internet/11every.html.

Johnston, Chris. 2014. "The 'Snappening': Explicit Snapchat Images Leaked via Third Party, Reports Say." *The Guardian*. October 11. www.theguardian.com/technology/2014/oct/11/snapchat-snappening-explicit-images-leak.

Jordan, Tim. 2015. *Information Politics: Liberation and Exploitation in the Digital Society.* London: Pluto Press.

Jurgenson, Nathan. 2013. "The Liquid Self." *Snapchat*. September 23. http://snapchat-blog. com/post/61770468323/the-liquid-self.

Juul, Jesper. 2010. *A Casual Revolution: Reinventing Video Games and Their Players*. Cambridge: MIT Press.

Juzwiak, Rich. 2010. "My Own Private *Untucked*: Behind the Scenes at the *RuPaul's Drag Race* Season 2 Reunion." *VH1 Blog*. April 28. http://blog.vh1.com/2010-04-28/my-own-private-untucked-behind-the-scenes-at-the-rupauls-drag-race-season-2-reunion.

Kaldrack, Irina, and Martina Leeker. 2015. Introduction to *No Software Just Services*, edited by Irina Kaldrack and Martina Leeker, 9–20 Lüneburg: Meson Press.

Kalish, Rachel, and Kimmel, Michael. 2011. "Hooking Up." *Australian Feminist Studies* 26, no. 67: 137–51. doi: 10.1080/08164649.2011.546333.

Kantar Worldpanel. 2017. "Android 87% Share in China; More Brands Competing." *Kantar WorldPanel*. www.kantarworldpanel.com/global/News/Android-87-Share-in-China-More-Brands-Competing.

Kaplan, Katharine A. 2003. "Facemash Creator Survives Ad Board." *Harvard Crimson*. www.thecrimson.com/article/2003/11/19/facemash-creator-survives-ad-board-the.

Karpf, David. 2017. "Digital Politics after Trump." *Annals of the International Communication Association* 41, no. 2: 198–207. doi: 10.1080/23808985.2017.1316675.

Keane, Michael. 2015. *The Chinese Television Industry*. London: Palgrave.

Kearney, Mary Celeste. 2010. "Pink Technology: Mediamaking Gear for Girls." *Camera Obscura* 25, no. 2: 1–39.

Kearney, Mary Celeste. 2015. "Sparkle: Luminosity and Post–Girl Power Media." *Continuum: Journal of Media and Cultural Studies* 29, no. 2: 263–73.

Keating, Lauren. 2016. "Gaming On-The-Go: Why 'Kendall & Kylie' Will Be More Successful Than Kim." *Tech Times*. March 7. www.techtimes.com/articles/138076/20160307/gaming-go-heres-why-kylie-kendalls-app-will-more-successful.htm.

Keightley, Keir. 1996. "'Turn It Down!' She Shrieked: Gender, Domestic Space, and High Fidelity, 1948–59." *Popular Music* 15, no. 2:149–77.

Kelty, Christopher. 2008. *Two Bits: The Cultural Significance of Free Software*. Durham, NC: Duke University Press.

Kennedy, Helen, et al. 2016. "The Work That Visualisation Conventions Do." *Information, Communication & Society* 19, no. 6: 715–35.

Keqiang, Li. March 5, 2015. "Report on the Work of the Government." http://www.gov.cn/guowuyuan/2015-03/16/content_2835101.htm.

Kerr, Ian. 2010. "The Devil Is in the Defaults." *Ottawa Citizen*. May 29. http://iankerr.ca/content/2010/05/29/the-devil-is-in-the-defaults.

Kim, Dorothy, and Eunsong Kim. 2014. "The #TwitterEthics Manifesto." *Model View Culture*. April 7. https://modelviewculture.com/pieces/the-twitterethics-manifesto.

Kindy, Kimberly, Marc Fisher, Julie Tate, and Jennifer Jenkins. 2015. "A Year of Reckoning: Police Fatally Shoot Nearly 1,000." *Washington Post*. December 26. www.washingtonpost.com/sf/investigative/wp/2015/12/26/2015/12/26/a-year-of-reckoning-police-fatally-shoot-nearly-1000.

Kitchin, Rob, and Martin Dodge. 2011. *Code/Space: Software and Everyday Life*. Software Studies. Cambridge: MIT Press.

Kjellberg, Hans, and Claes-Fredrik Helgesson. 2007. "On the Nature of Markets and their Practices." *Marketing Theory* 7, no. 2: 137–62.

Kleeman, Sophie. 2016. "Tinder's Newest Feature Will Make It Easier to Find Orgies Near You." *Gizmodo*. http://gizmodo.com/tinders-newest-feature-will-make-it-easier-to-find-orgi-1773339524.

Klein, Naomi. 2017. "W.W.E. the People." *Harper's Magazine*. September. https://harpers.org/archive/2017/09/w-w-e-the-people.

Klosowski, Thorin. 2013. "The Most Sadistic Apps That Force You to Get Stuff Done" *Lifehacker.com*. January 29. http://lifehacker.com/5979811/the-most-sadistic-apps-that-force-you-to-get-stuff-done.

Kofoed, Jette, and Malene Charlotte Larsen. 2016. "A Snap of Intimacy: Photo-Sharing Practices among Young People on Social Media." *First Monday* 21, no. 11. https://firstmonday.org/ojs/index.php/fm/article/view/6905.

Kreiss, Daniel. 2016. *Prototype Politics: Technology-Intense Campaigning and the Data of Democracy*. Oxford Studies in Digital Politics. New York: Oxford University Press.

Kreiss, Daniel, Joshua O. Barker, and Shannon Zenner. 2017. "Trump Gave Them Hope: Studying the Strangers in Their Own Land." *Political Communication* 34, no. 3: 470–78. doi: 10.1080/10584609.2017.1330076.

Kreiss, Daniel, and Shannon C. McGregor. 2017. "Technology Firms Shape Political Communication: The Work of Microsoft, Facebook, Twitter, and Google With Campaigns During the 2016 U.S. Presidential Cycle." *Political Communication* 35, no. 2, 155–77. https://doi.org/10.1080/10584609.2017.1364814.

Kumura, Hugh. 2014. "25 Top IOS Apps and Their Version Update Frequencies." *SensorTower*. April 15. https://sensortower.com/blog/25-top-ios-apps-and-their-version-update-frequencies.

Kuriyama, Shigehisa. 2008. "The Forgotten Fear of Excrement." *Journal of Medieval and Early Modern Studies* 38, no. 3: 413–42.

Kwinter, Sanford. 2008. *Far from Equilibrium: Essays on Technology and Design Culture*. Barcelona: Actar.

Laja, Peep. 2017. "Intuitive Web Design: How to Make Your Website Intuitive to Use." *CXL*. https://conversionxl.com/blog/intuitive-web-design-how-to-make-your-website-intuitive-to-use.

Lam, Bourree. 2015. "Tinder's New Profiles Make It Easier for Privileged People to Date Each Other." *The Atlantic*. www.theatlantic.com/business/archive/2015/11/tinder-adds-job-and-education-profiles/415584.

Lamont, Tom. 2016. Life after the Ashley Madison Affair. *The Guardian*. February 28. www.theguardian.com/technology/2016/feb/28/what-happened-after-ashley-madison-was-hacked.

Landau, Neil. 2016. *TV Outside the Box: Trailblazing in the Digital Television Revolution*. New York: Focal Press.

Lang, Arne. 2016. *Sports Betting and Bookmaking: An American History*. Lanham, MD: Rowman and Littlefield.

Lang, Nico. 2017. "Sorry, Gay Guys, I'm Not Here for Your Casual Misogyny." *New Now Next*. March 8. www.newnownext.com/casual-misogyny/03/2017.

laplacian. 2009. "How Shazam Works." *Free Won't*. January 10. https://laplacian.wordpress.com/2009/01/10/how-shazam-works.

Lapowsky, Issie. 2016. "For Philando Castile, Social Media Was the Only 911." *Wired*. July 7. www.wired.com/2016/07/philando-castile-social-media-911.

Larkin, Brian. 2004. "Degraded Images, Distorted Sounds: Nigerian Video and the Infrastructure of Piracy." *Public Culture* 16, no. 2: 289–314.

Larson, Selena. 2014. "The Era of the Check-In Is Over, and Foursquare Is Moving On." *Readwrite*. June 4. http://readwrite.com/2014/06/04/check-in-foursquare-dennis-crowley-readwritemix.

Lash, Scott. 2000 "Risk Culture." In *The Risk Society and Beyond*, edited by Barbara Adam, Ulrich Beck, and Joost Van Loon, 47–62. London: Sage.

Latour, Bruno. 2005. *Reassembling the Social: An Introduction to Actor Network Theory*. Oxford: Oxford University Press.

Law Office of the Los Angeles County Public Defender. n.d. "Frequently Asked Questions: Three (3) Strikes." http://pd.co.la.ca.us/faqs_3strikes.html.

Lee, Jasmine (2015) "Interview with Leah Busque, TaskRabbit CEO Founder." *Medium*. https://medium.com/@jasminelee/interview-with-leah-busque-taskrabbit-ceo-founder-41e5e1cce402.

Lee, Seung Hwan, and K. Douglas Hoffman. 2015. "Learning the Shamwow: Creating Infomercials to Teach the Aida Model." *Marketing Education Review* 25, no. 1: 9–14. doi: 10.1080/10528008.2015.999586.

Lefebvre, Henri, John Moore, and Michel Trebishsch. 2008. *Critique of Everyday Life*. London: Verso.

Lessig, Lawrence. 2006. *Code, Version 2.0*. New York: Basic Books.

Leurs, Koen. 2015. *Digital Passages: Migrant Youth 2.0: Diaspora, Gender, and Youth Cultural Intersections*. MediaMatters. Amsterdam: Amsterdam University Press.

Levenson, Jill S., Yolanda N. Brannon, Timothy Fortney, and Juanita Baker. 2007. "Public Perceptions about Sex Offenders and Community Protection Policies." In *Analyses of Social Issues and Public Policy* 7, no. 1: 1–25.

Levine, E. 2015. *Cupcakes, Pinterest, and Ladyporn: Feminized Popular Culture in the Early Twenty-First Century*. Urbana: University of Illinois Press.

Lewis, Clayton, Peter G. Polson, Cathleen Wharton, and John Rieman. 1990. "Testing a Walkthrough Methodology for Theory-Based Design of Walk-up-and-Use Interfaces." Proceedings of the SIGCHI Conference on Human Factors in Computing Systems, Seattle, Washington.

Leyda J., and D. Negra, eds. 2015. *Extreme Weather and Global Media*. Abingdon, Oxon: Routledge.

Licoppe, Christian, Carole Anne Riviere, and Julien Morel. 2015. "Grindr Casual Hook-Ups As Interactional Achievements." *New Media & Society* 18, no. 11: 2540–58. doi: 10.1177/1461444815589702.

Light, Ben. 2016a. "Producing Sexual Cultures and Pseudonymous Publics with Digital Networks." In *Race and Gender in Electronic Media: Challenges and Opportunities*, edited by Rebecca. A. Lind, 231–46 London: Routledge.

Light, Ben. 2016b. "The Rise of Speculative Devices: Hooking Up with the Bots of Ashley Madison. *First Monday* 21, no. 6. https://firstmonday.org/ojs/index.php/fm/article/view/6426.

Light, Ben, Jean Burgess and Stefanie Duguay 2018. "The Walkthrough Method: An Approach to the Study of Apps." *New Media & Society* 20, no. 3: 881-900. doi: 10.1177/1461444816675438.

Lin, Trisha T. C., and Chanansara Oranop. 2016. "Responding to Media Convergence: Regulating Multi-Screen Television Services in Thailand." *Telematics and Informatics* 33, no. 2: 722–32.

Lindqvist, Janne, Justin Cranshaw, Jason Wiese, Jason Hong, and John Zimmerman. 2011. "I'm the Mayor of My House: Examining Why People Use Foursquare—A Social-Driven Location Sharing Application." *Proceedings of the SIGCHI Conference on Human Factors in Computing Systems (CHI '11)*, ACM, New York. 2409–18.

Ling, Rich. 2012. *Taken for Grantedness: The Embedding of Mobile Communication into Society*. Cambridge: MIT Press.

Lingel, Jessa, and Bradley Wade Bishop. 2014. "The Geo-Web and Everyday Life: An Analysis of Spatial Tactics and Volunteered Geographic Information." *First Monday* 19, no. 7. http://firstmonday.org/ojs/index.php/fm/article/view/5316/4095.

Lobinger, Katharina. 2016. "Photographs as Things—Photographs of Things. A Texto-Material Perspective on Photo-Sharing Practices." *Information, Communication & Society* 19, no. 4: 475–88. doi: 10.1080/1369118X.2015.1077262.

Lopez, Michael J. 2016. "The Industry Leading Resource fo App Market Data Was Already Making Waves in a Crowded App Space." www.michealjlopez.com/app-annie.

Lotan, Gilad. 2015a. "Apple, Apps, and Algorithmic Glitches." *Medium*. https://medium.com/message/apple-apps-and-algorithmic-glitches-f7bc8dd2cda6.

Lotan, Gilad. 2015b. "Apple's App Charts: 2015 Data and Trends . . . or How Much Harder It Is to Get into the Top Charts." *Medium*, December 31. https://medium.com/i-data/apple-s-app-charts-2015-data-and-trends-abb95300df57.

Lotz, Amanda D. 2014. *The Television Will Be Revolutionized*. 2nd ed. New York: New York University Press.

Lovely, Stephen. 2016. "Review of Tubi TV." *Cordcutting.com*. July 11. http://cordcutting.com/review-of-tubi-tv.

Loza, Susana. 2014. "Hashtag Feminism, #SolidarityIsForWhiteWomen, and the Other #FemFuture." *Ada: A Journal of Gender, New Media, and Technology* 5. http://adanewmedia.org/2014/07/issue5-loza.

Luckerson, Victor. 2015. "How Snapchat Built Its Most Addictive Feature." *Time*. September 25. http://time.com/4049026/snapchat-live-stories.

Lupton, Deborah. 2013. "Quantifying the Body: Monitoring and Measuring Health in the Age of mHealth Technologies." *Critical Public Health* 23, no. 4: 393–403.

Lupton, Deborah. 2015. Health Promotion in the Digital Era: A Critical Commentary. *Health Promotion International* 30, no. 1: 174–83.

Lupton, Deborah. 2016. "Personal Data Practices in the Age of Lively Data." In *Digital Sociologies*, edited by Jessie Daniels, Karen Gregory, and Tressie McMillan Cottom, 339–354. Bristol, UK: Polity Press.

Luu, Chieu, et al. 2015. "Neko Atsume Is the Addicting New App Where You Feed Stray Cats." *CNN.com*. July 7. www.cnn.com/2015/07/03/asia/neko-atsume-japan-cat-game.

Lyon, David. 2003. *Surveillance as Social Sorting: Privacy Risk and Digital Discrimination*. London: Routledge.

Macari, Matt. 2012. "Apple Gets Yet Another, Broader Slide-to-Unlock Patent." *The Verge*. www.theverge.com/2012/10/10/3479550/apple-expands-patent-coverage-on-slide-to-unlock-feature.

Macauley, Eyabo. 2015. "The Evolution of Shazam." Presentation at the IAB Digital Summit. February 19.

MacKinnon, Rebecca. 2012. *Consent of the Networked: The Worldwide Struggle for Internet Freedom*. New York: Basic Books.

Magnet Shoshana, and Tara Rodgers. 2012. "Stripping for the State." *Feminist Media Studies* 12, no. 1: 101–18.

Mahan, Joseph, Joris Drayer, and Emily Sparvero. 2012. "Gambling and Fantasy: An Examination of the Influence of Money on Fan Attitudes and Behaviors." *Sports Marketing Quarterly* 21: 159–69.

Mahler, Jonathan. 2015. "Who Spewed That Abuse? Anonymous Yik Yak App Isn't Telling." *New York Times*. March 8. www.nytimes.com/2015/03/09/technology/popular-yik-yak-app-confers-anonymity-and-delivers-abuse.html?_r=1.

Manjoo, Farhad. 2009. "That Tune, Named." *Slate*. October 19. www.slate.com/articles/technology/technology/2009/10/that_tune_named.html.

Mann, Stephen L. 2011. "Drag Queens' Use of Language and the Performance of Blurred Gendered and Racial Identities." *Journal of Homosexuality* 58, nos. 6–7: 793–811.

Mannie, Sierra. 2014. "Dear White Gays: Stop Stealing Black Female Culture." *Time*. July 9. http://time.com/2969951/dear-white-gays-stop-stealing-black-female-culture.

Manovich, Lev. 2001. *The Language of New Media*. Cambridge: MIT Press.

Manovich, Lev. 2013. *Software Takes Command: Extending the Language of New Media*. International Texts in Critical Media Aesthetics. New York: Bloomsbury.

Mao, Qiying. 2015. "Yingyong fenfa baidu tengxun 360 zhan 84%; zhongxiao yingyong shangdian tuichu jingzheng" [Baidu Tencent 360 Take 84 Percent Share in App Distribution; Small Firms Withdraw from the Competition]. *Sohu.com*. http://it.sohu.com/20150802/n418009158.shtml.

Marwick, Alice, and boyd, danah. 2011. "I Tweet Honestly, I Tweet Passionately: Twitter Users, Context Collapse, and the Imagined Audience." *New Media & Society* 13, no. 1: 114–33.

Mary H. 2015. "Shazam Has Really Gone Downhill." *SiteJabber*. November 11. www.sitejabber.com/reviews/www.shazam.com.

Massey, Doreen. 2013/1994. *Space, Place, and Gender*. Minneapolis: University of Minnesota Press.

Massumi, Brian. 2010. "The Future Birth of the Affective Fact: The Political Ontology of Threat." In *The Affect Theory Reader*, edited by Melissa Gregg and Gregory J. Seigworth, 52–70. Durham, NC: Duke University Press.

Matney, Lucas. 2016. "Apple Shows Soft China Revenues at \$18.37B in Q1, up 14% YoY." *Techcrunch.com*. http://techcrunch.com/2016/01/26/apple-shows-q1-china-revenues-of-18-37b-up-14-yoy.

Matviyenko, Svitlana, and Paul D. Miller. 2014. *The Imaginary App*. Software Studies. Cambridge: MIT Press.

Maurer, Bill, and Lana Swartz. 2015. "Wild, Wild West: A View from Two Californian Schoolmarms." In *The Moneylab Reader*, edited by Geert Lovink, , Nathaniel Tkacz, and Patricia de Vries, 221–29. Amsterdam: Institute of Network Cultures.

Maureira, Marcello Gómez. 2014. "Tender—It's How People Meat." *Vimeo*. https://vimeo.com/111997940.

May, Emily. 2011. "Hollaback!'s 'I've Got Your Back Campaign.'" *Vimeo*. https://vimeo.com/24870641.

May, Emily, and Samuel Carter. 2016. "Hollaback! You Have the Power to End Street Harassment." In *Gender, Sex and Politics: In the Streets and Between the Sheets in the 21st Century*, edited by Shira Tarrant, 11–21. New York: Routledge.

Mayer, Vicki. 2011. *Below the Line: Producers and Production Studies in the New Television Economy*. Durham, NC: Duke University Press.

McCormick, Rich. 2016. "Hillary Clinton's new mobile app offers real-world rewards to her biggest supporters." *TheVerge.com*. Retrieved August 25, 2016, from http://www.theverge.com/2016/7/25/12268528/hillary-clinton-mobile-app-supporter-reward.

McGarry, Caitlin. 2015. "Forget Television: The Revolution Will Be Live-Streamed." *MacWorld*. August 19. www.macworld.com/article/2972892/social-media/forget-television-the-revolution-will-be-live-streamed.html.

McGonigal, J. 2011. *Reality Is Broken: Why Games Make Us Better and How They Can Change the World*. New York: Penguin Books.

McGuigan, Jim. 2010. "Creative Labour, Cultural Work, and Individualisation." In *International Journal of Cultural Policy* 16, no. 3: 323–35.

McGuigan, Jim. 2014. "The Neoliberal Self." *Culture Unbound* 6: 223–40.

McGuigan, Lee. 2015. "Direct Marketing and the Productive Capacity of Commercial Television: T-Commerce, Advanced Advertising, and the Audience Product." *Television & New Media* 16, no. 2: 196–214.

McKelvey, Fenwick. 2016. "No More Magic Algorithms: Cultural Policy in an Era of Discoverability." *Data & Society: Points*. May 9. https://points.datasociety.net/no-more-magic-algorithms-cultural-policy-in-an-era-of-discoverability-6ba07eda2b89#.x39cxvk5n.

McKelvey, Fenwick, and Jill Piebiak. 2016. "Porting the Political Campaign: The Nation-Builder Platform and the Global Flows of Political Technology." *New Media & Society*. December. doi: 10.1177/1461444816675439.

Medtronic. 2016. "Fitbit and Medtronic Provide Key Data for Those Living with Diabetes." *Medtronic*. www.medtronic.com/uk-en/about/news/news-fitbit-medtronic-partnership.html.

Meehan, Eileen. 1986. "Conceptualizing Culture as Commodity: The Problem of Television." *Critical Studies in Mass Communication* 3, no. 4: 448–57.

Meese, James, Rowan Wilken, Bjorn Nansen, and Michael Arnold. 2015. "Entering the Graveyard Shift: Disassembling the Australian TiVo." *Television & New Media* 16, no. 2: 165–79.

Mellamphy, Nandita Biswas, Nick Dyer-Witheford, Alison Hearn, Svitlana Matviyenko, and Andrew Murphie. 2015. "Introduction to Special Issue: Apps and Affect." *Fibreculture* 25: 1–9.

Menegus, Bryan. 2016. "Hillary Clinton Just Released a Truly Joyless Mobile Game." *Giz-

modo. July 25. http://gizmodo.com/hillary-clinton-just-released-a-truly-joyless-mobile-ga-1784239852.

Metz, Christian. 1974. *Film Language: A Semiotics of the Cinema*. University of Chicago Press.

Metz, Christian. 1982. *The Imaginary Signifier: Psychoanalysis and the Cinema*. Bloomington: Indiana University Press.

Miller, Sean. 2016. "NGP VAN Unveils Volunteer Tools That Counter Non-Partisan Competition." *Campaigns & Elections*. June 30. www.campaignsandelections.com/campaign-insider/ngp-van-unveils-volunteer-tools-that-counter-non-partisan-competition.

Miller, Tessa. 2013. "I Am Leah Busque, Founder of TaskRabbit and This Is How I Work." *Lifehacker.com*. http://lifehacker.com/im-leah-busque-founder-of-taskrabbit-and-this-is-how-496031842.

Miller, Vincent. 2008. "New Media, Networking, and Phatic Culture." *Convergence: The International Journal of Research into New Media Technologies* 14, no. 4: 387–400. doi: 10.1177/1354856508094659.

Millward, Steven. 2014. "What China's Netizens Want: Building Tech Ecosystems in the World's Toughest Market." In *The Imaginary App*, edited by Paul D. Miller and Svitlana Matviyenko, 179–86. Cambridge : MIT Press.

Miltner, Kate. 2015. "From #feels to Structure of Feeling: The Challenges of Defining 'Meme Culture.'" *Culture Digitally*. October 29. http://culturedigitally.org/2015/10/memeology-festival-02-from-feels-to-structure-of-feeling-the-challenges-of-defining-meme-culture.

Morozov, Evgeny. 2013. "The Perils of Perfection." *New York Times*. March 2. www.nytimes.com/2013/03/03/opinion/sunday/the-perils-of-perfection.html.

Morris, Jeremy W. 2013. "Non-Practical Entities: Business Method Patents and the Digitization of Culture." *Critical Studies in Media Communication* 31, no. 3: 212–29.

Morris, Jeremy W. 2015. *Selling Digital Music, Formatting Culture*. Berkeley: University of California Press.

Morris, Jeremy W., and Evan Elkins. 2015. "There's a History for That: Apps and Mundane Software as Commodity." *Fibreculture* 25: 62–87. doi: 10.15307/fcj.25.181.2015.

Mowlabocus, Sharif. 2016. "The 'Mastery' of the Swipe: Smartphones, Transitional Objects and Interstitial Time." *First Monday* 21, no. 10. http://firstmonday.org/ojs/index.php/fm/article/view/6950/5630.

Mullaney, Tim. 2015. "Why I Hate the Fitbit IPO (and You Should, Too)." *Market Watch*. www.marketwatch.com/story/why-i-hate-the-fitbit-ipo-and-you-should-too-2015-06-18.

Munt, Sally. 2008. *Queer Attachments: The Cultural Politics of Shame*. Farnham, UK: Ashgate Publishing, Ltd.

Murphy, Sean. 2012. "Many States Fall Short of Federal Sex Offender Law." *CNSnews.com*. October 4. www.cnsnews.com/news/article/many-states-fall-short-federal-sex-offender-law.

Murphy, Tim. 2014. "Gay Men and Misogyny: Rose McGowan's Half-Right." *New York Magazine*. November 9. http://nymag.com/thecut/2014/11/gay-men-and-misogyny-rose-mcgowans-half-right.html.

Muston, Samuel. 2013. "Meet Tinder—aka the 'Shy Grindr.'" *The Independent*. www.independent.co.uk/news/media/online/meet-tinder-aka-the-shy-grindr-8460863.html.

My Mobile Witness. n.d. "Technology." www.mymobilewitness.com/technology.

Myers, David G. *Intuition: Its Powers and Perils*. New Haven, CT: Yale University Press, 2002.

Nagy, Peter, and Gina Neff. 2015. "Imagined Affordance: Reconstructing a Keyword for Communication Theory." *Social Media + Society* 1, no. 2: 1–9. doi: 10.1177/2056305115603385.

Nakamura, Lisa. 2009. *Digitizing Race: Visual Cultures of the Internet*. Visual Studies. Minneapolis: University of Minnesota Press.

Nakamura, Lisa. 2013. *Cybertypes: Race, Ethnicity, & Identity on the Internet*. New York: Routledge.

Nakamura, Lisa. 2014. "'I will do everything that I am asked': Scambaiting, Digital Show-Space, and the Racial Violence of Social Media." *Journal of Visual Culture* 13, no. 3: 257–74.

Napoli, Philip M. 2012. *Audience Economics: Media Institutions and the Audience Marketplace*. New York: Columbia University Press.

Neate, Rupert. 2016. "Fitbit Stock Sinks After Company Warns Shareholders Over Profits." *The Guardian*, US Markets, February 23.

Neff, Gina, and David Stark. 2002/2004. "Permanently Beta: Responsive Organization in the Internet Era." In *Society Online: The Internet in Context*, edited by Phillip N. Howard and Steven Jones, 173–88. Thousand Oaks, CA: Sage.

Neves, Joshua. 2015. "The Long Commute: Mobile Television and the Seamless Social." In *Chinese Television in the Twenty-First Century: Entertaining the Nation*, edited by Ruoyun Bai and Geng Song, 51–66. London: Routledge.

Newitz Anna L. 2015. "Ashley Madison Code Shows More Women, and More Bots." *Gizmodo*. August 31. http://gizmodo.com/ashley-madison-code-shows-more-women-and-more-bots-1727613924.

Newman, Michael. 2017. *Atari Age: The Emergence of Video Games in America*. Cambridge: MIT Press.

Newman, Tony. 2002. "Promoting Resilience: A Review of Effective Strategies for Child Care Services." www.nursingacademy.com/uploads/6/4/8/8/6488931/promotingresiliencenewman.pdf .

Newport, Frank. 2015. "Most U.S. Smartphone Owners Check Phone at Least Hourly." *Gallup.com*. July 9. www.gallup.com/poll/184046/smartphone-owners-check-phone-least-hourly.aspx.

Newton, Esther. 1972. *Mother Camp: Female Impersonators in America*. Chicago: University of Chicago Press.

New York State. Homeland Security and Emergency Services. n.d. "Counter Terrorism." www.dhses.ny.gov/oct/safeguardNY.

New York Times. 1989. "Campus Life: Syracuse; In Wake of Rapes, Center for Victims to Open by July." October 22. www.nytimes.com/1989/10/22/style/campus-life-syracuse-in-wake-of-rapes-center-for-victims-to-open-by-july.html.

Nextdoor. 2016. "Guidelines." https://nextdoor.com/neighborhood_guidelines/#guidelines.

Ng, Eve. 2013. "A "Post-Gay" Era? Media Gaystreaming, Homonormativity, and the Politics of LGBT Integration." *Communication, Culture & Critique* 6, no. 2: 258–83.

NGP VAN Next. 2014. YouTube. www.youtube.com/watch?v=dlYoMgxtqMk&feature=youtube_gdata_player.

Nieborg, David. B. 2015. "Crushing Candy: The Free-to-Play Game in Its Connective Commodity Form." *Social Media + Society* 1, no. 2: 1–12. doi: 10.1177/2056305115621932.

Nieborg, David. 2016. "From Premium to Freemium: The Political Economy of the App." In *Social, Casual and Mobile Games: The Changing Gaming Landscape*, edited by Tamar Leaver and M. Willson. London: Bloomsbury Academic, 225–40.

Nielsen. 2015. "So Many Apps, So Much More Time for Entertainment." *Nielsen Research*. www.nielsen.com/us/en/insights/news/2015/so-many-apps-so-much-more-time-for-entertainment.html.

Nielsen. 2017a. "Millennials on Millennials." *Nielsen Research*. www.nielsen.com/content/dam/corporate/us/en/product%20info/millennials-on-millennials-one-sheet.pdf.

Nielsen. 2017b. "Millennials on Millennials: A Look at Viewing Behavior, Distraction and Social Media Stars." *Nielsen Research*. March 2. www.nielsen.com/us/en/insights/news/2017/millennials-on-millennials-a-look-at-viewing-behavior-distraction-social-media-stars.html.

Nielsen, Rasmus Kleis. 2012. *Ground Wars: Personalized Communication in Political Campaigns*. Princeton, NJ: Princeton University Press.

Norman, Donald A. 1988. *The Design of Everyday Things*. New York: Doubleday.

Norman, Donald A. 2004. *Emotional Design: Why We Love (or Hate) Everyday Things*. New York: Basic/Civitas Books.

Novas, Carlos, and Rose, Nikolas. 2010. "Genetic Risk and the Birth of the Somatic Individual." *Economy and Society* 29, no. 4: 485–513.

Noyes, Dan. 2010. "Jessica's Law Dilemma: Homeless Sex Offenders." *ABC7*. http://abclocal.go.com/kgo/story?section=news/iteam&id=7245220.

Ohikuare, Judith. 2013. "An app to Help Women Avoid Street Harassment." *The Atlantic*. September 13. www.theatlantic.com/technology/archive/2013/09/an-app-to-help-women-avoid-street-harassment/279642.

Olma, Sebastian. 2014. "Never Mind the Sharing Economy: Here's Platform Capitalism." *INC Blog*. October 16. http://networkcultures.org/mycreativity/2014/10/16/never-mind-the-sharing-economy-heres-platform-capitalism.

Ong, Aihwa. 2006. *Neoliberalism as Exception: Mutations in Citizenship and Sovereignty*. Durham, NC: Duke University Press.

Ong, Walter J. 1982. *Orality and Literacy: The Technologizing of the Word*. London: Routledge.

O'Reilly. 2010. Where 2.0 2010: Dennis Crowley. "Adventures in Mobile Social 2.0: Twelve Months of Foursquare." Where 2.0 Conference. www.youtube.com/watch?v=LcaJGE-GQJNU.

Orth, Samuel Peter. 1919. *The Boss and the Machine: A Chronicle of the Politicians and Party Organization*. New Haven, CT: Yale University Press.

Oswaks, Molly. 2015. "'This.' Has People Clamoring for an Invite." *New York Times*. January 28. www.nytimes.com/2015/01/29/fashion/this-has-people-clamoring-for-an-invite.html.

Ouellette, Laurie, and Julie Wilson. 2011. "Women's Work: Affective Labour and Convergence Culture." *Cultural Studies* 25, nos. 4–5: 548–65.

Papacharissi, Zizi. 2015. *Affective Publics: Sentiment, Technology, and Politics.* Oxford: Oxford University Press.

Parks, Lisa. 2010. "Around the Antenna Tree: The Politics of Infrastructural Visibility." *Flow.* March 5. http://flowtv.org/2010/03/flow-favorites-around-the-antenna-tree-the-politics-of-infrastructural-visibilitylisa-parks-uc-santa-barbara.

Parsell, Cameron, and Greg Marston. 2012. "Beyond the 'At Risk' Individual: Housing and the Eradication of Poverty to Prevent Homelessness." *Australian Journal of Public Administration* 71, no. 1: 33–44.

Peck, Jamie, and Adam Tickell. 2002. "Neoliberalizing Space." *Antipode* 34, no. 3: 380–404.

Pedwell, Carolyn. 2014. *Affective Relations: The Transnational Politics of Empathy.* Basingstoke: Palgrave Macmillan. Kindle Edition.

Perez, Sarah. 2016. "Yik Yak Tries to Make a Comeback with Launch of Private Chat." *TechCrunch.* April 25. https://techcrunch.com/2016/04/25/yik-yak-tries-to-make-a-comeback-with-launch-of-private-chat.

Periscope.tv. "About Us." Periscope. www.periscope.tv/about.

Pertierra, Anna Cristina, and Graeme Turner. 2013. *Locating Television: Zones of Consumption.* London: Routledge.

Peters, John Durham. 2001. "Witnessing." *Media, Culture & Society* 23, no. 6: 707–723.

Peters, Thomas. 2016. "Trump and Brexit Used a New Digital Organizing Tool to Help Achieve Their Surprise Victories." *Medium.* December 20. https://medium.com/@uCampaignapp/how-trump-and-brexit-used-a-new-digital-organizing-tool-to-win-their-surprise-victories-ceca7c720b3.

Peterson, Alan, and Deborah Lupton. 2000. *The New Public Health: Health and Self in the Age of Risk.* London: Sage.

Phillips, Andrea. 2010. "The Gamification of Politics." *Deus Ex Machinatio.* November 3. www.deusexmachinatio.com/blog/2010/11/3/the-gamification-of-politics.html.

Phillips, Whitney. 2015. *This Is Why We Can't Have Nice Things: The Origins, Evolution, and Cultural Embeddedness of Online Trolling.* Cambridge: MIT Press.

Phillips, Whitney, and Ryan Milner. 2017. *The Ambivalent Internet: Mischief, Oddity, and Antagonism Online.* Hoboken, NJ: Wiley.

Pierce, David. 2015. "Twitter's Periscope App Lets You Livestream Your World." *Wired.* March 26. www.wired.com/2015/03/periscope.

Pierce, David. 2018. "Inside the Downfall of Doppler Labs." *Wired.* November 1, 2017. https://www.wired.com/story/inside-the-downfall-of-doppler-labs/.

Pinch, Trevor J., and Wiebe E. BijkerE. 1984. "The Social Construction of Facts and Artefacts: Or How the Sociology of Science and the Sociology of Technology Might Benefit Each Other." *Social Studies of Science* 14, no. 3: 399–441. doi: 10.1177/030631284014003004.

Piwek, Lukasz, and Adam Joinson. 2016. "'What Do They Snapchat About?' Patterns of Use in Time-Limited Instant Messaging Service." *Computers in Human Behavior* 54 (January): 358–67. doi: 10.1016/j.chb.2015.08.026.

PocketGamer. 2016. "Count of Active Applications in the App Store." PocketGamer.biz. www.pocketgamer.biz/metrics/app-store/app-count.

Polanyi, Karl. 1957. *The Great Transformation.* Boston: Beacon Press

Politico. 2016. "Digital Campaigns: How Can the Democrats Win with Tech?" *POLITICO.* July 26. www.politico.com/video/2016/07/digital-campaigns-how-can-the-democrats-win-with-tech-060102.

Pon, Brian. 2015. "Locating Digital Production: How Platforms Shape Participation in the Global App Economy." *Caribou Digital.* http://cariboudigital.net/wp-content/up loads/2015/04/Pon-AAG-Platforms-and-app-economy.pdf.

Popper, Ben. 2014. "Meet Swarm: Foursquare's Ambitious Plan to Split Its App in Two." *The Verge.* May 1. www.theverge.com/2014/5/1/5666062/foursquare-swarm-new-app.

Powers, Devon. 2017. "First! Cultural Circulation in the Age of Recursivity." *New Media & Society* 19, no. 2: 165–80.

Pratt, Ythan. 2013. Google Play Rankings Explained by the NativeX Games Task Force. *NativeX* Blog. http://nativex.com/blog/google-play-rankings-explained-by-the-nativex-games-task-force.

Pressberg, Matt. 2017. "Tubi TV Is Bringing Traditional Television Economics to Streaming." *The Wrap.* June 13. www.thewrap.com/tubi-tv-bringing-traditional-television-economics-streaming.

Pretz, Jean E., and Kathryn Sentman Totz. 2007. "Measuring Individual Differences in Affective, Heuristic, and Holistic Intuition." *Personality and Individual Differences* 43: 1247–57.

Probyn, Elspeth. 2004. "Everyday Shame." *Cultural Studies* 18, nos. 2–3: 328–49.

Pugh, E. W. 2002. "Origins of Software Bundling." *Annals of the History of Computing, IEEE* 24, no. 1: 57–58. doi: 10.1109/85.988580.

Qu, Zhongfang, and Chenguang Wu. 2013. "Tengxun kaifang pingtai dazao yidong hulianwang shengtaiquan" [Tencent's Open Platform Aims to Build the Mobile Internet Ecosystem]. *163.com.* http://money.163.com/13/1120/00/9E38UCM800253B0H.html.

Quintas-Froufe, Natalia, and Ana González-Neira. 2015. "A New Challenge for Advertising on Mobile Devices: Social TV." *Revista Icono* 13, no. 1: 52–75.

Race, Kane. 2014. "Speculative Pragmatism and Intimate Arrangements: Online Hook-Up Devices in Gay Life." *Culture, Health & Sexuality* 17, no. 4: 37–41. doi: 10.1080/13691058.2014.930181.

Radway, Janice A. 1984. *Reading the Romance: Women, Patriarchy, and Popular Literature.* Chapel Hill: University of North Carolina Press.

Raile, Dan. 2014. "Deezer Partners with Vodafone in South Africa." *Billboard.* October 9. www.billboard.com/articles/business/6281301/deezer-south-africa-vodafone.

Ramirez, Fanny. 2015. "Affect and Social Value in Freemium Games." In *Social, Casual, and Mobile Games: The Changing Gaming Landscape,* edited by Tama Leaver and Michele Willson, 117–32. New York: Bloomsbury Academic.

Randle, Quint, and Rob Nyland. 2008. "Participation in Internet Fantasy Sports Leagues and Mass Media Use." *Journal of Website Promotion* 3, nos. 3/4: 143–52.

Ranzini, Giulia, and Lutz, Christoph. 2016. "Love at First Swipe? Explaining Tinder Self-Presentation and Motives." *Mobile Media & Communication* 5, no. 1: 80–101. doi: 10.1177/2050157916664559.

Raphel, Adrienne. 2014. "TaskRabbit Redux." *New Yorker*. July 22. www.newyorker.com/business/currency/taskrabbit-redux.

Razack, Sherene H. 2008. *Casting Out: The Eviction of Muslims from Western Law and Politics*. Toronto: University of Toronto Press.

Recode. 2016. "Full Video: Tinder CEO Sean Rad at Code 2016." www.recode.net/2016/6/17/11966748/sean-rad-tinder-full-video-code.

Rees, Alex. 2016. "Here's What Happens When You Play Kendall and Kylie Jenner's New App, Which Is Delightful." *Cosmopolitan.com*. February 17. www.cosmopolitan.com/entertainment/celebs/news/g5425/kendall-kylie-jenner-app-game-review-walkthrough.

Reeves, Joshua. 2012. "If You See Something, Say Something: Lateral Surveillance and the Uses of Responsibility." *Surveillance & Society* 10, nos. 3/4: 235–48.

Reisinger, Don. 2016. "Kendall and Kylie Have Another Mobile App Hit." *Fortune*. http://fortune.com/2016/02/17/kendall-kylie-mobile-app.

Renbourn, Edward T. 1960. "The Natural History of Insensible Perspiration: A Forgotten Doctrine of Health and Disease." *Medical History* 4, no. 2: 135–52.

Rentschler, Carrie. 2004. "Witnessing: US Citizenship and the Vicarious Experience of Suffering." *Media, Culture & Society* 26, no. 2: 296–304.

Rentschler, Carrie. 2014. "Rape Culture and the Feminist Politics of Social Media," *Girlhood Studies* 7, no. 1: 65–82.

Rentschler, Carrie. 2015. "Technologies of Bystanding: Learning to See Like a Bystander." In *Shaping Inquiry in Culture, Communication, and Media Studies*, edited by Sharrona Pearl, 15–40. London: Routledge.

Rettberg, Jill Walker. 2014a. *Blogging*. 2nd ed. Cambridge: Polity Press.

Rettberg, Jill Walker. 2014b. *Seeing Ourselves Through Technology: How We Use Selfies, Blogs and Wearable Devices to See and Shape Ourselves*. Basingbroke: Palgrave.

Rettberg, Jill Walker. 2017a. "Hand Signs for Lip-Syncing: The Emergence of a Gestural Language on Musical.ly as a Video-Based Equivalent to Emoji." *Social Media + Society*.

Rettberg, Jill Walker. 2017b. "Self-Representation in Social Media." In *SAGE Handbook of Social Media*, edited by Jean Burgess, Alice Marwick, and Thomas Poell, 429-443. London: SAGE Publications.

Rettberg, Jill Walker. Forthcoming. "Online Diaries and Blogs." In *The Diary*, edited by Batsheva Ben-Amos and Dan Ben-Amos. Bloomington: Indiana University Press.

Rettberg, Scott. 2011. "All Together Now: Collective Knowledge, Collective. Narratives, and Architectures of Participation." In *New Narratives: Stories and Storytelling in the Digital Age*, edited by Ruth Page and Bronwen Thomas, 187–204. Lincoln: University of Nebraska Press.

Rhodes S. 2015. "Companion App Promises You'll Never Walk Home Alone Again." *USA Today College*. http://college.usatoday.com/2015/09/11/companion-app-promises-youll-never-walk-home-alone-again.

Riaz, Saleha. 2014. "Tencent Reports 110M Daily Downloads from Myapp Store." *MobileWorldLive.com*. www.mobileworldlive.com/asia/asia-news/tencent-reports-110m-daily-downloads-myapp-store.

Robinson B. 1991. "Lifeline Services—An Industry Still Defining Its Own Identity." *Ageing International* 18, no. 1: 42–47.

Rogers, Kevin. 2013. "Jailbroken: Examining the Policy and Legal Implications of iPhone Jailbreaking." *Pittsburgh Journal of Technology Law and Policy* 13, no. 2:, 1–13.

Ronson, Jon. *So You've Been Publicly Shamed*. New York: Riverhead Books, 2016.

Rosenberg, Scott. 2008. *Dreaming in Code: Two Dozen Programmers, Three Years, 4,732 Bugs, and One Quest for Transcendent Software*. New York: Crown.

Rosoff, Matt. 2016. "The App Explosion Is Over." *Business Insider*. www.businessinsider.com/average-number-of-apps-vs-time-spent-2016–5.

Rossiter, Ned. 2016. *Software, Infrastructure, Labor: A Media Theory of Logistical Nightmares*. New York: Routledge.

Rotogrinders. 2016. "The Evolution of the Daily Fantasy Sports Industry." *Daily Fantasy Timeline: A History of the DFS Industry*. https://rotogrinders.com/static/daily-fantasy-sports-timeline.

Ruckenstein, Minna. 2017. "Keeping Data Alive: Talking DTC Genetic Testing." *Information, Communication & Society* 20, no. 7: 1024–39.

Rushkoff, Douglas. 2016. *Throwing Rocks at the Google Bus: How Growth Became the Enemy of Prosperity*. New York: Penguin Books.

Rutherford, Alexandra. 2009. *Beyond the Box: B.F. Skinner's Technology of Behavior from Laboratory to Life, 1950s-1970s*. Toronto: University of Toronto Press.

rVotes. 2016. "rStory." *rVotes Campaign and Election System*. www.rvotes.com/?page_id=272.

Ryan, Peter. 2016. "Mobile Platforms: The Medium and Rhetoric of the 2015 Canadian Federal Election Manifestos." Presented at the Canadian Political Science Association, Calgary, Alberta, June 2.

Sager, Ira. 2012. "Before IPhone and Android Came Simon, the First Smartphone." *Bloomberg*. www.bloomberg.com/news/articles/2012–06–29/before-iphone-and-android-came-simon-the-first-smartphone.

Sales, Nancy Jo. 2015. "Tinder and the Dawn of the 'Dating Apocalypse.'" *Vanity Fair*. www.vanityfair.com/culture/2015/08/tinder-hook-up-culture-end-of-dating.

Sandvig, Christian. 2013. "The Internet as Infrastructure." In *The Oxford Handbook of Internet Studies*, edited by William Dutton, 86–107. Oxford: Oxford University Press.

Sasson, E. 2014. "Does Gay Male Culture Have A Misogyny Problem?" *The New Republic*. December 15. www.newrepublic.com/article/120565/mcgowan-hilton-banks-controversies-are-gay-men-misogynistic.

Savage, Mike. 2013. "The 'Social Life of Methods': A Critical Introduction." *Theory, Culture & Society* 30, no. 4: 3–21.

Scannell, Paddy. 2000 "For-Anyone-As-Someone Structures." *Media, Culture & Society* 22, no. 1: 5–24.

Schichor, David. 1995. *Punishment for Profit: Private Prisons / Public Concerns*. Thousand Oaks, CA: Sage.

Schmidt, Leigh Eric. 2000. *Hearing Things: Religion, Illusion, and the American Enlightenment*. Cambridge, MA: Harvard University Press.

Scholz, Trebor (2016) *Uberworked and Underpaid: How Workers Are Disrupting the Digital Economy*. Cambridge, MA: Polity Press.

Schüll, Natasha. 2005. "Digital Gambling: The Coincidence of Desire and Design." *Annals of the American Academy of Political and Social Science* 597: 65–81.

Schüll, Natasha. 2012. *Addiction by Design: Machine Gambling in Las Vegas*. Princeton, NJ: Princeton University Press.

Schüll, Natasha. 2016. "Data for Life: Wearable Technology and the Design of Self-Care." *BioSocieties* 11, no. 3: 317–33.

Schwartz, Hillel. 1986. *Never Satisfied: A Cultural History of Diets, Fantasies, and Fat*. Anchor Books, 1986.

Segrave, Kerry. 2002. *Vending Machines: An American Social History*. Jefferson, NC: McFarland., 2002.

Seigworth, G. and M. Gregg. 2010. *The Affect Theory Reader*. Durham, NC: Duke University Press.

Sender, Katherine. 2007. Dualcasting: Bravo's Gay Programming and the Quest for Women Audiences. *Cable Visions: Television Beyond Broadcasting*. New York: NYU Press: 302–18.

Sender, Katherine. 2012. *The Makeover: Reality Television and Reflexive Audiences*. New York: NYU Press.

Sennett, Richard. 2008. *The Craftsman*. New Haven, CT: Yale University Press.

Shazam. 2009. "Shazam Celebrates 15 Million New Users in Just Six Months." *Shazam Blog*. February 13. http://news.shazam.com/pressreleases/shazam-celebrates-15-million-new-users-in-just-six-months-890474.

Shazam. 2016. "Shazam Expands Partnership with AdVine to Offer Shazam Advertising Solutions in South Africa." *Shazam Blog*. November 7. http://news.shazam.com/press releases/shazam-expands-partnership-with-advine-to-offer-shazam-advertising-solutions-in-south-africa-925908.

Sherburne, Philip. 2013. "Beatmatching: Shazam Adds 1.5 Million Dance Tracks to Database." *Spin*. February 7. www.spin.com/2013/02/shazam-adds-15-million-dance-tracks-database.

Shifman, Limor. 2013. *Memes in Digital Culture*. Cambridge: MIT Press.

Shim, Hongjin, Kyung Han, You, Jeong Kyu Lee, and Eun Go. 2015. "Why Do People Access News with Mobile Devices? Exploring the Role of Suitability Perception and Motives on Mobile News Use." *Telematics and Informatics* 32, no. 1: 108–17.

Shimkovitz, Brian. 2012. "Collateral Damage." *Wire*. May. www.thewire.co.uk/in-writing/essays/collateral-damage_awesome-tapes-from-africa_s-brian-shimkovitz.

Shinkle, Eugénie. 2008. "Video Games, Emotion, and the Six Senses." *Media, Culture & Society* 30, no. 6: 907–15.

Shirky, Clay. 2015. *Little Rice: Smartphones, Xiaomi, and the Chinese Dream*. New York: Columbia Global Reports.

Shu, Catherine. 2014. "Wandoujia, One Of China's Leading App Stores, Lands $120M In New Funding Led By SoftBank." *TechCrunch*. https://techcrunch.com/2014/01/12/wandoujia-120m.

Silver, Nate. 2011. "After 'Black Friday,' American Poker Faces Cloudy Future." *FiveThirtyEight*. https://fivethirtyeight.blogs.nytimes.com/2011/04/20/after-black-friday-american- poker-faces-cloudy-future.

Silverstone, Roger. 1994. *Television and Everyday Life*. London: Routledge.

Singer, Natasha. 2014. "In the Sharing Economy Workers Find Both Freedom and Uncertainty." *New York Times.* August 16. www.nytimes.com/2014/08/17/technology/in-the-sharing-economy-workers-find-both-freedom-and-uncertainty.html?_r=0.

Singer, Natasha. 2015. The War on Campus Sexual Assault Goes Digital. *New York Times.* November 13. www.nytimes.com/2015/11/15/technology/the-war-on-campus-sexual-assault-goes-digital.html.

Smith, Aaron. 2016. "15% of American Adults Have Used Online Dating Sites or Mobile Dating Apps." *Pew Research Center.* www.pewinternet.org/2016/02/11/15-percent-of-american-adults-have-used-online-dating-sites-or-mobile-dating-apps/.

Smith, Dave. 2014. "Chart of the Day: Most People Use Only 4 Apps." *Business Insider.* www.businessinsider.com/chart-of-the-day-most-people-use-only-4-apps-2014-9.

Smith D. L., and S. J. Notaro. 2015. Is Emergency Preparedness a 'Disaster' for People with Disabilities in the US? Results from the 2006–2012 Behavioral Risk Factor Surveillance System (BRFSS). *Disability & Society* 30, no. 3: 401–18.

SN, Vikas. 2013. "Shazam Ties up with Saavn to Identify Hindi & Regional Music; Implications." *MediaNama.* April 4. www.medianama.com/2013/04/223-shazam-saavn-tieup.

Snap Inc. 2017. "Q2 2017 Earnings Report." *Snap.com.* August 10. https://investor.snap.com/-/media/Files/S/Snap-IR/reports-and-presentations/q2–17-earnings-slides.pdf.

Snapchat. 2016. "Stories." *Snapchat.com.* https://support.snapchat.com/en-US/ca/stories.

"Snaps Solutions—Social Messaging." 2015. Snaps. https://snaps.io/custom-emoji-keyboard/.

Snickars, Pelle. 2012. "A Walled Garden Turned into a Rain Forest." In *Moving Data: The iPhone and the Future of Media*, edited Pelle Snickars and Patrick Vonderau, 155–70. New York: Columbia University Press.

Snickars, Pelle, and Patrick Vonderau. 2012. *Moving Data: The iPhone and the Future of Media.* New York: Columbia University Press.

Sparf J. 2016. "Disability and Vulnerability: Interpretations of Risk in Everyday Life." *Journal of Contingencies and Crisis Management* 24, no. 4: 244–5.

Spigel, Lynn. 1992. *Make Room for TV: Television and the Family Ideal in Postwar America.* Chicago: University of Chicago Press.

Stallabrass, Julian. 1996. *Garguantua: Manufactured Mass Culture.* London: Verso.

Stampler, Laura. 2014. "Inside Tinder: Meet the Guys Who Turned Dating into an Addiction." *Time.* http://time.com/4837/tinder-meet-the-guys-who-turned-dating-into-an-addiction.

Standing, Guy. 2011. *The Precariat: The New Dangerous Classes.* London: Bloomsbury.

Star, Susan Leigh. 1999. "The Ethnography of Infrastructure." *American Behavioral Scientist* 43, 377–91.

Stark, Luke, and Kate Crawford. 2014. "The Conservatism of Emoji." *New Inquiry.* August 20. http://thenewinquiry.com/essays/the-conservatism-of-emoji/.

Statista. 2015. "Average Number of Apps Used Per Day According to Smartphone Users in the United States as of August 2015." www.statista.com/statistics/473831/number-of-daily-smartphone-apps-used-usa.

Statista. 2016a. Number of Apps Available in Leading App Stores as of June 2016. www.statista.com/statistics/276623/number-of-apps-available-in-leading-app-stores.

Statista. 2016b. Smartphone User Penetration as Percentage of Total Global Population from 2014 to 2020. www.statista.com/statistics/203734/global-smartphone-penetra-tion-per-capita-since-2005.

Stein, Louisa. 2009. "Playing Dress Up: Digital Fashion and Gamic Extensions of Televisu-al Experience in *Gossip Girl*'s *Second Life*." *Cinema Journal* 48, no. 3: 116–22.

Stelter, Brian. 2016. "Philando Castile and the Power of Facebook Live." *CNN Money*. July 7.

Sterne, Jonathan. 2003. *The Audible Past: Cultural Origins of Sound Reproduction*. Durham, NC: Duke University Press.

Sterne, Jonathan. 2007. "Out with the Trash: On the Future of New Media." In *Residual Media,* edited by Charles Acland, 16–31. Minneapolis: University of Minnesota Press.

Sterne, Jonathan. 2012. *MP3: The Meaning of a Format*. Durham, NC: Duke University Press.

Sterne, Peter. 2015. "Now Spun off from Atlantic Media, 'This.' Raises $610,000." *Politico*. August 11. www.politico.com/media/story/2015/08/now-spun-off-from-atlantic-media-this-raises-610–000–004038.

Stewart, Kathleen. 2007. *Ordinary Affects*. Durham, NC: Duke University Press.

Streitfeld, David. 2017. "'The Internet Is Broken': @ev Is Trying to Salvage It, *New York Times*. May 20. www.nytimes.com/2017/05/20/technology/evan-williams-medi um-twitter-internet.html.

Straw, Will. 2004. "Cultural Scenes." *Loisir et Société/Society and Leisure* 27, no. 2: 411–22.

Striphas, Ted. 2009. *The Late Age of Print: Everyday Book Culture from Consumerism to Control*. New York: Columbia University Press.

Summers, Nick. 2014. "The Truth about Tinder and Women Is Even Worse Than You Think." *Bloomberg Businessweek*. www.businessweek.com/articles/2014–07–02/tinders-forgotten-woman-whitney-wolfe-sexism-and-startup-creation-myths.

Sun, Bing. 2014. "Juhe qida pingtai 'qiang rukou'; 'yingyong bao' cheng tengxun zhanlue hexin" [Unifying Tencent's 7 platforms; MyApp Becoming the Strategic Core]. *Sohu. com*. http://business.sohu.com/20140225/n395597952.shtml.

Surdu, Nicolae. 2011. "How Does Shazam Work to Recognize a Song?" *So, You Code?*. January 20. www.soyoucode.com/2011/how-does-shazam-recognize-song.

Sutton, Candace. 2015. "Accused Tinder Killer Gable Tostee Posts a Picture . . ." *Daily Mail*. www.dailymail.co.uk/news/article-3278700/Accused-balcony-killer-Gable-Tostee-pub-lishes-picture-playing-GOLF-describes-witch-witch-hunt-bizarre-Facebook-post.html.

Swindle, Monica. 2011. "Feeling Girl, Girling Feeling: An Examination of 'Girl' as Affect." *Rhizomes* 22. www.rhizomes.net/issue22/swindle.html.

Szubiak, Ali. 2016. "Kendall and Kylie Jenner Launch Weirdly Addictive Mobile Game." *PopCrush*, February 17. http://popcrush.com/kendall-jenner-kylie-jenner-mobile-game.

Tabberer, Jamie 2017. "Gay Men Like Me Need to Start Acknowledging Our Misogyny Problem." *The Independent*. August 27. www.independent.co.uk/voices/gay-men-lgbt-50th-anniversary-misogyny-rupaul-drag-race-fishy-queen-lesbians-a7862516.html.

Taylor, Astra. 2014. *The People's Platform: Taking Back Power and Culture in the Digital Age*. New York: Metropolitan.

Taylor, Timothy D. 2014. *Strange Sounds: Music, Technology and Culture*. New York: Rout-ledge.

Taylor, T. L. 2012. *Raising the Stakes: E-Sports and the Professionalization of Computer Gaming*. Cambridge: MIT Press.

Tech, Robin P. G., Jan-Peter Ferdinand, and Martina Dopfer. 2016. "Open Source Hardware Startups and Their Communities." In *The Decentralized and Networked Future of Value Creation*, edited by J. P. Ferdinand, U. Petschow, and S. Dickel, 129–46 (Heidelberg: Springer 2016).

Theberge, Paul. 1997. *Any Sound You Can Imagine Making Music/Consuming Technology*. Middleton, CT: Wesleyan University Press.

Thierer, A. 2008. Review of Zittrain's *Future of the Internet*. *The Technology Liberation Front*, March 23. https://techliberation.com/2008/03/23/review-of-zittrains-future-of-the-internet.

Think Gaming. 2016. Kendall & Kylie. https://thinkgaming.com/app-sales-data/94583/kendall-and-kylie.

"This. Editorial Policy." n.d. https://this.cm/editorial.

Thompson, Derek. 2014. "The Shazam Effect." *Atlantic*. December. www.theatlantic.com/magazine/archive/2014/12/the-shazam-effect/382237.

Thrift, Nigel. 2004. "Remembering the Technological Unconscious by Foregrounding Knowledges of Position." *Environment and Planning D: Society and Space* 22, no. 1: 175–90. doi.org/ 10.1068/d321t.

Tinder. 2013a. "Tinder CEO Sean Rad, during Keynote at the Glimpse Conference." www.youtube.com/watch?v=rj9sysMTqUE.

Tinder. 2015a. "It's about Meeting . . ." https://twitter.com/Tinder/status/631249399591473153?ref_src=twsrc%5Etfw.

Tinder. 2015b. "Keeping Tinder Real." http://blog.gotinder.com/keeping-tinder-real.

Tinder. 2015c. "We Just Revamped Your Tinder Profile." http://blog.gotinder.com/we-just-revamped-your-tinder-profile.

Tinder. 2016. "Introducing Tinder Social." http://blog.gotinder.com/introducing-tinder-social.

Tiwana, Amrit. 2014. *Platform Ecosystems: Aligning Architecture, Governance, and Strategy*. Amsterdam: Elsevier, 2014.

Tiwana, Amrit, Benn Konsynsky, and Ashley Bush. 2010. "Platform Evolution: Coevolution of Platform Architecture, Governance, and Environmental Dynamics." *Information Systems Research* 21, no. 4: 675–87.

Tompkins, S., E. K. Sedgwick, and A. Frank. 1995. *Shame and Its Sisters: A Silvan Tomkins Reader*. Durham, N.C.: Duke University Press.

Totodee. 2017. Comment on Thread: "Tubi TV Is Bringing Traditional Economics to Streaming." *Reddit*. June 16. https://redd.it/6hlky0.

Truong, Alice. 2016. "Periscope's Lead Designer Explains the Painstaking Labor behind Its Fluttering Hearts." *Quartz*. February 27. https://qz.com/626443/periscopes-lead-designer-explains-the-painstaking-labor-behind-its-fluttering-hearts.

Trustify. 2016. Suspect Your Spouse Is on Ashley Madison? *Trustify*. www.trustify.info/landing/catch-a-cheating-spouse-ashley-madison.

Tryon, C. 2013. *On-Demand Culture: Digital Delivery and the Future of Movies*. New Brunswick, NJ: Rutgers University Press.

Tsing, Anna L. 2005. *Friction: An Ethnography of Global Connection.* Princeton, NJ: Princeton University Press.

Tubi TV. 2017. "Tubi Announces $20M in Funding." Media release, May 9. http://tubitv. com/static/press/tubitv_funding.

Turkle, Sherry. 1995. *Life on the Screen: Identity in the Age of the Internet.* New York: Simon and Schuster.

Turton, William. 2015. "Most Young Adults Watched the GOP Debate on Snapchat, Not TV." *Daily Dot.* August 14. www.dailydot.com/layer8/teens-love-politics-but-on-snapchat.

Tyson, Gareth, Perta, Vasile C., Haddadi, Hamed, and Seto, Michael C. 2016. "A First Look at User Activity on Tinder." *arXiv*, 1–8. http://arxiv.org/abs/1607.01952.

Uricchio, William. 2004. "Storage, Simultaneity, and the Media Technologies of Modernity." In *Allegories of Communication: Intermedial Concerns from Cinema to the Digital*, edited by John Fullerton and Jan Olsson, 123–38. (Bloomington: Indiana University Press.

US Department of Homeland Security. "If You See Something, Say Something." www.dhs. gov/see-something-say-something/about-campaign.

Vaidhyanathan S. 2011. *The Googlization of Everything.* Berkeley: University of California Press.

Vallury, R. 2013. "Thought-Perception beyond Form or, the Logic of Shame." *New Formations: A Journal of Culture/Theory/Politics* 80/81: 230–33.

van Dijck, José. 2011. "Tracing Twitter: The Rise of a Microblogging Platform." *International Journal of Media & Cultural Politics* 7, no. 3: 333–48. doi: 10.1386/macp.7.3.333_1

van Dijck, José. 2013. *The Culture of Connectivity: A Critical History of Social Media.* Oxford: Oxford University Press.

van Dijck, José, and Thomas Poell. 2013. "Understanding Social Media Logic." *Media and Communication* 1, no. 1: 2–14.

van Dijk, Elisabeth T., et al. 2015. "Unintended Effects of Self-Tracking: Beyond Personal Informatics." *Designing for Experiences of Data. CHI'15.* April 18–April 23, 2015, Seoul, South Korea.

van Natta, Don. 2016. "Welcome to the Big Time." *ESPN.com.* www.espn.com/espn/feature/story/_/id/17374929/otl-investigates-implosion-daily-fantasy-sports-leaders-draftkings-fanduel.

Vanderhoef, John. 2013. "Casual Threats: The Feminization of Casual Video Games." *Ada: A Journal of Gender, New Media, and Technology*, no. 2. http://adanewmedia. org/2013/06/issue2-vanderhoef.

van Rijn, Roy. 2010. "Patent, Publicity." *royvanrijn.com.* July 9. http://royvanrijn.com/ blog/2010/07/patent-publicity.

Vara, Vauhini. 2005. "Song-Recognition Firms Offer to Name That Tune, for a Price." *Wall Street Journal.* August 4.

Varan, Duane, Jamie Murphy, Charles F. Hofacker, Jennifer A. Robinson, Robert F. Potter, and Steven Bellman. 2013. "What Works Best When Combining Television Sets, PCs, Tablets, or Mobile Phones? How Synergies across Devices Result from Cross-Device Effects and Cross-Format Synergies." *Journal of Advertising Research* 53, no. 2: 212–20.

Vester, Katharina. 2011. "Regime Change: Gender, Class, and the Invention of Dieting in Post-Bellum America." *Journal of Social History* 44, no. 1: 39–70.

Vonderau, Patrick. 2015. "The Politics of Content Aggregation." *Television & New Media* 16, no. 8: 717–33.

Vonderau, Patrick. 2017. "The Spotify Effect: Digital Distribution and Financial Growth." *Television & New Media.November 21.* https://doi.org/10.1177/1527476417741200.

Voyce, Malcolm. 2006. "Shopping Malls in Australia: The End of Public Space and the Rise of 'Consumerist Citizenship'?" *Journal of Sociology* 42, no. 3: 269–86.

Wacquant, Loïc. 2009. *Punishing the Poor: The Neoliberal Government of Social Insecurity.* Durham, NC: Duke University Press.

Waggoner, Glen. 1984. *Rotisserie League Baseball: The Greatest Game for Baseball Fans since Baseball.* New York: Bantam Books.

Walker, S. 2006. *Fantasyland.* New York: Penguin.

Wang, Avery. 2003. "An Industrial Strength Audio Search Algorithm." In *Proceedings of the Fourth International Society for Music Information Retrieval Conference,* 7–13. Baltimore, MD. www.ee.columbia.edu/~dpwe/papers/Wang03-shazam.pdf.

Wang, Avery. 2006. "The Shazam Music Recognition Service." *Communications of the ACM* 49, no. 8: 44–48.

Wang, Shawn. 2016. "This.cm Wants to Deliver the Only Links You'll Really Read Each Evening." *Nieman Lab.* March 4. www.niemanlab.org/2016/03/this-cm-wants-to-deliver-the-only-links-youll-really-read-each-evening.

Wardrip-Fruin, Noah. 2009. *Expressive Processing: Digital Fictions, Computer Games, and Software Studies.* Software Studies. Cambridge: MIT Press.

Wardrip-Fruin, Noah, and Nick Montfort. 2003. *The New Media Reader.* Cambridge: MIT Press.

Wee, Willis. 2013. "The Story of App Annie: Building a Company for the Mobile App Economy." *Tech in Asia.* May17. www.techinasia.com/app-annie-mobile-app-data-intelligence.

Weeks, Kathi. 2011. *The Problem with Work: Feminism, Marxism, Antiwork Politics, and Postwork Imaginaries.* Durham, NC: Duke University Press.

Wen, Ting. 2014. "Tengxun gongbu kaifang yidonghua celue" [Tencent Announced Its Mobile Open Platform Strategies]. *Shanghai Securities.* http://money.163.com/14/0117/05/9IP3BFP700253B0H.html.

Werning, Stefan. 2015. "Swipe to Unlock." *Digital Culture & Society* 1, no. 1: 53–69. doi: 10.14361/dcs-2015-0105.

White, Michele. 2011. "Engaged with eBay." *Feminist Media Studies* 11, no. 3: 303–19.

Wiley, Stephen B. Crofts, Tabitha Moreno Becerra, and Daniel M. Sutko. 2012. "Subjects, Networks, Assemblages: A Materialist Approach to the Production of Social Space." In *Communication Matters: Materialist Approaches to Media, Mobility, and Networks,* edited by Jeremy Packer and Stephen B. Crofts Wiley, 183–95. Abington, UK: Routledge.

Wilken, Rowan, and Gerard Goggin, eds. 2015. *Locative Media.* New York: Routledge.

Willett, Rebekah. 2007. "Consuming Fashion and Producing Meaning through Online Paper Dolls." In *Growing Up Online: Children and Technology,* edited by Sandra Weber and Shanly Dixon, 170–92. London: Palgrave.

Willett, Rebekah. 2008. "'What You Wear Tells a Lot about You:' Girls Dress Up Online." *Gender & Education* 20, no. 5: 421–34.

Williams, Raymond. 1974/2003. *Television: Technology and Cultural Form.* London: Routledge.

Williams, Raymond. 1989. "Culture Is Ordinary [1958]." In *Resources of Hope: Culture, Democracy, Socialism*, 3–14. London: Verso.

Wilson, Matthew W. 2012. "Location-Based Services, Conspicuous Mobility, and the Location-Aware Future." *Geoforum*. Themed issue: *Spatialities of Ageing* 43, no. 6: 1266–75, doi: 10.1016/j.geoforum.2012.03.014.

Wordsworth, Dorothy. 1798. *Dorothy Wordsworth's Journal, Written at Alfoxden 20th January to 22nd May*. (Vol. 1, edited by W. Night). New York: Macmillan.

Wu, Wenting. 2015. "Meiyou yuzhuang ruanjian, shoujishang kao shenme huiben" [How Do Phone Manufacturers Profit without Preinstalled Apps?]. *China Business News*. http://tech.sina.com.cn/t/2015-12-06/doc-ifxmihae9089843.shtml.

@Wyntonav. 2016. "When you Shazam African music." *Twitter*. July 17. https://mobile.twitter.com/Wyntonav/status/754612077696839680.

Xiang, Tracey. 2014. "Tencent's Mobile Open Strategy even including Developer Incubation." *Technode.com*. http://technode.com/2014/01/16/tencent-mobile-open-strategy/.

Yarrow, J. 2014. *Business Insider*. February 20. www.businessinsider.com/WhatsApp-note-2014-2.

Yee, Lee Chyen, and Paul Carsten. 2013. "Baidu to Buy Chinese App Store for $1.9 Billion in Mobile Push." *Reuters*. http://uk.reuters.com/article/uk-baidu-netdragon-idUK BRE96F01Q20130716.

Yik Yak. 2016. "Privacy Policy." August 16. www.yikyak.com/privacy.

Yip, Patrick. 2015. "The Ultimate Cheat Sheet to China's Android App Stores." *OneSky* Blog. www.oneskyapp.com/blog/chinese-app-stores/.

Young, Liam. 2013. "Un–Black Boxing the List: Knowledge, Materiality, and Form." *Canadian Journal of Communication* 38, no. 4: 497–516.

Yu, Hong. 2017. *Networking China: The Digital Transformation of the Chinese Economy*. Urbana: University of Illinois Press.

Zelizer, Viviana. 1997. *The Social Meaning of Money: Pin Money, Paychecks, Poor Relief, and Other Currencies*. Princeton, NJ: Princeton University Press.

Zelizer, Viviana. 2016. "My Money Obsession." In "Twenty Years after *The Social Meaning of Money*." *Books&Ideas*. www.booksandideas.net/Twenty-Years-After-The-Social-Meaning-of-Money.html.

Zimmerman, Jackie. 2015. "This Is What It Takes to Make 2000 Working for TaskRabbit." *Money*. March 12. http://time.com/money/3714829/working-for-taskrabbit.

Zittrain, Jonathan. 2008. *The Future of the Internet—And How to Stop It*. New Haven, CT: Yale University Press.

Zittrain, Jonathan. 2011. "*The Future of the Internet—And How to Stop It*: A Dialog with Jonathan Zittrain Updating His 2008 Book." *John Battell's SearchBlog*. http://battelle media.com/archives/2011/08/the_future_of_the_internet_and_how_to_stop_it_-_a_dia log_with_jonathan_zittrain_updating_his_2008_book.php.

Zittrain, Jonathan. 2012. "Protecting the Internet without Wrecking It." *BostonReview*. http://bostonreview.net/forum/protecting-internet-without-wrecking-it/jona than-zittrain-responds.

Zomorodi, Manoush. 2016. "Your Quantified Body, Your Quantified Self." *Note to Self*. Podcast audio. March 16. www.wnyc.org/story/quantified-bodies.

Contributors

Megan Sapnar Ankerson is Associate Professor of Communication Studies at the University of Michigan. Her research examines the histories and politics of digital media and communication technologies, focusing on the intersection of design, experience, and cultural imagination. She is the author of *Dot-com Design: The Rise of a Usable, Social, Commercial Web* (2018) and co-editor of the international journal *Internet Histories: Digital Technology, Culture, Society.*

Finn Brunton is an assistant professor in the Department of Media, Culture, and Communication at NYU. He is the author of *Spam: A Shadow History of the Internet* (MIT, 2013); *Obfuscation: A User's Guide for Privacy and Protest* with Helen Nissenbaum (MIT, 2015); *Communication* with Mercedes Bunz and Paula Bialski (Minnesota, 2018); and *Digital Cash: The Unknown History of the Anarchists, Immortalists, and Utopians Who Built Cryptocurrency* (Princeton, 2019).

Shira Chess (PhD 2009, Rensselaer Polytechnic Institute) is an Assistant Professor of Entertainment and Media Studies in the Grady College of Journalism and Mass Communication at the University of Georgia. She is the author of *Ready Player Two: Women Gamers and Designed Identity*(University of Minnesota Press, 2017) and co-author of *Folklore, Horror Stories, and the Slender Man: The Development* of an Internet Mythology (Palgrave Pivot, 2015). Her research has been published in *Critical Studies in Media Communication*; *The Journal of Broadcasting and Electronic Media*; *Feminist Media Studies*; *New Media & Society*; *Games and Culture*; and *Information, Communication & Society* as well as several essay collections.

Christopher Cwynar is an Assistant Professor of Communication Studies at Defiance College. His research is primarily concerned with the evolution of public media during periods of intensive cultural, technological, and institutional change. His scholarly work has appeared in a number of journals including *Media, Culture & Society*, *The International Communication Gazette*, *The Journal of Radio and Audio Media*, and *The Radio Journal: International Studies in Broadcast & Audio Media*.

Stefanie Duguay is Assistant Professor in the Department of Communication Studies at Concordia University, Montreal, Canada. Her research focuses on the influence of digital media technologies in everyday life, with particular attention to sexual identity, gender and social media. Stefanie's research has been published in *New Media & Society*, *Information, Communication & Society*, and other international journals.

Elizabeth Ellcessor is an assistant professor of media studies at the University of Virginia. She is the author of Restricted Access: Media, Disability, and the Politics of Participation (NYU Press, 2016) and co-editor, with Bill Kirkpatrick, of Disability Media Studies (NYU Press, 2017). Her work has been published in New Media & Society, Cinema Journal, and Television & New Media, among other venues.

Greg Elmer is Professor of Communication and Culture, Ryerson University. Greg has published widely on questions of media, technology and politics with an emphasis on theories of surveillance, social control and protest, including *The Permanent Campaign: New Media, New Politics* with Ganele Langlois & Fenwick McKelvey (Peter Lang, 2012) and *Infrastructure Critical: Sacrifice at Toronto's G8/20 Summit*, with Alessandra Renzi (ARP, 2014).

Radhika Gajjala is Professor of Media and Communication and American Culture Studies at Bowling Green State University, USA. She was Fulbright Professor in Digital Culture at University of Bergen, Norway in 2015–16 and previously Senior Fulbright scholar at Soegijapranata Catholic University in 2012. Her work engages themes related to globalization, digital labor, feminism and social justice. Published books include *Online Philanthropy: Connecting, Microfinancing, and Gaming for Change* (Lexington Press, 2017), *Cyberculture and the Subaltern* (Lexington Press, 2012), and *Cyberselves: Feminist Ethnographies of South Asian Women* (Altamira, 2004). Co-edited collections include *Cyberfeminism 2.0* (2012), *Global Media Culture and Identity* (2011), *South Asian Technospaces* (2008), and *Webbing Cyberfeminist Practice* (2008). She is currently working on book length projects on *Digital Diasporas: Labor* *and Affect in Gendered Indian Digital Publics* (forthcoming 2019): Rowman and Littlefield International, UK.

Tarleton Gillespie is a principal researcher at Microsoft Research, an affiliated associate professor in the Department of Communication and Department of Information Science at Cornell University, co-founder of the blog Culture Digitally, co-editor of *Media Technologies: Essays on Communication, Materiality, and Society* (MIT, 2014), and author of *Wired Shut: Copyright and the Shape of Digital Culture* (MIT, 2007) and *Custodians of the Internet: Platforms, Content Moderation, and the Hidden Decisions that Shape Social* Media (Yale, 2018).

Gerard Goggin is Professor of Media and Communications at the University of Sydney. He is widely published on social, cultural, and policy aspects of digital technologies, especially in relation to mobile media and communication. Gerard's books include *Locative Technologies in International Contexts* (2019), *Routledge Companion to Global Internet Histories* (2017), *Disability and the Media* (2015), *Global Mobile Media* (2011), *Cell Phone Culture* (2006), and *Digital Disability* (2003).

Mack Hagood is Robert H. and Nancy J. Blayney Assistant Professor of Comparative Media Studies at Miami University, Ohio, where he does ethnographic and archival research in media, sound, and popular music. He is particularly interested in how people use audio media to control their spatial surroundings, social interactions, and sense of self. He is author of *Hush: Media and Sonic Self-Control* (Duke University Press, 2019) and producer/co-host of the sound studies podcast *Phantom Power*.

Germaine R. Halegoua is an Associate Professor in Film & Media Studies at University of Kansas. Her research focuses on digital media and place, urban informatics, and cultural geographies of digital media. Her work is published in several journals including*New Media & Society, International Journal of Cultural Studies, Social Media + Society, Journal of Urban Technology* as well as anthologies and online venues. She is also the co-editor of the book, *Locating Emerging Media*.

Alison Harvey is Lecturer in Media and Communication at the University of Leicester. Her research focuses on issues of inclusivity and accessibility in digital culture, with an emphasis on games. She is the author of *Gender, Age, and*

Digital Games in the Domestic Context. Her work has also appeared in a range of interdisciplinary journals, including *International Journal of Cultural Studies, Feminist Media Studies, Games & Culture, Information, Communication & Society Social Media & Society,* and *Studies in Social Justice.*

Jessalynn Keller is an Assistant Professor in the Department of Communication, Media and Film at the University of Calgary. Her research focuses on feminist digital cultures, girls' media, and mediated identities. Jessalynn is author of *Girls' Feminist Blogging in a Postfeminist Age* (Routledge 2015) and co-editor of *Emergent Feminisms: Complicating a Postfeminist Media Culture* (Routledge 2018). Her research has also been published in journals including *Feminist Media Studies, Continuum: Journal of Media and Cultural Studies, Journal of Gender Studies, Information, Communication & Society,* and *Social Media + Society,* as well as in several edited anthologies.

Luzhou Nina Li is a lecturer in cultural studies in the Institute for Advanced Studies in the Humanities at the University of Queensland, Australia. Her research focuses on digital culture, global media and cultural industries, media history, Chinese media, among others. Her work has appeared in *Television & New Media, International Journal of Cultural Studies* and *Media, Culture & Society.*

Ben Light is Professor of Digital Society at the University of Salford. His research is concerned with understanding people's everyday experiences of digital media. Light engages science and technology studies bringing it into dialogue with questions of (non)consumption practices, digital methods, gender and sexuality. He is author of *Disconnecting with Social Networking Sites* (Palgrave 2014). His latest book with Susanna Paasonen and Kylie Jarrett explores the phenomenon of not safe for work (MIT Press 2019).

Jason Kido Lopez is a Visiting Assistant Professor in the University of Wisconsin-Madison's Department of Communication Arts. His research and teaching interests include sports media broadly construed, fantasy sports and gambling, media ethics, and race. He is the author of *Self-Deception's Puzzles and Processes: A Return to a Sartrean View.*

Fenwick McKelvey is an Associate Professor in Information and Communication Technology Policy in the Department of Communication Studies at Concordia University. He studies the digital politics and policy. He is the author of *Internet Daemons: Digital Communications Possessed* (University of Minnesota Press, 2018) and co-author of *The Permanent Campaign: New Media, New Politics* (Peter Lang, 2012) with Greg Elmer and Ganaele Langlois.

Kate M. Miltner is a PhD Candidate at the USC Annenberg School for Communication and Journalism and a Joint Fellow at the UC Berkeley Center for Technology, Society and Policy and the UC Berkeley Center for Long-Term Cybersecurity. She also holds an MSc in Media and Communications from the London School of Economics and Political Science. Her research agenda centers on the intersections between digital technology, identity, and power.

Jeremy Morris is an Associate Professor in Media and Cultural Studies in the Department of Communication Arts at the University of Wisconsin-Madison. He is the author of *Selling Digital Music, Formatting Culture* (University of California Press, 2015) and has published widely on digital culture, music technologies and podcasting. He is the founder of http://podcastre.org and is currently working on a manuscript about the history of mundane software and the software commodity.

Sharif Mowlabocus is a Senior Lecturer in Digital Media at the University of Sussex, where his teaching and research explores the relationship between human sexuality and technology. He published *Gaydar Culture* (Routledge) in 2010, which documented the first ten years of British gay male digital culture. He has since worked with NGOs and commercial organisations on a variety of projects and has held visiting positions at Microsoft Research and Stanford University.

Sarah Murray is Assistant Professor in the Department of Film, Television, and Media and affiliate faculty in the Digital Studies Institute at the University of Michigan. She uses histories and theories of emerging media to study media's role in self-actualization, data-driven technological intimacies, mobile media cultures, and the consumer tech industry. She is published in *Critical Studies in Media Communication, Feminist Media Histories*, and *International Journal of Cultural Studies*.

Bahar Nasirzadeh is PhD candidate in the Communication and Culture program at York University, Canada. Her research and teaching interests focus on media theory, critical theories, surveillance studies, and affect theory. Her research explores the relationship between governance and technology, with a particular focus on automation and production of subjectivity. She is interested in understanding the implications of automated decision-making in allocating access to options and chances. Nasirzadeh also teaches design at Centennial College.

Kate O'Riordan is Professor of Digital Culture and Head of the School of Media, Film and Music, University of Sussex. She works on media and cultural studies of science and technology, including public engagement with emerging technologies and their intersections with gender and sexuality. Books include: *Unreal Objects* (Pluto Press, 2017); *The Genome Incorporated* (Ashgate, 2010); *Human Cloning and the Media*(Routledge, 2007); *Queer Online* (Peter Lang, 2007).

Devon Powers is Associate Professor in the Department of Advertising and Public Relations, Temple University. She is the author of *Writing the Record: The* Village Voice *and the Birth of Rock Criticism* (University of Massachusetts Press, 2013), and co-editor of *Blowing Up the Brand: Critical Perspectives on Promotional Culture* (Peter Lang, 2010). Her research explores historical and contemporary consumer culture and the dynamics of cultural intermediation, circulation, and promotion. Recent work has appeared in *Journal of Consumer Culture, New Media & Society*, and *Journal of Historical Research in Marketing*. Her next book, *On Trends*, will publish in 2019.

Elena Razlogova is Associate Professor of History at Concordia University. She is the author of *The Listener's Voice: Early Radio and the American Public* (2011) and co-editor of the "Radical Histories in Digital Culture" issue of *Radical History Review* (2013). She has published on U.S. radio history, digital music recommendation and recognition algorithms, and film festivals and film translation in the Soviet Union. Her chapter on freeform DJs, automated music recommendation, and the WFMU appeared in *Radio's New Wave* (2013). She is finishing a book on the uneasy relationship of the freeform radio tradition and the AI-driven music culture.

Carrie Rentschler's research examines feminist activism, media making, gender violence and the politics of witnessing. She is author of *Second Wounds: Victims' Rights and the Media in the U.S.* and co-editor of *Girls and the Politics* *of Place*, and is currently writing a history of contemporary bystander culture. She teaches in the Department of Art History and Communication Studies and is associate faculty in the Institute for Gender, Sexuality, and Feminist Studies at McGill University.

Jill Walker Rettberg is Professor of Digital Culture at the University of Bergen. Her books include *Seeing Ourselves Through Technology: How We Use Selfies, Blogs and Wearable Devices to See and Shape Ourselves* (Palgrave 2014) and *Blogging* (Polity Press, 2008, 2014). She is the Principal Investigator of MACHINE VISION, a €2,000,000 ERC-funded project running from 2018–23 that investigates the cultural effects of algorithmic visual technologies. She can be found in social media as @jilltxt.

Natasha D. Schüll is a cultural anthropologist in NYU's department of Media, Culture, and Communication. She is author of *Addiction by Design* (2012), an ethnographic exploration of the relationship between gambling technology and the experience of addiction. Her next book, *Keeping Track* (forthcoming), concerns the rise of digital self-tracking technologies and the new modes of self-regulation they engender. Her research has been featured in such media venues as *60 Minutes, The New York Times, The Economist,* and *The Atlantic.*

Sarah Sharma is Director of the McLuhan Centre for Culture and Technology at the University of Toronto. She holds her faculty appointment at the ICCIT and the Faculty of Information. She is the author of In the Meantime: Temporality and Cultural Politics (Duke UP, 2014). She is currently working on a new monograph on technology and the gendered politics of exit and refusal.

Tamara Shepherd is an Assistant Professor in the Department of Communication, Media and Film at the University of Calgary. She studies the feminist political economy of digital culture, looking at policy, labor, and literacy in social media, mobile technologies, and digital games. She is an editorial board member of *Social Media + Society*, and her work has been published in *Convergence, First Monday, International Journal of Cultural Studies,* and the *Canadian Journal of Communication.*

Victoria Simon is Visiting Assistant Professor of Media Studies at Pitzer College. She earned her PhD in Communication Studies from McGill University in 2018. She researches the history and cultural politics of touchscreens for music production and the social implications of algorithms in user interface design. She is published in *Communication, Culture and Critique* and *Television and New Media*.

Tarishi Verma is a PhD student at the School of Media and Communication, Bowling Green State University, Ohio. She comes from a varying background of journalism and audio-visual production. She graduated in journalism from the University of Delhi, and completed her Master's degree in Media and Cultural Studies from the Tata Institute of Social Sciences in Mumbai, India. She has directed three documentaries, been an assistant director for one, and has produced several radio shows. She has also interned with Indian news organizations like *The Deccan Herald, The Week*, and *Hindustan Times*. She has also worked as a journalist with The Indian Express for a little more than a year writing on politics, gender, film, among other topics. Her research interests lie in new/digital media and feminist activism online, gendered spaces, digital labors, and representations of gender.

Patrick Vonderau is Professor at the Department of Media Studies at Stockholm University. His most recent book publication is the co-authored *Spotify Teardown: Inside the Black Box of Music Streaming* (MIT Press, forthcoming 2018). Together with social anthropologist Johan Lindquist he is currently leading a three-year research project entitled *Shadow Economies of the Internet: An Ethnography of Click Farming*. Patrick is on the editorial board of *Convergence, Media Industries Journal*, and *Montage AV*.

Index

Page numbers in *italics* indicate illustrations.